A·N·N·U·A·L E·D·I·T·I·O·N·S

Developing World *06/07*

Sixteenth Edition

EDITOR

Robert J. Griffiths

University of North Carolina at Greensboro

Robert J. Griffiths is associate professor of political science and director of the International Studies Program at the University of North Carolina at Greensboro. His teaching and research interests are in the fields of comparative and international politics, and he teaches courses on the politics of development, African politics, international law and organization, and international political economy. His publications include articles on South African civil/military relations, democratic consolidation in South Africa, and developing countries and global commons negotiations.

Contemporary Learning Series

2460 Kerper Blvd., Dubuque, IA 52001

Visit us on the Internet
http://www.mhcls.com

Credits

1. **Understanding the Developing World**
   Unit photo—USMC photo by Staff Sgt. Jim Goodwin
2. **Political Economy and the Developing World**
   Unit photo—© Getty Images/John A. Rizzo
3. **Conflict and Instability**
   Unit photo— Photo courtesy of USAID
4. **Political Change in the Developing World**
   Unit photo—© Getty Images/PhotoLink/Jack Star
5. **Population, Development, Environment, & Health**
   Unit photo—© Photo courtesy of USAID
6. **Women and Development**
   Unit photo—© *Saudi Aramco World*/PADIA/Brynn Bruijn

Copyright

Cataloging in Publication Data
Main entry under title: Annual Editions: Developing World. 2006/2007.
1. Developing World—Periodicals. I. Griffiths, Robert J., *comp.* II. Title: Developing World.
ISBN 0–07–3209643   658'.05   ISSN 1096–4215

Sixteenth Edition

Cover image © Digital Vision/PunchStock and Corbis/Royalty Free
Printed in the United States of America   1234567890QPDQPD98765   Printed on Recycled Paper

Editors/Advisory Board

Members of the Advisory Board are instrumental in the final selection of articles for each edition of ANNUAL EDITIONS. Their review of articles for content, level, currentness, and appropriateness provides critical direction to the editor and staff. We think that you will find their careful consideration well reflected in this volume.

EDITOR

Robert J. Griffiths
University of North Carolina

Staff

Preface

In publishing ANNUAL EDITIONS we recognize the enormous role played by the magazines, newspapers, and journals of the public press in providing current, first-rate educational information in a broad spectrum of interest areas. Many of these articles are appropriate for students, researchers, and professionals seeking accurate, current material to help bridge the gap between principles and theories and the real world. These articles, however, become more useful for study when those of lasting value are carefully collected, organized, indexed, and reproduced in a low-cost format, which provides easy and permanent access when the material is needed. That is the role played by ANNUAL EDITIONS.

The developing world is home to the vast majority of the world's population. Because of its large population, increasing role in the international economy, frequent conflicts and humanitarian crises, and importance to environmental preservation, the developing world continues to be a growing focus of international concern. The developing world has also figured prominently in the recent protests over globalization and the September 11, 2001 terrorist attacks against the United States are tied to certain circumstances in developing countries.

Developing countries demonstrate considerable ethnic, cultural, political, and economic diversity, making generalizations about them difficult. Increasing differentiation further complicates our ability to understand developing countries and to comprehend the challenges of modernization, development, and globalization that they face. Comprehending these challenges must take into account the combination of internal and external factors that contribute to the current circumstances throughout the developing world. Issues of peace and security, international trade and finance, debt, poverty, the environment, human rights, and gender illustrate the effects of growing globalization and interdependence. These trends emphasize the importance of understanding how these issues intersect and suggest the need for a cooperative approach to dealing with them. There is significant debate regarding the best way to address the developing world's problems. Moreover, the developing world's needs compete for attention on an international agenda that is often dominated by relations between the industrialized nations and more recently by the war on terrorism. Domestic concerns within the industrial nations also continue to overshadow the plight of the developing world.

This sixteenth edition of *Annual Editions: Developing World* seeks to provide students with an understanding of the diversity and complexity of the developing world and to acquaint them with the challenges that these nations confront. I am convinced that there remains a need for greater awareness of the problems that confront the developing world and that the international community must make a commitment to effectively address these issues—especially in an international environment dominated by globalization and the war on terrorism. I hope that this volume contributes to students' knowledge and understanding of current trends and their implications and serves as a catalyst for further discussion.

Over two-thirds of the articles in this edition are new. I chose articles that I hope are both interesting and informative and that can serve as a basis for further student research and discussion. The units deal with what I regard as the major issues facing the developing world. In addition, I have attempted to suggest similarities and differences between developing countries, the nature of their relationships with the industrialized nations, and the differences in perspective regarding the causes of and approaches to the issues.

I would again like to thank McGraw-Hill/Contemporary Learning Series for the opportunity to put together a reader on a subject that is the focus of my teaching and research. I would also like to thank those who have sent in the response forms with their comments and suggestions. I have tried to take these into account in preparing the current volume. No book on a topic as broad as the developing world can be comprehensive. There are certainly additional and alternative readings that might be included. Any suggestions for improvement are welcome. Please complete and return the postage-paid article rating form at the end of the book with your comments.

Robert J. Griffiths
Editor

Contents

UNIT 1
Understanding the Developing World

Unit Overview   xviii

The concepts in bold italics are developed in the article. For further expansion, please refer to the Topic Guide and the Index.

UNIT 2
Political Economy and the Developing World

The concepts in bold italics are developed in the article. For further expansion, please refer to the Topic Guide and the Index.

The concepts in bold italics are developed in the article. For further expansion, please refer to the Topic Guide and the Index.

UNIT 3
Conflict and Instability

The concepts in bold italics are developed in the article. For further expansion, please refer to the Topic Guide and the Index.

UNIT 4
Political Change in the Developing World

The concepts in bold italics are developed in the article. For further expansion, please refer to the Topic Guide and the Index.

The concepts in bold italics are developed in the article. For further expansion, please refer to the Topic Guide and the Index.

UNIT 5
Population, Development, Environment, and Health

The concepts in bold italics are developed in the article. For further expansion, please refer to the Topic Guide and the Index.

UNIT 6
Women and Development

The concepts in bold italics are developed in the article. For further expansion, please refer to the Topic Guide and the Index.

Topic Guide

This topic guide suggests how the selections in this book relate to the subjects covered in your course. You may want to use the topics listed on these pages to search the Web more easily.

On the following pages a number of Web sites have been gathered specifically for this book. They are arranged to reflect the units of this *Annual Edition.* You can link to these sites by going to the student online support site at *http://www.mhcls.com/online/.*

ALL THE ARTICLES THAT RELATE TO EACH TOPIC ARE LISTED BELOW THE BOLD-FACED TERM.

Africa
26. Africa's Unmended Heart
27. Sudan's Darfur: Is it Genocide?
35. Africa's Democratization: A Work in Progress
36. The New South Africa, a Decade Later

Agriculture
8. International Trade

AIDS
36. The New South Africa, a Decade Later
40. Malaria, The Child Killer
43. The Payoff From Women's Rights

Civil War
22. The Market for Civil War

Communal conflict
22. The Market for Civil War
26. Africa's Unmended Heart
27. Sudan's Darfur: Is it Genocide?
28. Blaming the Victim: Refugees and Global Security

Corruption
16. A Regime Changes
37. Latin America's Populist Turn

Culture
5. Development as Poison
6. Selling to the Poor

Debt
2. The Development Challenge
11. Why Developing Countries Need a Stronger Voice
15. How to Run the International Monetary Fund
18. Calculating the Benefits of Debt Relief

Democracy
13. Without Consent: Global Capital Mobility and Democracy
16. A Regime Changes
30. The Democratic Mosaic
31. Keep the Faith
32. Voices within Islam: Four Perspectives on Tolerance and Diversity
33. First Steps: The Afghan Elections
34. The Syrian Dilemma
35. Africa's Democratization: A Work in Progress
36. The New South Africa, a Decade Later
37. Latin America's Populist Turn
44. Iraq's Excluded Women

Development
2. The Development Challenge
4. Institutions Matter, but Not for Everything
5. Development as Poison
14. Making the WTO More Supportive of Development
17. Ranking the Rich 2004

20. Microfinance and the Poor

Doha Development Round
10. Why Prospects for Trade Talks are Not Bright
12. Why Should Small Developing Countries Engage in the Global Trading System

Economic growth
11. Why Developing Countries Need a Stronger Voice
21. The Real Digital Divide
29. National Income and Liberty
36. The New South Africa, a Decade Later
39. Development in Africa: The Good, the Bad, the Ugly

Environment
17. Ranking the Rich 2004
42. Why We Owe So Much to Victims of Disaster

Ethnic conflict
22. The Market for Civil War
23. The End of War?
26. Africa's Unmended Heart
27. Sudan's Darfur: Is it Genocide?
28. Blaming the Victim: Refugees and Global Security

Failing states
24. The Failed States Index

Foreign aid
2. The Development Challenge
15. How to Run the International Monetary Fund
17. Ranking the Rich 2004
19. Recasting the Case for Aid

Gender issues
43. The Payoff From Women's Rights
44. Iraq's Excluded Women
45. Ten Years' Hard Labour

Globalization
1. More or Less Equal?
6. Selling to the Poor

Governance
3. Strengthening Governance: Ranking Countries Would Help
19. Recasting the Case for Aid
39. Development in Africa: The Good, the Bad, the Ugly

Health
3. Strengthening Governance: Ranking Countries Would Help
40. Malaria, The Child Killer
41. The Price of Life

HIV-AIDS
36. The New South Africa, a Decade Later
40. Malaria, The Child Killer
41. The Price of Life

Internet References

The following internet sites have been carefully researched and selected to support the articles found in this reader. The easiest way to access these selected sites is to go to our student online support site at *http://www.mhcls.com/online/*.

AE: Developing World 06/07

The following sites were available at the time of publication. Visit our Web site—we update our student online support site regularly to reflect any changes.

General Sources

Foreign Policy in Focus (FPIF): Progressive Response Index
http://fpif.org/progresp/index_body.html

This index is produced weekly by FPIF, a "think tank without walls," which is an international network of analysts and activists dedicated to "making the U.S. a more responsible global leader and partner by advancing citizen movements and agendas." This index lists volume and issue numbers, dates, and topics covered by the articles.

People & Planet
http://www.peopleandplanet.org

People & Planet is an organization of student groups at universities and colleges across the United Kingdom. Organized in 1969 by students at Oxford University, it is now an independent pressure group campaigning on world poverty, human rights, and the environment.

United Nations System Web Locator
http://www.unsystem.org

This is the Web site for all the organizations in the United Nations family. According to its brief overview, the United Nations, an organization of sovereign nations, provides the machinery to help find solutions to international problems or disputes and to deal with pressing concerns that face people everywhere, including the problems of the developing world, through the UN Development Program at *http://www.undp.org* and UNAIDS at *http://www.unaids.org*.

United States Census Bureau: International Summary Demographic Data
http://www.census.gov/ipc/www/idbsum.html

The International Data Base (IDB) is a computerized data bank containing statistical tables of demographic and socioeconomic data for all countries of the world.

World Health Organization (WHO)
http://www.who.ch

The WHO's objective, according to its Web site, is the attainment by all peoples of the highest possible level of health. Health, as defined in the WHO constitution, is a state of complete physical, mental, and social well-being and not merely the absence of disease or infirmity.

UNIT 1: Understanding the Developing World

Africa Index on Africa
http://www.afrika.no/index/

A complete reference source on Africa is available on this Web site.

African Studies WWW (U. Penn)
http://www.sas.upenn.edu/African_Studies/AS.html

The African Studies Center at the University of Pennsylvania supports this ongoing project that lists online resources related to African Studies.

UNIT 2: Political Economy and the Developing World

Center for Third World Organizing
http://www.ctwo.org/

The Center for Third World Organizing (CTWO, pronounced "C-2") is a racial justice organization dedicated to building a social justice movement led by people of color. CTWO is a 20-year-old training and resource center that promotes and sustains direct action organizing in communities of color in the United States.

ENTERWeb
http://www.enterweb.org

ENTERWeb is an annotated meta-index and information clearinghouse on enterprise development, business, finance, international trade, and the economy in this age of cyberspace and globalization. The main focus is on micro-, small-, and medium-scale enterprises, cooperatives, and community economic development both in developed and developing countries.

International Monetary Fund (IMF)
http://www.imf.org

The IMF was created to promote international monetary cooperation, to facilitate the expansion and balanced growth of international trade, to promote exchange stability, to assist in the establishment of a multilateral system of payments, to make its general resources temporarily available under adequate safeguards to its members experiencing balance-of-payments difficulties, and to shorten the duration and lessen the degree of disequilibrium in the international balances of payments of members.

TWN (Third World Network)
http://www.twnside.org.sg/

The Third World Network is an independent, nonprofit international network of organizations and individuals involved in issues relating to development, the Third World, and North-South issues.

U.S. Agency for International Development (USAID)
http://www.info.usaid.gov

USAID is an independent government agency that provides economic development and humanitarian assistance to advance U.S. economic and political interests overseas.

The World Bank
http://www.worldbank.org

The International Bank for Reconstruction and Development, frequently called the World Bank, was established in July 1944 at the UN Monetary and Financial Conference in Bretton Woods, New Hampshire. The World Bank's goal is to reduce poverty and improve living standards by promoting sustainable growth and investment in people. The bank provides loans, technical assistance, and policy guidance to developing country members to achieve this objective.

UNIT 3: Conflict and Instability

The Carter Center
http://www.cartercenter.org

The Carter Center is dedicated to fighting disease, hunger, poverty, conflict, and oppression through collaborative initiatives in the areas of democratization and development, global health, and urban revitalization.

Center for Strategic and International Studies (CSIS)
http://www.csis.org/

For four decades, the Center for Strategic and International Studies (CSIS) has been dedicated to providing world leaders with strategic insights on, and policy solutions to, current and emerging global issues.

Conflict Research Consortium
http://www.Colorado.EDU/conflict/

The site offers links to conflict- and peace-related Internet sites.

Institute for Security Studies
http://www.iss.co.za

This site is South Africa's premier source for information related to African security studies.

PeaceNet
http://www.igc.org/peacenet/

PeaceNet promotes dialogue and sharing of information to encourage appropriate dispute resolution, highlights the work of practitioners and organizations, and is a proving ground for ideas and proposals across the range of disciplines within the conflict-resolution field.

Refugees International
http://www.refintl.org

Refugees International provides early warning in crises of mass exodus. It seeks to serve as the advocate of the unrepresented—the refugee. In recent years, Refugees International has moved from its initial focus on Indochinese refugees to global coverage, conducting almost 30 emergency missions in the last 4 years.

UNIT 4: Political Change in the Developing World

Latin American Network Information Center—LANIC
http://www.lanic.utexas.edu

According to *Latin Trade,* LANIC is "a good clearing house for Internet-accessible information on Latin America."

ReliefWeb
http://www.reliefweb.int/w/rwb.nsf

ReliefWeb is the UN's Department of Humanitarian Affairs clearinghouse for international humanitarian emergencies.

World Trade Organization (WTO)
http://www.wto.org

The WTO is promoted as the only international body dealing with the rules of trade between nations. At its heart are the WTO agreements, the legal ground rules for international commerce and for trade policy.

UNIT 5: Population, Development, Environment, and Health

Earth Pledge Foundation
http://www.earthpledge.org

The Earth Pledge Foundation promotes the principles and practices of sustainable development—the need to balance the desire for economic growth with the necessity of environmental protection.

EnviroLink
http://envirolink.org

EnviroLink is committed to promoting a sustainable society by connecting individuals and organizations through the use of the World Wide Web.

Greenpeace
http://www.greenpeace.org

Greenpeace is an international NGO (nongovernmental organization) that is devoted to environmental protection.

Linkages on Environmental Issues and Development
http://www.iisd.ca/linkages/

Linkages is a site provided by the International Institute for Sustainable Development. It is designed to be an electronic clearinghouse for information on past and upcoming international meetings related to both environmental issues and economic development in the developing world.

Population Action International

http://www.populationaction.org

According to its mission statement, Population Action International is dedicated to advancing policies and programs that slow population growth in order to enhance the quality of life for all people.

The Worldwatch Institute

http://www.worldwatch.org

The Worldwatch Institute advocates environmental protection and sustainable development.

UNIT 6: Women and Development

WIDNET: Women in Development NETwork

http://www.focusintl.com/widnet.htm

This site provides a wealth of information about women in development, including the Beijing '95 Conference, WIDNET statistics, and women's studies.

WomenWatch/Regional and Country Information

http://www.un.org/womenwatch/

The UN Internet Gateway on the Advancement and Empowerment of Women provides a rich mine of information.

We highly recommend that you review our Web site for expanded information and our other product lines. We are continually updating and adding links to our Web site in order to offer you the most usable and useful information that will support and expand the value of your Annual Editions. You can reach us at: *http://www.mhcls.com/annualeditions/*.

UNIT 1

Understanding the Developing World

Unit Selections

1. **More or Less Equal?**, The Economist
2. **The Development Challenge**, Jeffrey D. Sachs
3. **Strengthening Governance: Ranking Countries Would Help**, Robert I. Rotberg
4. **Institutions Matter, but Not for Everything**, Jeffrey D. Sachs
5. **Development as Poison**, Stephen A. Marglin
6. **Selling to the Poor**, Allen L. Hammond and C.K. Prahalad
7. **The Challenge of Worldwide Migration**, Michael W. Doyle

Key Points to Consider

- Has the number of poor people declined worldwide?

- What accounts for discrepancies in the estimated number of the world's poor?

- What factors might speed up the achievement of the Millennium Development Goals?

- Besides institutions, what other factors are crucial to development?

- What constitutes the Western model of development?

- Is the Western model of development transferable to the developing world?

- What are the advantages and disadvantages of marketing to the world's poor?

- Why does hunger persist in the developing world?

- What factors are shaping international migration patterns?

Student Website

www.mhcls.com/online

Internet References

Further information regarding these websites may be found in this book's preface or online.

Africa Index on Africa
    http://www.afrika.no/index/
African Studies WWW (U. Penn)
    http://www.sas.upenn.edu/African_Studies/AS.html

It has never been easy to characterize and understand the diverse countries that make up the developing world and the task has become even more difficult as further differentiation among these countries has occurred. "Developing world" is a catch-all term that lacks precision and explanatory power. It is used to describe a wide range of societies from traditional to modernizing. To complicate things even further, there is also controversy over what actually constitutes development. For some, it is economic growth or progress towards democracy; for others it involves greater empowerment and dignity. There are also differing views on why progress toward development has been uneven. The West tends to see the problem as stemming from poor governance, institutional weakness and failure to embrace free-market principles. Critics from the developing world cite the legacy of colonialism and the nature of the international political and economic structures as the reasons for a lack of development. In any case, lumping together the 100-plus nations that make up the developing world obscures the disparities in size, population, resources, forms of government, industrialization, distribution of wealth, ethnic diversity, and a host of other indicators that make it difficult to categorize and generalize about this large, diverse group of countries.

Despite their diversity, most nations of the developing world share some characteristics. Developing countries often have large populations with annual growth rates between 2 and 4 percent. Although there has been some improvement, poverty is widespread in both rural and urban areas, with rural areas often containing the poorest of the poor. While the majority of the developing world's inhabitants continue to live in the countryside, there is a massive rural-to-urban migration under way, and cities are growing rapidly. Wealth is unevenly distributed, making education, employment opportunities, and access to health care luxuries that few can enjoy. Corruption and mismanagement are common. With very few exceptions, these nations share a colonial past that has affected them both politically and economically. Moreover, critics charge that the neocolonial structure of the international economy and the West's political, military, and cultural links with the developing world amount to continued domination.

Developing countries continue to struggle to improve their citizens' living standards. Despite the economic success in some areas, poverty remains prevalent, and over a billion people live on less than a dollar a day. There is some indication, depending on how the data is measured, that the number of poor worldwide is declining but poverty is still a major feature of the developing world. There is also growing economic inequality between the industrial countries and the developing world. This is especially true of the poorest countries that have become further marginalized due to their fading strategic importance since the end of the cold war and their limited participation in the global economy. Inequality is also growing within developing countries where elite access to education, capital, and technology has significantly widened the gap between rich and poor.

Although the gap between rich and poor nations persists, some emerging markets saw significant growth during the 1990s. However, even these countries experienced the harsh realities of the global economy. The 1997 Asian financial crisis demonstrated the potential consequences of global finance and investment, and investors remain wary of investing in all but a small number of developing countries. Focus on the relatively few countries that are making progress further marginalizes the large number of developing countries that have not experienced economic growth. The reasons for poor economic performance in many developing countries are complex. Both internal and external factors play a role in the inability of some developing countries to make economic progress. Among the factors are the colonial legacy, continued reliance on the export of primary products, stagnating or declining terms of trade for those primary products, protectionism in the industrialized countries, inadequate foreign aid contributions, debt, as well as weak governance and a lack of regime transparency and accountability. Sluggish growth in the United States and lagging economies in the rest of the industrialized world has also had a significant impact on the economic prospects of the developing world.

Shortly after independence a divide between developing and industrialized countries emerged. The basis for this division was the view that the industrialized world continued to dominate and exploit the developing countries. This viewpoint encouraged efforts to alter the international economic order during the 1970s. While the New International Economic Order (NIEO) succumbed to neoliberalism in the 1980s, developing countries still seek solidarity in their interactions with the West. The efforts to extract concessions from the industrialized countries in the negotiations on the Doha trade round illustrate this effort. Moreover, developing countries still view Western prescriptions for development skeptically and chafe under the Washington Consensus that dictates terms for access to funds from international financial institutions and foreign aid. Some critics suggest that Western development models are detrimental and result in inequitable development and cultural imperialism. In contrast to the developing world's criticism of the West, industrial countries continue to maintain the importance of institution-building and following the Western model that emphasizes a market-oriented approach to development. There is clearly a divergence of opinion between the industrialized countries and the developed world on issues ranging from economic development to governance and human rights. The effects of globalization may strengthen the criticism of Western views as more and more poor people in the developing world come into contact with the international economy through the efforts of corporations seeking to sell their products in the untapped markets of the developing world. Globalization may have the additional effect of increasing the migration of poor people seeking to take advantage of economic opportunities that elude them at home.

More or less equal?

Is economic inequality around the world getting better or worse?

CRITICS of capitalism are convinced that the gap between rich and poor is widening across the world. For them, the claim amounts almost to an article of faith: worsening inequality is a sure sign of the moral bankruptcy of "the system". Whether rising inequality should in fact be seen as condemning capitalism in this way is a question worth addressing in its own right. There are reasons to doubt it. But it would also be interesting to know the answer to the narrow factual question. Is the familiar claim that capitalism makes global inequality worse actually true?

Unfortunately, this apparently straightforward question turns out to be harder to answer than one might suppose. There are three broad areas of difficulty. The first is measuring what people, especially the poorest people in developing countries, consume. The second is valuing consumption in a way that allows useful comparisons to be made across countries and over time. And the third, in effect, is settling on an appropriate basis of comparison. Which matters more, for instance: whether inequality is widening among nations, or whether inequality is widening among all the people of the world, regardless of which country they happen to live in? Judging any claim about global inequality is impossible without a clear understanding of how the researchers concerned have dealt with all three questions.

The third deserves to be emphasised at the outset. A thought-experiment reveals how easy it is to get muddled. Suppose it is true that inequality measured across countries is widening. (In other words, the gap between average incomes in the richest countries and average incomes in the poorest countries, measured without regard to changes in population, is growing.) Also suppose that inequality is worsening within every individual country. Given that cross-country inequality is widening, and that within-country inequality is getting worse as well, it would have to follow that global inequality, measured across all the world's individuals, is rising too, would it not? Actually, no. Even if those first two assumptions were true, global inequality measured across all the world's individuals might well be falling.

How so? Simply add a third assumption: namely, that a group of poor countries accounting for a big share of all the poor people in the world was growing very rapidly. Suppose, for instance, that average incomes in India and China were growing much faster than average incomes in the rich industrial economies. Then it could be true that inequality was widening within every country, including within China and India themselves; and also that the gap between the very poorest countries (of sub-Saharan Africa) and the richest (Europe and the United States) was widening; and yet, at the same time, that inequality measured across all the individuals in the world was falling fast, because average incomes in the two most populous poor countries were rapidly going up.

It so happens that average incomes in India and China are going up extremely rapidly. Without knowing anything else, one should therefore be sceptical about all the claims that are so confidently made about rising "global inequality".

Measuring matters

Much of the frequently acrimonious debate among economists about global poverty and inequality turns out to revolve around a single technical issue: is it better to measure consumption (and hence living standards) using data drawn from national accounts or data drawn from household surveys? The two sources ought to marry up. In fact they differ systematically, and by a wide margin. Worse, growth in consumption, not merely levels of consumption, differs persistently according to which source is used. National-accounts data tend nearly always to give a much more optimistic view of trends in poverty than do household-survey data.

Accordingly, in a recent review of the literature* by Angus Deaton of Princeton University (Mr Deaton is perhaps the only economist at work in this area who is acknowledged by all sides both as author-

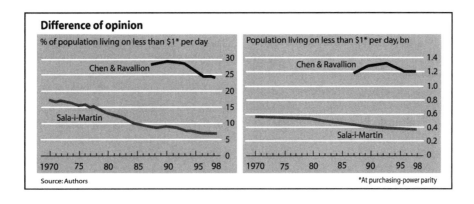

Difference of opinion

% of population living on less than $1* per day

Chen & Ravallion

Sala-i-Martin

1970   75   80   85   90   95 98

Source: Authors

Population living on less than $1* per day, bn

Chen & Ravallion

Sala-i-Martin

1970   75   80   85   90   95 98

*At purchasing-power parity

itative and as having no ideological axe to grind) two sets of studies are contrasted. The first draws mainly on national-accounts data, the second on household surveys. Their results are at odds.

Work by Surjit Bhalla, by Xavier Sala-i-Martin, and by Francis Bourguignon and Christian Morrisson shows rapid—indeed historically unprecedented—falls in poverty during the 1980s and 1990s, the new golden age of global capitalism. According to these papers, the proportion of the world's people living on less than a dollar a day (inflation adjusted) has fallen so quickly that the decline has been enough to offset rising population in the developing countries. In other words, the number of people in poverty has been falling not only as a share of the world's population but also, remarkably, in absolute terms.

Mr Sala-i-Martin's calculations, for instance, show that the proportion of the world's people living in acute poverty (on less than a dollar a day) fell from 17% in 1970 to 7% in 1998; the proportion living on less than $2 a day fell from 41% to 19%. The absolute headcount of global $1-a-day poverty fell, according to the same estimates, by 200m (see chart); and the count of $2-a-day poverty fell by 350m. Mr Bhalla, who finds the sharpest drop in poverty of these authors, wryly states that in 2000 when the United Nations (UN) announced its Millennium Development Goal on poverty—to bring the number of people living on less than a dollar a day in 2015 down to half

the level in 1990—the goal had already been achieved.

But this is not at all the picture that emerges from the second, and far more widely cited, set of estimates. Calculations by the World Bank, using direct surveys of households, carry the official imprimatur of the UN, which uses them in monitoring progress towards its Millennium Development Goal on poverty. And they seem to show relatively little reduction in poverty over recent decades.

A paper by Shaohua Chen and Martin Ravallion of the Bank lays out the thinking behind the Bank's estimates. The authors put the proportion of people living on less than a dollar a day at 28% in 1987—far higher than the corresponding figure according to Mr Sala-i-Martin's work. By 1998, the proportion in poverty had in fact fallen (something which you might not guess if you listened only to those who deplore the wickedness of global capitalism), but only to 24%. Compare that with Mr Sala-i-Martin's estimate of just 7%.

That discrepancy draws attention to the danger of focusing too much on the dollar-a-day threshold. That is a crowded part of the global income distribution. For this reason alone, switching from one data source to another, or moving the official poverty line from one level to another, is apt to have a large effect on the figures. This underlines the importance of not regarding any of these numbers as holy writ.

Still, the question remains, why are the differences so big? Several

factors are at work. The World Bank attempts to measure "consumption poverty", as opposed to "income poverty". To the extent that poor people manage to save, their consumption will be less than their income, and so there will be more poor people on the Bank's definition. Also, the Bank expresses its poverty ratios as proportions of population in the developing countries; Mr Sala-i-Martin, for instance, uses global population; the effect is to make Mr Sala-i-Martin's estimates, other things equal, smaller than the Bank's. Country samples also vary from study to study. And on top of all this comes the effect of basing estimates on national accounts rather than on household surveys.

It is revealing to consider why, according to the Chen-Ravallion study, poverty fell relatively slowly on their household-survey measure. It was not because of an increase in within-country inequality—in other words, brisk growth in average consumption was not being hogged by the better off. It was because growth in average consumption was slower than what growth in national incomes, as measured in the national accounts, would lead you to expect. Growth in consumption per head across the countries in the Chen-Ravallion sample was less than 1% a year between 1987 and 1998, according to the household surveys. Growth in consumption per head according to the national accounts was more than 3% a year.

Mr Deaton notes that a plethora of new data has so far failed to resolve this issue "because the new sources are mutually contradictory". Summing up, he states: "If the surveys are wrong, and the national accounts right, either inequality has been widening in ways that our data do not appear to show, or poverty has been falling more rapidly than shown by the dollar-a-day counts. If the surveys are right, there has been less growth in the world in the 1990s than we are used to thinking."

It would be a mistake to presume that either source of data is better in principle. Surveys are famously prone to error because of bad or fluctuating design, discrepancies in samples and poor execution. But national accounts have drawbacks as well, especially in poor countries. For instance, they fail to capture some sorts of non-market income and consumption. This makes them prone to understate the consumption of the poor, but also to overstate the growth of consumption of the poor as incomes rise and as more activities fall within the scope of market transactions.

Take your pick

Still, most of the discrepancy between the survey estimates and the national-accounts estimates—with the surveys persistently pessimistic on trends in poverty—is probably due to the fact that as people get better off, they are less likely to respond (accurately, or at all) to surveys. As a result, as countries get richer, the ratio of "survey consumption" to "national-accounts consumption" is usually found to fall. Consistent with this, the ratio of the two measures is highest in the poorest countries.

Mr Deaton argues that both sources ought to be used, though combining them properly raises a host of difficult technical issues.

Meanwhile, the truth about global poverty and inequality presumably lies somewhere between the extremes suggested by the two methodologies.

One can at least conclude that the official World Bank data, used by the UN and other agencies, are too pessimistic: poverty has most likely fallen faster than these widely cited figures suggest, and possibly fast enough to reduce the global head-count of those living on less than a dollar a day, even as population rises. More accurate answers will require more work to be done. In the meantime, however, the official position on global poverty ought to start, at a minimum, to acknowledge the uncertainty surrounding the figures and, further, to concede that the truth is likely to be better than the official figures say.

Beside the point?

But what of the fear that global capitalism is making progress at the expense of the poor? The true figures would probably be quite reassuring on this—but even if the more pessimistic official figures were correct, it would be worth questioning the conclusions that the anti-globalists draw from them. If poverty was proving as tenacious in the face of growth as the Bank's estimates say, would it make sense to blame global capitalism for that?

Hardly. On any estimate, poverty is at its most impervious in sub-Saharan Africa. These are not just the poorest countries in the world, but also the slowest-growing. Can it be plausibly claimed that these countries are the victims of globalisation? That would be an odd conclusion, given that sub-Saharan Africa's economies are so comparatively isolated from the rest of the world econ-

omy—by force of history, circumstance and, to a large extent, the policies of their own and other governments. Sub-Saharan Africa plainly suffers not from globalisation, but from lack of it. The focus of attention should be on how to extend the benefits of international economic linkages to the region. Removing every rich-country barrier to trade with these countries would be an excellent place to start.

By contrast, India and China are showing how great the benefits of international economic integration can be. Neither country is an exemplar of free-market capitalism—far from it. But it is undeniable that both countries have consciously chosen to seize the opportunities afforded by the global economy, through both trade and foreign investment. As incomes surge, while the living standards of the poorest improve more modestly, if at all, inequality within both countries may well be rising. The gaps between urban and rural incomes, especially, have widened lately.

This may prove a temporary phenomenon. But suppose otherwise; suppose the problem persists. Would any such worsening of inequality entitle one to conclude that India and China had taken a wrong turn these past 20 years? Of course not. Look at Africa to understand that there are worse things than inequality.

"Measuring poverty in a growing world (or measuring growth in a poor world)", revised February 2004, by Angus Deaton. May be downloaded from http://www.wws.princeton.edu/deaton/working.htm

Papers referred to in this article are listed on http://www.economist.com/inequalitypapers

The Development Challenge

Jeffrey D. Sachs

PROMISES, PROMISES

As a MATTER of stated policy, there is no doubt that Washington is committed to supporting economic development in impoverished countries. In September 2000, it joined the UN in issuing the Millennium Declaration, in which the world pledged to cut extreme poverty in half and reduce child mortality by two-thirds within the next 15 years (aims later formalized as part of the Millennium Development Goals). In March 2002, the United States and the international community adopted the Monterrey Consensus, which laid out a multifaceted strategy to achieve these objectives by promoting the private sector in developing countries, opening trade with them, and increasing official development assistance (ODA). That year, the U.S. National Security Strategy promised to "secure public health," "emphasize education," and "continue to aid agricultural development" in low-income countries. The United States and other developed countries, the document asserted, "should set an ambitious and specific target: to double the size of the world's poorest economies within a decade."

Most Americans, and perhaps most senior U.S. government officials, believe that the United States has been following through on such commitments. The U.S. response, public and private, to December's Indian Ocean tsunami has seemed to confirm the nation's generous engagement with those in need. Ironically, though, this outpouring of concern may obscure rather than clarify a deeper truth. Other than in response to disasters—famines, floods, and earthquakes—U.S. assistance for the world's poorest countries is utterly inadequate. It falls far short of meeting the needs of recipient countries, fails to tap into the vast U.S. capacity for providing aid, does not fulfill Washington's many promises to fund development adequately, and is a small fraction of what Americans believe the U.S. actually provides.

Without a new approach, Washington risks undermining the most important international development goals that the world has accepted—and plunging the international community into a maelstrom of recrimination.

Without dramatic reform, the United States will increasingly be tarred as a country ready to invest in war, and perhaps in emergency relief, but not in peaceful development. The number of failed states will increase, spreading disorder and threatening global and national security.

Only a new U.S. international development strategy can avoid this outcome by achieving the objectives set forth in the Millennium Declaration, the Monterrey Consensus, and the National Security Strategy. Embarking on a new, practical course of assistance will have short-term costs, but the long-term benefits will far outweigh them. Continued failure, on the other hand, will be far too expensive to bear.

THE DEVELOPMENT MIRAGE

THE DEVELOPMENT ASSISTANCE COMMITTEE (DAC) of the Organization for Economic Cooperation and Development defines official development assistance as the sum of grants and sub-market-rate loans made to developing countries to promote economic development and welfare. Military aid is not counted, nor is aid to high-income countries such as Israel. Even with these parameters, the DAC definition is too expansive to measure real assistance for economic development, but it nonetheless gives a systematic measure of U.S. foreign assistance.

By the DAC's definition, in 2003 (the most recent year for which comprehensive international data are available), the United States gave $16.3 billion in net ODA. Of that amount, $1.7 billion went to multilateral organizations such as the World Bank, which in turn grant or lend the money to developing countries. Washington distributed the remaining $14.6 billion bilaterally, directly targeting recipient nations. Together, the multilateral and bilateral aid represented 0.15 percent of the $11 trillion gross national income (GNI) of the United States in 2003. In the 2004 U.S. budget—which totaled $2.3 trillion—development assistance represented just 0.7 percent of budgetary expenditures.

These sums are vastly smaller than the American people think they are. In a 2001 survey, the Program on International Policy Attitudes (PIPA) at the University of Maryland found that Americans, on average, believe that foreign aid accounts for 20 percent of the federal budget, around 30 times the actual figure. PIPA surveys in the mid-1990s came up with essentially the same result.

U.S. assistance for the world's poorest countries is utterly inadequate.

Many Americans and senior government officials also mistakenly believe that private giving represents a substantial amount of U.S. aid to developing countries. *The Wall Street Journal* and others, citing an earlier study by the U.S. Agency for International Development (USAID), have even reported that private giving significantly exceeds official giving, but their estimates erroneously include $18 billion in private remittances, which are not development aid but income transfers between family members in the United States and abroad. The DAC estimates that U.S. assistance from private voluntary agencies in 2003 amounted to about $6.3 billion. Even if one added to this figure a high-end estimate of $4 billion in other giving from private foundations, corporate philanthropy, and other organizations, the sum of U.S. public and private financial contributions to international development would amount to around $26.6 billion, or just 0.25 percent of GNI.

Not only are the figures for official development assistance themselves not nearly as large as most Americans believe, but the DAC'S estimate of $16 billion also overstates U.S. official aid for economic development by including a considerable amount of assistance that contributes little or nothing to long-term development. In a recent white paper, USAID makes the point by distinguishing between five operational goals for foreign aid: promoting transformational development, supporting strategic states, strengthening fragile states, providing humanitarian relief, and addressing global challenges such as the HIV/AIDS epidemic and climate change.

All five operational goals make sense from a foreign policy standpoint, but only the first directly targets economic development. Aid intended for transformational development aims to support long-term economic change by helping a country achieve structural transformations that should allow it ultimately to escape dependence on outside aid. Assistance to strategic countries focuses on nations that have geopolitical importance, such as Colombia, Egypt, Iraq, and Afghanistan, and helps fight terrorism, strengthen alliances, or reduce narcotics trafficking. Aid to fragile states is designed to head off conflict or help countries recover from it. Lastly, humanitarian assistance is earmarked for relief following natural disasters and often takes the form of U.S. grain deliveries. A surprisingly small proportion of U.S. bilateral assistance is directed at transformational development, and only a small part of that actually transforms the economies of developing countries.

Washington's own aid accounting (as opposed to the DAC'S) makes a key distinction between developmental assistance and geopolitical aid, which is distributed to strategic countries mostly as Economic Support Funds (ESF). In 2004, the ESS program provided $3.3 billion to 42 countries. Economic development is a side effect, not a basic objective, of such aid. Strategic states—defined here as developing countries receiving more than half of their U.S. assistance from the ESF or similar funding (such as the Iraqi Relief and Reconstruction Fund, the Andean Counterdrug Initiative, and the Emergency Response Fund for Afghanistan)—include several countries in the Middle East, Asia, Europe, and Latin America[1]. These 16 countries, plus the West Bank and Gaza Strip, received about 45 percent of total U.S. bilateral assistance to developing countries in 2003, even though they accounted for only 11 percent of the population of developing countries receiving U.S. aid. Many of the strategic states are in fact middle-income countries that are not high development priorities. In many cases, ESF supports corruption or allows a government to reduce its own development spending to free up funds for its military.

Meanwhile, according to the DAC database, very little of the $6.1 billion the United States spent on bilateral assistance to "nonstrategic" developing countries in 2003 actually reached the ground as long-term investment in transformational development. For one thing, $2 billion was distributed as emergency assistance or as non-emergency food aid. Emergency assistance, although salutary, merely addresses immediate crises. Food aid for nonemergency purposes, moreover, has enormously high transaction costs and distorts the local economy by depressing the prices local farmers receive for their goods. Amazingly, nearly half of the money spent on U.S. food aid in 2004 went to cover transport costs rather than the food itself.

In 2003, only $4.1 billion in U.S. bilateral aid was not spent on strategic countries, food aid, or other emergency aid. Of that total, moreover, $1.3 billion took the form of "debt forgiveness grants"—the cancellation of old debts, not the granting of new money. Furthermore, recipient countries saved even less than that amount in actual debt payments in 2003, because the cash-flow savings that year were only a small fraction of the debt that was erased from the books. In some countries, debt payments have actually risen thanks to debt cancellation. (After reaching agreements with creditor nations, countries that had previously been servicing none of their debt burden have had to resume payments on a smaller base.)

Of the remaining $2.8 billion in 2003 U.S. bilateral aid, very little actually funded investments in transformational development. According to the DAC, the entire sum went toward technical cooperation: payments made primarily to U.S. entities—consultants from government agencies or nongovernmental organizations (NGOS)—for assignments in recipient nations. These missions may be useful, but the expenditures are not long-term investments in local clinics, schools, power plants, sanitation, or other infrastructure.

Washington gave very little money directly to nonstrategic developing countries to support specific investments in transformational development. Poor countries have proposed sound plans to build schools and clinics and pay the salaries of teachers and doctors, but the United States virtually never funds such programs directly, sending its own consultants instead. In doing so, Washington contributes to an unworkable proliferation of donor-country pet projects, rather than to an integrated strategy adopted by the recipient country and supported by the donors. A balanced and judicious aid program would provide both technical cooperation and budgetary support to countries that could use the money effectively.

The case of sub-Saharan Africa—the poorest region of the world—shows how dangerously skewed U.S. aid priorities are. The prevailing image in the United States is that Washington gives Africa vast sums of money, which corrupt officials there then fritter away or stash in offshore accounts. But this image, fueled by inaccurate stereotypes, badly misconstrues the truth.

Washington's aid to developing countries is vastly smaller than Americans think.

In fact, in 2003, the United States gave $4.7 billion to sub-Saharan Africa in net bilateral ODA. Of that sum, $0.2 billion went to a handful of middle-income countries, especially South Africa. Of the remaining $4.5 billion, $1.5 billion was apportioned for emergency aid and $0.3 billion for non-emergency food aid. Another $1.3 billion was designated for debt forgiveness grants, and $1.4 billion went to technical assistance. This distribution left only $118 million for U.S. in-country operations and direct support for programs run by African governments and communities—just 18 cents for each of the nearly 650 million people in low-income sub-Saharan Africa. This figure represents the total U.S. bilateral support, beyond aid in the form of technical cooperation, for investments in health, education, roads, power, water and sanitation, and democratic institutions in the region that year. The next time U.S. officials visit Africa and wonder aloud where the "trillions and trillions" of dollars went, they should be reminded of how small those trillions actually are.

Two recent U.S. initiatives will modestly improve the picture, but so far their impact remains very limited. Announced in 2002, the new Millennium Challenge Account (MCA) is designed to give grants to low-income countries that demonstrate good governance. For the current fiscal year, $1.5 billion has been appropriated for the MCA, and in 2002 the Bush administration promised to request $5 billion per year in 2006 and beyond. The MCA is a highly meritorious new approach. No funds, however, have yet been disbursed. The President's Emergency Plan for AIDS Relief (PEPFAR) is also budgeted to give an average of $3 billion per year to certain African and Caribbean countries. In 2004, roughly $2.4 billion was disbursed. This important initiative should be increased in scale significantly, with more of the funding disbursed with other donors through the Global Fund to Fight AIDS, Tuberculosis, and Malaria. Such an approach would better leverage U.S. funding and allow recipient countries to pursue a more integrated approach to fighting AIDS.

FLUNKING

TO EVALUATE AND IMPROVE the current U.S. foreign aid system, a three-prong test should be used: How much foreign assistance is needed and can be used effectively to achieve transformational development? What is the U.S. capacity to give? And—most important for the United States' international image—how does U.S. aid stack up against Washington's promises to poor countries?

The answer to the first question can be derived from careful studies conducted to determine the amount of worldwide ODA needed to achieve the goals of the Millennium Declaration. Most recently, the UN Millennium Project undertook the most extensive analysis of this question ever performed and determined that the developing world will require an additional $70 billion in aid over current levels by 2006, rising to $130 billion over current levels by 2015 (in constant U.S. dollars at 2003 prices). With these added funds, total projected aid in 2006 would represent 0.44 percent of total projected donor GNI that year, increasing to 0.54 percent in 2015. Assuming that all 22 DAC donor countries contribute an equal percentage of their national income and that the U.S. economy grows at an average of three percent a year, the $16 billion contributed by Washington in 2003 would have to increase to $51 billion in 2006 and to $74 billion in 2015. Thus, even the full funding currently promised for the MCA and PEPFAR ($5 billion and $3 billion per year, respectively) would leave the United States far short of doing its part to help poor countries meet the Millennium Development Goals.

These calculations are based on a transparent and rigorous economic analysis (albeit one subject to uncertainty). The basic idea is straight-forward. To achieve the Millennium Development Goals, each impoverished country must make specific, identifiable investments in health, education, and basic infrastructure such as roads, electricity, water, and sanitation. Of course, since these countries are impoverished, their own financial means are limited; most of their current income must be used simply to stay alive rather than to invest in the future. Thus, the poorest nations are caught in a poverty trap. They are poor because they lack the basic necessities of health, education, and infrastructure, and because they are poor, they cannot invest in these basic necessities on the scale necessary to achieve the Millennium Development Goals.

Development assistance can close this financing gap. In the typical African country, total investment needs are about $111 per person per year. Assuming a substantial increase in domestic resource mobilization, around $10 per person can be financed by local households, and another $35 by governments in the low-income countries. The balance of $65 per person is the financing gap that should be covered by donors. Global estimates of necessary ODA are reached by projecting this level of aid for all low-income countries in need, with some further adjustments.

Such an approach is built on the principles of private-sector-led, market-based economic growth. After all, private-sector-led growth depends on adequate infrastructure (roads, power, ports, water, and sanitation) and human capital (a healthy population with adequate levels of literacy, education, and job skills). Domestic and foreign investors will shun a developing country without those prerequisites.

The second standard—U.S. capacity to provide foreign assistance is far from being fully realized. The small sums that Washington gives in ODA are driven by political considerations, not by economic need. To ensure that the Millennium Development Goals are met, the UN Millennium Project calls for ODA from each donor country to rise to at least 0.44 percent of GNI in 2006, and then to continue to increase to 0.54 percent of GNI by 2015. To meet needs beyond these goals—geopolitical and humanitarian needs, for example—the project recommends that each donor country actually reach 0.7 percent of GNI in development assistance by 2015.

These sums are small not only relative to GNI, but also relative to recent changes in the U.S. budget. Since 2001, defense spending has expanded by about 1.7 percent of GNI, and tax revenues have declined by 3.3 percent of GNI, due mainly to tax cuts. In the same period, U.S. official development aid rose by only 0.04 percent of GNI. The United States has, in short, chosen to spend its money on priorities other than development assistance, yet such assistance is just as fundamental to national security as the military. When ODA is measured as a share of GNI, each of the other 21 donor countries contributes more than the United States, and most by a wide margin. The United States ranks second from last (slightly ahead of Italy) when NGO development assistance is added to ODA.

Development aid is just as fundamental as military spending to U.S. national security.

U.S. political leaders could mobilize public support for increased ODA. President George W. Bush's emergency AIDS program is enormously popular, and he is rightly proud of having launched the effort. Moreover, the public repeatedly supports the purposes of development assistance, especially when convinced that the money can be effectively used. Indeed, as mentioned earlier, Americans also believe that U.S. assistance is more than one order of magnitude greater than is really the case.

As for the final test of measuring U.S. development assistance against U.S. commitments, the enormous gap between promise and performance has been weighing heavily on Washington's foreign policy for many years. The U.S. political leadership repeatedly emphasizes that it is party to the Monterrey Consensus, and Bush traveled personally to Monterrey to make that case. Yet the administration has failed to follow through.

In paragraph 42 of the Monterrey Consensus, the signatories commit themselves to making "concrete efforts towards the target of 0.7 percent of gross national product" in official development aid, a commitment that dates back to a vote of the UN General Assembly in 1970. Although some U.S. government officials have long expressed resistance to that international target, the U.S. government has in fact repeatedly signed onto it, not only in Monterrey but on other occasions as well. Meanwhile, five countries—Denmark, Luxembourg, the Netherlands, Norway, and Sweden have already met the goal (and in fact did so many years ago). Six more countries have recently set a timetable to reach it before 2015: Belgium, Finland, France, Ireland, Spain, and the United Kingdom. This year, other countries are likely to join the pledge with specific schedules of their own. The gap between Washington and the other donors is growing.

Washington's palpable shortfall has become a pervasive source of friction in U.S. relations with low-income countries. The United States regularly asks these nations for help in the war on terrorism, only to plead its own "poverty" when asked for more development aid—even for areas such as health, education, and agriculture, which are focal points of Washington's national security doctrine. In one striking example, the United States contributed a meager $4 million to Ethiopia in 2002 to raise its agricultural output—and then gave $500 million in emergency food aid when famine predictably hit the country a year later. Low-income nations are painfully aware of the truth: the United States can be counted on to respond to emergencies, but not to help them break free of poverty.

MAKE OR BREAK

THE MILLENNIUM DEVELOPMENT GOALS were announced in 2000. Five years later, and with just a decade until the target date of 2015, dozens of the world's poorest and most desperate countries remain tragically off course—often well governed but too impoverished to make the investments in infrastructure and human capital necessary to meet the goals. And despite repeated commitments to help such nations free themselves from this poverty trap, the amounts of U.S. aid contributed continue to fall woefully short of what is needed. Privately, many U.S. officials around the world are wringing their hands.

Most of the world has recognized that 2005 is a make-or-break year. The United Kingdom has already charted a path to success, albeit one that the United States has still not embraced. The United Kingdom will chair both the EU and the G-8 group of highly industrialized nations plus Russia this year, and in its capacity as host of the G-8's July summit has promised to put development assistance at the center of the agenda. U.K. Chancellor of the Exchequer Gordon Brown has put on the table an important proposal for an International Finance Facility IFF), backed by donor countries, to double development aid during the next decade. There is widespread European and developing-country support for the IFF. Yet, unwisely, the United States has not only given the proposal a cold shoulder but also failed to explain how the donor nations should otherwise meet the needs of the poorest countries.

In September, international leaders will convene on the millennium summit's fifth anniversary to take stock of the world's progress—or lack thereof—in meeting the Millennium Development Goals. They will also consider reforming the UN to support greater global security. The summit's participants must realize that the two agendas—development assistance and global security—are inexorably linked, not only in substance but in process. Countries such as Germany and Japan that aspire to permanent seats on an expanded Security Council will find that low-income states have aspirations of their own. The poorest nations will hold their wealthy counterparts accountable for what they have and have not done to enable the indigent to overcome early death, mass hunger, disease, and extreme poverty.

The United States itself faces a widely unappreciated risk: the risk of business as usual, with the world wanting to discuss development while Washington focuses on the war on terrorism. The emotional, geopolitical, and operational divide between the United States and the rest of the international community could very well widen markedly, and dangerously. If their life-and-death needs are not met, impoverished countries may be much more reluctant to support Washington and its many security concerns. The United States should not be complacent about the growing perception of this country as one that cares solely about itself, even to the neglect of its own pronouncements.

It is not too late to rectify this situation, but it is growing more critical by the day. By taking four fundamental steps, Washington can turn this crisis in international development into an opportunity for the United States to reassert its moral and political authority as a world leader.

First, President Bush and congressional leaders should explain to the American people that U.S. development assistance is far less than what they believe it to be—and far

less than what is needed, affordable, and already promised by Washington. The government must explain that aid delivered to well-governed countries for well-targeted purposes—health care, schools, roads, power, sanitation—would not only save millions of lives each year but also help countries break free of the poverty trap that binds them. And Washington must explain that expanded aid is a bipartisan, indeed nonpartisan, cause that is crucial to U.S. national security because impoverished countries are vulnerable to becoming failed states and even havens for terrorists.

The giant gap between the United States and other donors threatens to keep growing.

Next, the president and Congress should commit to fulfilling the financing needs that have been identified and internationally agreed on: U.S. development assistance should be increased to an average of 0.5 percent of the nation's GNI during the coming decade, reaching 0.7 percent of GNI by 2015. By 2006, U.S. support for the Millennium Development Goals should rise by another $40 billion per year above the current $16 billion already designated for development aid. An increase on this scale is affordable; in fact, it equals approximately half of Washington's annual spending on Iraq and Afghanistan, yet it would help more than a billion people in low-income countries, in addition to saving millions of lives each year.

Third, Washington should spend the new total of $56 billion in aid through a number of reliable channels. Plausibly, the government could distribute $3 billion to the Global Fund to Fight AIDS, Tuberculosis, and Malaria (partly for AIDS in countries not covered by PEPFAR); $5 billion to PEPFAR programs; $10 billion to the Millennium Challenge Account; $8 billion to the International Development Agency of the World Bank, as well as to regional development banks; $20 billion to USAID in non-MCA countries for transformational development; and $10 billion to strategic states (up from the current $5 billion). Washington's agencies and programs should continue to engage with and support the work of U.S. NGOs, while also funneling a large part of their new funding directly to the budgets of the recipient countries.

Finally, the president and Congress should overhaul the structure of U.S. development assistance programs to enable development to play the strategic role required for national security. USAID should be raised to the rank of a cabinet department, with MCA and other agencies housed under the same roof, similar to the United Kingdom's Department for International Development. The new department must be invested with the analytical capacity and political clout to ensure that the United States becomes a true leader of the global effort to fight poverty and achieve transformational development.

By reassessing its priorities and implementing these sweeping reforms, the United States can seize the initiative in international development. Last year at the UN, President Bush declared, "Our wider goal is to promote hope and progress as the alternatives to hatred and violence. Our great purpose is to build a better world beyond the war on terror." The safety of the nation requires nothing less.

Notes

1. The states or territories classified as "strategic" by this definition are, by region, Egypt, Iraq, Jordan, Lebanon, West Bank/Gaza, and Yemen; Afghanistan, East Timor, Mongolia, Myanmar, Pakistan, and Sri Lanka; Turkey; Bolivia, Colombia, Ecuador, and Peru.

JEFFREY D. SACHS is Director of the Earth Institute at Columbia University and Director of the UN Millennium Project, which in January issued the report *Investing in Development: A Practical Plan to Achieve the Millennium Goals*. He is also the author of the new book *The End of Poverty*.

Strengthening Governance: Ranking Countries Would Help

Robert I. Rotberg

The demonstrated link among poor governance, poverty, and nation-state failure makes strengthening the quality of governance in the developing world an urgent task. In weak, troubled states, there is a strong likelihood that an excess of grievances will offer fertile ground for the nurturing of terrorism. Thus, improving the governance capabilities and effectiveness of developing countries is crucial not only to fostering their economic development, but also to reducing the potential for local and global conflict.

Governance is the term used to describe the tension-filled interaction between citizens and their rulers and the various means by which governments can either help or hinder their constituents' ability to achieve satisfaction and material prosperity. In developed countries, citizens often take it for granted that their leaders will help them meet their fundamental needs. Furthermore, they understand that they possess the tools to improve governance when they are dissatisfied: mobilizing interest groups, employing legal means, or acting at the ballot box.

Most of the world's inhabitants, however, are unable to hold their rulers accountable, to participate in or influence their governments, or to use electoral mechanisms to affect significant change. Governance thus becomes a capricious endeavor at best and, for so much of the developing world, especially the poorest countries and those ravaged by war and disease, a synonym for autocracy and despotism. It is the plight of many: three-fifths of the world's population lives in the developing world, and the vast majority suffer from being poorly governed. In those parts of the world, strengthening governance directly improves the lives of the governed.

Past efforts have demonstrated that no amount of exhortation alone from Washington, London, Brussels, or Tokyo will accelerate the practice of beneficial governance. Tying donor assistance to good governance conditionalities may help at the margin. Programs such as the Millennium Challenge Account (MCA), which aim to provide U.S. financial assistance only to those poorer nation-states making an effort to implement good governance, are well intentioned. In their assessment of governance quality, however, they rely more than they should on available indices that are inherently subjective (and thus prone to bias) and are therefore less than transparent. Moreover, even the well-intentioned and well-directed MCA (and similar attempts by other donors) are also fundamentally political, choosing their recipients on the basis of impressionistic more than rigorously objective criteria. Strengthening local civil societies in general could also add pressure to any internal national momentum for better governance.

There is abundant talk in diplomatic and assistance circles about the need to improve governance in the developing world. Yet, little has been done or accomplished, largely because there is no basis on which a nation-state's governance effectiveness and quality can be assessed objectively and meaningfully, compared to the standards or best practices in its neighborhood or region. It is therefore essential to introduce a new method that is rigorous, bias-free, and capable of distinguishing degrees of good governance among countries.

This new method should spotlight those nation-states that govern unusually poorly, clearly contrasting them with their neighbors or allies who govern more effectively and thus perform more creditably for their citizens. A new nongovernmental organization (NGO), funded broadly or by a range of developing-world nation-states themselves, would be able to issue an annual report card. It would reveal which countries are well or poorly governed and why, as well as suggest the specific areas in which each country needs improvement. Such a system will compel countries to recognize that governance counts internationally as well as locally; that good governance is measurable and bad governments can no longer hide; and, in so doing, provide both the carrot and the stick for positive change.

The Rationale for Ranking

No generally acceptable, objective governance-ranking system exists today, although several dozen partial schemes contain approaches that are useful. Rating systems for intrinsic components of governance such as corruption, freedom, competitiveness, trade openness, political risk, receptivity to private enterprise, and contract enforcement, among others, already exist. The most comprehensive of those currently available schemes is the World Bank's Governance Matters III: Governance Indicators for 1996-2002, but it explicitly refuses to rank countries and is itself a compilation of indices that are mostly subjective in origin. The UN Development Program's Human Development Index is another excellent source of comparative data, but only on attainments of development, such as health and education. It does not set out to rank nation-states according to their governmental effectiveness.

> # Improving governance is crucial to reducing the potential for nation-state failure.

Ranking can, however, make a difference. The efforts of NGOs in relevant fields demonstrate the potential for a governance ranking system to affect significant change. The most compelling example is Transparency International (TI), the Berlin-based NGO that ranks nation-states according to the perceived levels of their corruption. TI has managed to shame countries and rulers in Africa and Asia to reduce corruption at the national level, leading more nations to seek to be perceived as less corrupt. TI's system has also proved that outside rating systems can embolden and support civil society activists within afflicted countries, having done so demonstrably in many of the developing world's more graft-ridden states. Beyond issuing corruption rankings, TI has local chapters in many countries that focus on helping governments act against corrupt activities. Similarly, establishing a new governance NGO could help regimes adopt good governance practices and serve their citizens more effectively.

The experience of other existing ranking systems for nation-states, such as credit rating systems, confirms that a carefully detailed report-card system would catch the eye of government leaders, international organizations, investors, and donors and would lead to at least some of the desired improvements. When Moody's or Standard & Poor's downgrades a country's sovereign credit rating, investors and donors take notice; purchasers of bonds and securities know that they will need to absorb greater risk. So do even the least responsible rulers of the downgraded countries, largely because it is in their personal self-interest; their political futures and the incomes of their countries are affected. International organizations and transnational investors also pay attention to such increased risk, as it leads to lending becoming more problematic.

The MCA based its first round of selected recipient countries on data derived from the World Bank, Freedom House, the Heritage Foundation, and TI. Because economic growth is impossible in the absence of good gover-

nance (witness the levels of per capita gross domestic product (GDP) achieved in Botswana, Mauritius, Singapore, and Taiwan versus those of Angola, Burma, Congo, Nigeria, and Zimbabwe—all intrinsically wealthy nation-states), donors are likely to welcome a sophisticated, new report card system detailing countries' governance practices. Donor aid agencies and the international lending agencies need such a tool, if only to target those countries that, on close inspection, are less well governed than popularly assumed.

U.S. policymakers would also pay heed. A comprehensive report card would enable them to focus antiterrorism efforts on those countries ranked at the lower end of the governance scale, presumably where the absence of good governance predisposes local populations to ethnic violence and increases the likelihood of civil war. Evaluating governments, including Arab or Asian regimes, for their practice of good or weak governance could potentially be more influential than unilaterally asserting whether a government is democratic, particularly if anti-Americanism continues to grow overseas. This result comes about both because governance is perceived to be more directly in the interests of citizens overseas, not America's interests, and also because the rating agency proposed would be a nongovernmental agency, reducing the skepticism with which anything Washington does overseas is currently viewed.

Overall, rating governance would set in motion a virtuous, competitive cycle among neighbors and throughout the developing world of governance improvement, while simultaneously reducing the threat of terrorism. Using Moody's Investment grades, TI's Corruption Index, or Freedom House's Freedom in the World as models, a nongovernmental entity should be established to rank countries by governance and to advise interested governments on how to improve their own practices. The challenge is not as difficult as it may seem because many of the components, such as corruption, freedom, and trade openness, already exist. What is required is a more comprehensive, holistic, mostly objective, and quantifiable method of ranking developing countries according to their governmental performance accomplishments.

Political Goods Set the Standard

Governmental performance can be assessed, and thus countries can be ranked, by measuring how many or how few political goods a nation-state provides for its inhabitants; stronger states may be distinguished from weaker states according to the effectiveness of their delivery. The higher the quality and the greater the quantity of the political goods delivered, the better the level of governance. Delivery and performance are approximately synonymous in this context. If a government patches the streets or fixes broken stoplights, it delivers valuable political goods and performs well for its constituents.

Political goods are the somewhat intangible and difficult-to-quantify claims that citizens once made on sovereigns and now make on the governments of their nation-states. Political goods include indigenous expectations and, conceivably, state obligations; inform the local political culture; and collectively give substance to the social contract between ruler and ruled that is at the core of interactions between states and their citizenries. Indeed, governments (and nation-states) exist primarily to provide for their taxpayers or inhabitants, to perform for their citizens in areas and in ways that are more easily managed and organized by the overarching state than by private enterprises or collective civic bodies. The provision of physical security from outside attack and from crime is a prime example. Political goods include citizens' desires to be secure, to exist under a robust rule of law, to be free politically, to enjoy a stable economic environment, to have access to high quality educational and health services, and so on.

The most important political good is the supply of security, especially human security: freedom from crime and threats to the person. Groups of individuals can theoretically band together to purchase goods or services that provide substantial measures of security. Traditionally and most typically, however, individuals and groups are more securely protected by publicly provided security arrangements. Only when reasonable provisions for security exist within a country, especially in a fragile, newly reconstructed nation-state in the developing world, can governments deliver other desirable political goods.

After security, rule of law is of primary importance. Effective modern states provide predictable, recognizable, systematized methods of adjudicating disputes and regulating both the norms and the prevailing mores of a host society. The essentials of this political good are usually embodied in codes and procedures that include an enforceable body of law, security of property and contract, an independent and efficacious judicial system, and a set of norms that comprise the values contained in the local version of a legal system.

Donor aid agencies and the international lending agencies need such a tool.

Another key political good enables citizens to participate freely and fully in politics and the political process. This political good encompasses citizens' essential freedoms: the right to participate in politics and to compete for office; respect and support for national and regional political institutions; tolerance of dissent and difference; and the existence of fundamental civil liberties and human rights.

Among other basic political goods typically supplied by the state are medical and health care; schools and educational instruction; roads, railways, harbors, and airports—the physical arteries of commerce; communications networks; a money and banking system, usually presided over by a central bank and facilitated by a nationally or regionally created currency; a beneficent fiscal and institutional context within which citizens can pursue personal entrepreneurial goals and potentially prosper; a political and social atmosphere conducive to the emergence and sustainability of civil society; and a fully articulated system for regulating access to the environmental commons.

Rating would set in motion a virtuous, competitive cycle improving governance.

Each of these criteria can be measured by creating proxy indicators and subindicators. For example, the extent to which an economic environment is conducive to entrepreneurial initiative and the pursuit of prosperity is reflected by GDP growth rates, inflation levels, and gini coefficient spreads, all of which can be measured and ranked by country. Indicators appropriate for security, rule of law, and other criteria are also available. Essentially, putting numbers on performance tells us whether within a region or across regions a country is doing better than its neighbors or others. Those who are falling behind can catch up or at least be encouraged to do so. Is one country more or less secure than its neighbors? Does it have more or less rule of law? Is it politically freer? Are its citizens receiving more or less instructional quality and medical services than other countries in the developing world or even in its region? Is civil society empowered? These are some of the key inquiries. Only by answering them as objectively as possible are we able to answer the overall question: Is one country better or more poorly governed than its neighbors?

Measuring governmental performance requires measuring outcomes, not inputs. The proposed NGO must employ proxies that assess a government's delivery of political goods, not its budgetary allocations. Primarily, the goal is to discern what a particular government has actually accomplished with its appropriated funds, rather than simply to study its original intentions (as good as they may have been). For example, in a more corrupt country, appropriated funds may have been siphoned away from service delivery into individual pockets; the mere fact that a nation-state appropriates or expends more for health or education than its neighbors may in fact produce few quantifiable results.

Using proxy indicators as indirect measures for governance leaves a number of questions unanswered. Do results that are quantifiable really capture the essence of governance? For example, are we actually able to evaluate citizen satisfaction? Can governance and governmental capacity for good be separated from income level, that is, should we assess less well-endowed countries according to the same criteria as wealthy ones? Does the proposed method account for the difficult cases? Secure, authoritarian, but well-performing nation-states could and sometimes do score higher than democratic, insecure, less-effective deliverers of essential political goods. How quasi-democratic nation-states ultimately rank depends to some extent on how indicators for political freedom and rule of law are weighted against indicators for security, economic prosperity, infrastructural accomplishments, educational and medical delivery, and so on. Fortunately, in the real world, only those partially democratic states that supply high levels of political goods can rank with the fully democratic countries, providing the latter perform reasonably well for their citizens, especially in the security category. Despotisms such as North Korea, Turkmenistan, and Zimbabwe always fall to the bottom of the scale, even if they provide a certain type of tyrannical control and ostensible stability.

Additionally, data are unequal. Not all developing countries are represented in the available indices. The poorer and the most problematic are often missing from TI's lists, and the World Bank's World Development Indicators: Distribution of Income or Consumption numbers are unavailable for some developing countries and are not compiled every year. Even when data are available, they must be suspect. Yet, the proposed method presumes that proxy indicators can reliably capture the delivery of a political good by obtaining in-country quantitative measures and seeking to refine them with an emphasis on reliability and comparability with data from neighboring countries. Currently, proxies are the most effective and direct way of supplying information about government performance. The focus should be on reaching a consensus among experts about appropriate proxies and then fine-tuning the proxies while working toward the construction of a complete index. All of the questions raised are answerable given sufficient research time and attention; none presents an insurmountable obstacle.

Testing Potential Ranking Systems

Aggregate efforts by six years of graduate-student work at Harvard University's Kennedy School of Government illustrate that the proposed method of ranking can work and indeed already has worked in about a dozen separate iterations. Different student groups have experimented with various methods of developing proxy indicators and subindicators for good governance in the nation-states of the developing world. Using these indicators, they have been able to rank countries from best governed to least well governed and have done so across and within regions.

The experiments have collectively demonstrated that rule of law can be measured by assessing the effectiveness

and predictability of the judiciary, the number of judges per 1,000 people (the more judges, the less judicial delay), the number of political prisoners, the level of corruption, the extent of demonstrated respect for private property, and the ability to enforce contracts.

Pilot efforts demonstrate that ranking countries produces defensible results.

For political freedom, that is, for the existence of functioning participatory democratic institutions and the rights and freedoms that make such institutions viable, the subindicators utilized by the students were voice and accountability, political stability, press and media freedom (number of newspapers closed and journalists jailed), voter participation rates, political rights, civil liberties, female adult literacy rates, respect for human rights, and the existence (or not) of the death penalty for criminals.

Per capita GDP in constant dollars, inflation rates, foreign direct investment (FDI), and donor assistance as percentages of GDP all help to measure relative economic success. To distinguish qualities of governance more finely, it is also helpful to measure GDP growth per capita; national poverty/inequality based on gini coefficient scores; the percentages of populations that qualify as "poor" (using the official global definition of earning $1 a day); trade openness as a percentage of GDP; and gross FDI inflows as percentages of GDP. Deficits as percentages of GDP are also relevant. Inflation levels usually distinguish between well- and poorly run countries, as does the amount of contract-intensive money: the ratio of noncurrency money to the total money supply (roughly the amount of money held outside banks); the present value of debt as a percentage of GDP; fiscal balances; and the amounts of domestic credit available to the private sector. Relative levels of foreign currency reserves provide additional insight into the performance of a government in the economic sphere.

Economic performance is enhanced if a nation-state's infrastructure is sound. It is thus important to measure paved road miles or kilometers per capita and per area; numbers of airport arrivals and departures; harbor capacities for those countries that are not landlocked; and teledensity: telephone landlines and mobile users, telephone faults, internet usage, and personal computers per 1,000 people. Comparative electric power and natural gas power transmission and usage are also helpful statistics. Aggregating data for access to potable water and sanitation, although difficult to find and assess, reveals additional information about governmental performance.

Governments in the developing world are traditionally and almost without exception tasked by citizens to deliver the highest possible quality schooling and medi-

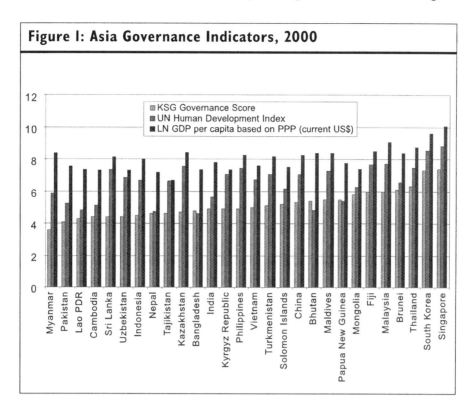

Figure I: Asia Governance Indicators, 2000

- KSG Governance Score
- UN Human Development Index
- LN GDP per capita based on PPP (current US$)

cal care that their treasuries can afford. Measuring those governance outputs seems straightforward, but devising appropriate and informative subindicators has proved challenging. Nevertheless, analyzing net enrollment ratios in primary education, adult illiteracy rates, literacy gender-parity indices, pupil-teacher ratios, the percentage of certified teachers, the average number of years of schooling (educational persistence), the number of years of compulsory schooling, and public expenditures on education as a percentage of total governmental expenditures demonstrates how well a nation-state provides for its citizens' education.

Ranking governance would provide both the carrot and the stick for positive change.

For health outputs, useful subindicators include life expectancy levels, infant mortality percentages as a proportion of 1,000 live births, the maternal mortality ratio per 100,000 live births, childhood immunization rates, HIV prevalence rates, the numbers of hospital beds per 1,000 people, and health expenditures as percentages of budgets and GDP. The underlying presumption is that good governance shows up as good delivery of health services. The most important subindicator, summing up mortality rates and life expectancies, is the World Health Organization's Health Adjusted Annual Life Expectancy

(HALE) Index. HALE measures life expectancy adjusted for morbidity and time spent in poor health. It provides the equivalent number of years in full health that a new-born can expect to live, based on mortality rates and prevailing health states. In some regions, additional subindicators representing the incidence and treatment of diseases such as malaria, tuberculosis, and dengue fever should be used.

When the indicators and subindicators are arranged and the individual indicator scores are summed, the results are overall objective national-governance scores which, by and large, seem to meet the test of reason and common sense. Although a few surprises have emerged, an example illustrates the possibilities in the method for producing valid results. In Asia, the best-governed states in a student-examined sample (based on data for 2000 only) were Singapore, South Korea, Thailand, Brunei, and Fiji; at the bottom of the same list were Myanmar (Burma), Pakistan, Laos, and Cambodia. China ranked 11th out of 29, and India ranked 18th.

Brunei's higher-than-expected placement is probably explained by a high rating (after top-rated Singapore) in the rule of law and infrastructure categories, as well as reasonably high ratings for security. Otherwise, the top- and bottom-ranked countries of the Asia table are consistent with common impressions. The fact that the snapshot year chosen for this particular experiment was 2000 also explains the high (12th-place) ranking of the Solomon Islands, which subsequently plunged into civil war. Using a single year's data may also largely explain why Sri Lanka, which was war-torn but otherwise (since 2000) has been prosperous and fast-growing, rated 24th, after Indonesia and Tajikistan and barely above Uzbekistan.

Singapore and Brunei rank near but not at the very top of the overall rankings for Asia because of their wealth and because they deliver many essential political goods to their citizens, particularly security. Both were downgraded somewhat in their total scores, however, because of deficiencies in providing political rights and freedoms. That downgrading illustrates the intrinsic strength and utility of the method (they ranked high despite some low-rated variables) and the critical importance of deciding how to weigh the value of each indicator. If political rights had been deemed less valuable to good governance than security, for example, both Singapore and Brunei would have fared more poorly in the final overall evaluation.

At a minimum, the student pilot efforts demonstrate that ranking countries of the developing world according to their qualities of governance using elaborate proxy indicators to evaluate levels of performance is plausible and can produce reasonable and defensible results. The mix of objectively and subjectively derived data represents the best efforts and combinations to date. To replace more detailed subjective numbers, for corruption and rule of law, for example, with truly objective quantifications will require extensive fieldwork and/or the development of new objective measures capable of being applied by data collectors within each country when sufficient funds become available. Absent such on-the-ground data collection and analysis, the trend of the quantified results prepared by the student pilot groups makes sense and verifies that this endeavor is indeed promising. What is required for the future, however, are more fully refined indicators and subindicators, each tested laboriously, and a statistically valid method for smoothing data and/or replacing missing data points.

Time for an Impartial Evaluation of Governance

Converting this proposal into a full-fledged governance ranking system will require creating a new, nonpartisan, nonprofit NGO and the investment of seed funding from foundations, a consortium of donors, or international organizations. It will also require the general idea to be accepted by many if not all of the prospective stakeholders: nation-states in the developing world and donors such as the Group of Eight. Any attempt to construct and then use a ranking system for governance quality is guaranteed to be controversial; those nation-states which rank low on the scales of governance may take exception and claim bias. There will thus need to be extensive additional work on and thorough testing of the hypotheses that underlie the proposal and the proxy indicator method to establish the system as comprehensive, independent, and responsible.

The aforementioned are necessary caveats, but as the threat of nation-state failure and the problem of governance quality in the developing world become increasingly daunting and well documented, the need to improve governance grows consistently more urgent. Existing efforts are too subjective, partial, and insufficient. Creating an effective, impartial good-governance ranking system and advisory service will help meet this critical challenge.

Robert I. Rotberg is director of the Kennedy School of Government's Program on Intrastate Conflict and Conflict Resolution at Harvard University and president of the World Peace Foundation.

Institutions Matter, but Not for Everything

The role of geography and resource endowments in development shouldn't be underestimated.

Jeffrey D. Sachs

THE DEBATE over the role of institutions in economic development has become dangerously simplified. The vague concept of "institutions" has become, almost tautologically, the intermediate target for all efforts to improve an economy. If an economy is malfunctioning, the reasoning goes, something must be wrong with its institutions. In fact, recent papers have argued that institutions explain nearly everything about a country's level of economic development and that resource constraints, physical geography, economic policies, geopolitics, and other aspects of internal social structure, such as gender roles and inequalities between ethnic groups, have little or no effect. These papers have been written by such respected economists as Daron Acemoglu, Simon Johnson, and James Robinson; Dani Rodrik, Arvind Subramanian, and Francesco Trebbi; and William Easterly and Ross Levine.

Indeed, a single-factor explanation of something as important as economic development can be alluring, and the institutions-only argument has special allure for two additional reasons. First, it attributes high income levels in the United States, Europe, and Japan to allegedly superior social institutions; it even asserts that when incomes rise in other regions, they do so mainly because of the Western messages of freedom, property rights, and markets carried there by intrepid missionaries intent on economic development. Second, according to the argument, the rich world has little, if any, financial responsibility for the poor because development failures are the result of institutional failures and not of a lack of resources.

The problem is that the evidence simply does not support those conclusions. Institutions may matter, but they don't matter exclusively. The barriers to economic development in the poorest countries today are far more complex than institutional shortcomings. Rather than focus on improving institutions in sub-Saharan Africa, it would be wise to devote more effort to fighting AIDS, tuberculosis, and malaria; addressing the depletion of soil nutrients; and building more roads to con-

nect remote populations to regional markets and coastal ports. In other words, sub-Saharan Africa and other regions struggling today for improved economic development require much more than lectures about good governance and institutions. They require direct interventions, backed by expanded donor assistance, to address disease, geographical isolation, low technological productivity, and resource limitations that trap them in poverty. Good governance and sound institutions would, no doubt, make such interventions more effective.

When economic growth fails

When Adam Smith, our profession's original and wisest champion of sound economic institutions, turned his eye to the poorest parts of the world in 1776, he did not so much as mention institutions in explaining their woes. It is worth quoting at length from Smith's *Wealth of Nations* on the plight of sub-Saharan Africa and central Asia, which remain the world's most troubled development hot spots:

> All the inland parts of Africa, and all that part of Asia which lies any considerable way north of the Euxine and Caspian seas, the ancient Scythia, the modern Tartary and Siberia, seem in all ages of the world to have been in the same barbarous and uncivilised state in which we find them at present. The Sea of Tartary is the frozen ocean which admits of no navigation, and though some of the greatest rivers in the world run through that country, they are at too great a distance from one another to carry commerce and communication through the greater part of it. There are in Africa none of those great inlets, such as the Baltic and Adriatic seas in Europe, the Mediterranean and Euxine seas in both Europe and Asia, and the gulfs of Arabia, Per-

sia, India, Bengal, and Siam, in Asia, to carry maritime commerce into the interior parts of that great continent: and the great rivers of Africa are at too great a distance from one another to give occasion to any considerable inland navigation. (Book I, Chapter III)

Smith's point is that Africa and central Asia could not effectively participate in international trade because transport costs were simply too high. And, without international trade, both regions were condemned to small internal markets, an inefficient division of labor, and continued poverty. These disadvantages of the hinterland exist to this day.

The ability of a disease to cut off economic development may seem surprising to some but reflects a lack of understanding of how disease can affect economic performance.

Smith couldn't know the half of it. The problems of African isolation went far beyond mere transport costs. Characterized by the most adverse malaria ecology in the world, Africa was as effectively cut off from global trade and investment by that killer disease. Although the disease ecology of malaria was not understood properly until two centuries after Adam Smith, what was known demonstrated that Africa's suffering was unique. It had a climate conducive to year-round transmission of malaria and was home to a species of mosquito ideally suited to transmitting malaria from person to person. When Acemoglu, Johnson, and Robinson find that the high mortality rates of British soldiers around 1820 in various parts of the world correlate well with the low levels of GNP per capita in the 1990s, they are discovering the pernicious effects of malaria in blocking long-term economic development.

The ability of a disease to cut off economic development may seem surprising to some but reflects a lack of understanding of how disease can affect economic performance. Thus, in writing that malaria has a limited impact in sub-Saharan Africa because most adults have some acquired immunity, Acemoglu, Johnson, and Robinson completely neglect the fact that the disease dramatically lowers the returns on foreign investments and raises the transaction costs of international trade, migration, and tourism in malarial regions. This is like claiming that the effects of the recent SARS (Severe Acute Respiratory Syndrome) outbreak in Hong Kong SAR can be measured by the number of deaths so far attributable to the disease rather than by the severe disruption in travel to and from Asia.

In an environment in which capital and people can move around with relative ease, the disadvantages of adverse geography—physical isolation, endemic disease, or other local problems (such as poor soil fertility)—are magnified. It is probably true that when human capital is high enough in any location,

physical capital will flow in as a complementary factor of production. Skilled workers can sell their outputs to world markets almost anywhere, over the Internet or by plane transport. Landlocked and at a high altitude, Denver can still serve as a high-tech hub of tourism, trade, and information technology. But when countries that are remote or have other problems related to their geography also have few skilled workers, these workers are much more likely to emigrate than to attract physical capital into the country. This is true even of geographically remote regions within countries. For example, China is having great difficulty attracting investments into its western provinces and is instead facing a massive shift of labor, including the west's few skilled workers, to the eastern and coastal provinces.

Recent history, then, confirms Smith's remarkable insights. Good institutions certainly matter, and bad institutions can sound the death knell of development even in favorable environments. But poor physical endowments may also hamper development. During the globalization of the past 20 years, economic performance has diverged markedly in the developing world, with countries falling into three broadly identifiable categories. First are the countries, and regions within countries, in which institutions, policies, and geography are all reasonably favorable. The coastal regions of east Asia (coastal China and essentially all of Korea, Taiwan Province of China, Hong Kong SAR, Singapore, Thailand, Malaysia, and Indonesia) have this beneficent combination and, as a result, have all become closely integrated with global production systems and benefited from large inflows of foreign capital.

Second are the regions that are relatively well endowed geographically but, for historical reasons, have had poor governance and institutions. These include the central European states, whose proximity to Western Europe brought them little benefit during the socialist regime. For such countries, institutional reforms are paramount. And, finally, there are impoverished regions with an unfavorable geography, such as most of sub-Saharan Africa, central Asia, large parts of the Andean region, and the highlands of Central America, where globalization has not succeeded in raising living standards and may, indeed, have accelerated the brain drain and capital outflows from the region. The countries that have experienced the severest economic failures in the recent past have all been characterized by initial low levels of income and small populations (and hence small internal markets) that live far from coasts and are burdened by disease, especially AIDS, tuberculosis, and malaria. These populations have essentially been trapped in poverty because of their inability to meet the market test for attracting private capital inflows.

When institutions *and* geography matter

It is a common mistake to believe—and a weak argument to make—that geography equals determinism. Even if good health is important to development, not all malarial regions are condemned to poverty. Rather, special investments are needed to fight malaria. Landlocked regions may be burdened by high transport costs but are not necessarily condemned to poverty.

Rather, special investments in roads, communications, rail, and other transport and communications facilities are even more important in those regions than elsewhere. Such regions may also require special help from the outside world to initiate self-sustaining growth.

A poor coastal region near a natural harbor may be able to initiate long-term growth precisely because few financial resources are needed to build roads and port facilities to get started. An equally poor landlocked region, however, may be stuck in poverty in the absence of outside help. A major project to construct roads and a port would most likely exceed local financing possibilities and may well have a rate of return far below the world market cost of capital. The market may be right: it is unlikely to pay a market return to develop the hinterland without some kind of subsidy from the rest of the world. Nor will institutional reforms alone get the goods to market.

In the short term, only three alternatives may exist for an isolated region: continued impoverishment of its population; migration of the population from the interior to the coast; or sufficient foreign assistance to build the infrastructure needed to link the region profitably with world markets. Migration would be the purest free market approach, yet the international system denies that option on a systematic basis; migration is systemically feasible only within countries. When populations do migrate from the hinterlands, the host country often experiences a political upheaval. The large migration from Burkina Faso to Côe d'Ivoire was one trigger of recent ethnic riots and civil violence.

A fourth and longer-term strategy that merits consideration is regional integration: a breaking down of artificial political barriers that limit the size of markets and condemn isolated countries to relative poverty. In this regard, the recent initiative to strengthen subregional and regional cooperation in Africa should certainly be supported. Yet, given political realities, this process will be too slow, by itself, to overcome the crisis of the poorest inland regions.

A good test of successful development strategy in these geographically disadvantaged regions is whether development efforts succeed in attracting new capital inflows. The structural adjustment era in sub-Saharan Africa, for example, was very disappointing in this dimension. Although the region focused on economic reforms for nearly two decades, it attracted very little foreign (or even domestic) investment, and what it did attract largely benefited the primary commodity sectors. Indeed, these economies remained almost completely dependent on a few primary commodity exports. The reform efforts did not solve the underlying fundamental problems of disease, geographical isolation, and poor infrastructure. The countries, unattractive to potential investors, could not break free from the poverty trap, and market-based infrastructure projects could not make up the difference.

Helping the poorest regions

Development thinking and policy must return to the basics: both institutions and resource endowments are critical, not just one or the other. That point was clear enough to Adam Smith but has been forgotten somewhere along the way. A crucial corollary is that poverty traps are real: countries can be too poor to find their own way out of poverty. That is, some locales are not favorable enough to attract investors under current technological conditions and need international help in even greater amounts than have been made available to them in recent decades.

An appropriate starting point for the international community would be to set actual developmental goals for such regions rather than "make do" with whatever economic results emerge. The best standards, by far, would be the Millennium Development Goals, derived from the international commitments to poverty alleviation adopted by all countries of the world at the UN Millennium Assembly of September 2000. The goals call for halving the 1990 rates of poverty and hunger by the year 2015 and reducing child mortality rates by two-thirds. Dozens of the poorest countries—those trapped in poverty—are too far off track to achieve these goals. Fortunately, at last year's UN Financing for Development Conference held in Monterrey, Mexico, and at the World Summit on Sustainable Development held in Johannesburg, South Africa, the industrial world reiterated its commitment to help those countries by increasing debt relief and official development assistance, including concrete steps toward the international target of 0.7 percent of donor GNP. The extra $125 billion a year that would become available if official development assistance were raised from the current 0.2 percent of GNP to 0.7 percent of GNP should easily be enough to enable all well-governed poor countries to achieve the Millennium Development Goals. Like official development assistance, debt-relief mechanisms have been wholly inadequate to date.

Armed with these goals and assurances of increased donor assistance, the international community, both donors and recipients, should be able to identify, for each country and in much greater detail than in the recent past, those obstacles—whether institutional, geographical, or other (including barriers to trade in the rich countries)—that are truly impeding economic development. For each of the Millennium Development Goals, detailed interventions—including their costs, organization, delivery mechanisms, and monitoring—can be assessed and agreed upon by stakeholders and donors. By freeing our thinking from one-factor explanations and understanding that poverty may have as much to do with malaria as with the exchange rate, we will become much more creative and expansive in our approach to the poorest countries. And, with this broader view, the international institutions can also be much more successful than past generations in helping to free these countries from their economic suffering.

References

Acemoglu, Daron, Simon Johnson, and James A. Robinson, 2001, "The Colonial Origins of Comparative Development: An Empirical Investigation," American Economic Review, Vol. 91 (December), pp. 1369–1401.

Bloom, David E., and Jeffrey D. Sachs, 1998, "Geography, Demography, and Economic Growth in Africa," Brookings Papers on Economic Activity: 2, Brookings Institution, pp.207–95.

Démurger, Sylvie, and others, 2002, "Geography, Economic Policy, and Regional Development in China," Asian Economic Papers, Vol. I (Winter), pp. 146–97.

Easterly, William, and Ross Levine, 2002, "Tropics, Germs and Crops: How Endowments Influence Economic Development," NBER Working Paper 9106 (Cambridge, Massachusetts: National Bureau of Economic Research).

Gallup, John Luke, and Jeffrey D. Sachs with Andrew D. Mellinger, 1998, "Geography and Economic Development," paper presented at the Annual World Bank Conference on Development Economics, Washington, D.C., April.

Rodrik, Dani, Arvind Subramanian, and Francesco Trebbi, 2002, "Institutions Rule: The Primacy of Institutions over Geography and Integration in Economic Development," NBER Working Paper 9305 (Cambridge, Massachusetts: National Bureau of Economic Research).

Sachs, Jeffrey D., 2002a , "A New Global Effort to Control Malaria," Science, Vol. 298 (October), pp. 122–24.

_____, 2002b, "Resolving the Debt Crisis of Low–Income Countries," Brookings Papers on Economic Activity: 1, Brookings Institution, pp. 257–86.

_____, 2003, "Institutions Don't Rule: Direct Effects of Geography on Per Capita Income," NBER Working Paper 9490 (Cambridge, Massachusetts: National Bureau of Economic Research).

_____ and Pia Malaney, 2002, "The Economic and Social Burden of Malaria," Nature Insight, Vol. 415 (February), pp. 680–85.

United Nations Development Program, 2003, Human Development Report (New York), forthcoming.

Jeffrey D. Sachs is Director of the Earth Institute at Columbia University and is a special adviser to the UN secretary-general on the UN Millennium Development Goals.

Development as Poison

Rethinking the Western Model of Modernity

STEPHEN A. MARGLIN

A*t the beginning of* Annie Hall, *Woody Allen tells a story about two women returning from a vacation in New York's Catskill Mountains. They meet a friend and immediately start complaining: "The food was terrible," the first woman says, "I think they were trying to poison us." The second adds, "Yes, and the portions were so small." That is my take on development: the portions are small, and they are poisonous. This is not to make light of the very* real gains that have come with development. In the past three decades, infant and child mortality have fallen by 66 percent in Indonesia and Peru, by 75 percent in Iran and Turkey, and by 80 percent in Arab oil-producing states. In most parts of the world, children not only have a greater probability of surviving into adulthood, they also have more to eat than their parents did—not to mention better access to schools and doctors and a prospect of work lives of considerably less drudgery.

Nonetheless, for those most in need, the portions are indeed small. Malnutrition and hunger persist alongside the tremendous riches that have come with development and globalization. In South Asia almost a quarter of the population is undernourished and in sub-Saharan Africa, more than a third. The outrage of anti-globalization protestors in Seattle, Genoa, Washington, and Prague was directed against the meagerness of the portions, and rightly so.

But more disturbing than the meagerness of development's portions is its deadliness. Whereas other critics highlight the distributional issues that compromise development, my emphasis is rather on the terms of the project itself, which involve the destruction of indigenous cultures and communities. This result is more than a side-effect of development; it is central to the underlying values and assumptions of the entire Western development enterprise.

The White Man's Burden

Along with the technologies of production, healthcare, and education, development has spread the culture of the modern West all over the world, and thereby undermined other ways of seeing, understanding, and being. By culture I mean something more than artistic sensibility or intellectual refinement. "Culture" is used here the way anthropologists understand the term, to mean the totality of patterns of behavior and belief that characterize a specific society. Outside the modern West, culture is sustained through community, the set of connections that bind people to one another economically, socially, politically, and spiritually. Traditional communities are not simply about shared spaces, but about shared participation and experience in producing and exchanging goods and services, in governing, entertaining and mourning, and in the physical, moral, and spiritual life of the community. The culture of the modern West, which values the market as the primary organizing principle of life, undermines these traditional communities just as it has undermined community in the West itself over the last 400 years.

The West thinks it does the world a favor by exporting its culture along with the technologies that the non-Western world wants and needs. This is not a recent idea. A century ago, Rudyard Kipling, the poet laureate of British imperialism, captured this sentiment in the phrase "White Man's burden," which portrayed imperialism as an altruistic effort to bring the benefits of Western rule to uncivilized peoples. Political imperialism died in the wake of World War II, but cultural imperialism is still alive and well. Neither practitioners nor theorists speak today of the white man's burden—no development expert of the 21st century hankers after clubs or golf courses that exclude local folk from membership. Expatriate development experts now work with local people, but their collaborators are themselves formed for the most part by Western culture and values and have more in common with the West than they do with their own people. Foreign advisers—along with their local collaborators—are still missionaries, missionaries for progress as the West defines the term. As our forbears saw imperialism, so we see development.

There are in fact two views of development and its relationship to culture, as seen from the vantage point of the modern West. In one, culture is only a thin veneer over a common, universal behavior based on rational calculation and maximization of individual self interest. On this view, which is probably the view of most economists, the Indian subsistence-oriented peasant is no less calculating, no less competitive, than the US commercial farmer.

> Cultural imperialism is still alive and well. ... Foreign advisers ... are still missionaries, missionaries for progress as the West defines the term. As our forebears saw imperialism, so we see development.

There is a second approach which, far from minimizing cultural differences, emphasizes them. Cultures, implicitly or explicitly, are ranked along with income and wealth on a linear scale. As the West is richer, Western culture is more progressive, more developed. Indeed, the process of development is seen as the transformation of backward, traditional, cultural practices into modern practice, the practice of the West, the better to facilitate the growth of production and income.

What these two views share is confidence in the cultural superiority of the modern West. The first, in the guise of denying culture, attributes to other cultures Western values and practices. The second, in the guise of affirming culture, posits an inclined plane of history (to use a favorite phrase of the Indian political psychologist Ashis Nandy) along which the rest of the world is, and ought to be, struggling to catch up with us. Both agree on the need for "development." In the first view, the Other is a miniature adult, and development means the tender nurturing by the market to form the miniature Indian or African into a full-size Westerner. In the second, the Other is a child who needs structural transformation and cultural improvement to become an adult.

Both conceptions of development make sense in the context of individual people precisely because there is an agreed-upon standard of adult behavior against which progress can be measured. Or at least there was until two decades ago when the psychologist Carol Gilligan challenged the conventional wisdom of a single standard of individual development. Gilligan's book *In A Different Voice* argued that the prevailing standards of personal development were male standards. According to these standards, personal development was measured by progress from intuitive, inarticulate, cooperative, contextual, and personal modes of behavior toward rational, principled, competitive, universal, and impersonal modes of behavior, that is, from "weak" modes generally regarded as feminine and based on experience to "strong" modes regarded as masculine and based on algorithm.

Drawing from Gilligan's study, it becomes clear that on an international level, the development of nation-states is seen the same way. What appear to be universally agreed upon guidelines to which developing societies must conform are actually impositions of Western standards through cultural imperialism. Gilligan did for the study of personal development what must be done for economic development: allowing for difference. Just as the development of individuals should be seen as the flowering of that which is special and unique within each of us—a process by which an acorn becomes an oak rather than being obliged to become a maple—so the development of peoples should be conceived as the flowering of what is special and unique within each culture. This is not to argue for a cultural relativism in which all beliefs and practices sanctioned by some culture are equally valid on a moral, aesthetic, or practical plane. But it is to reject the universality claimed by Western beliefs and practices.

Of course, some might ask what the loss of a culture here or there matters if it is the price of material progress, but there are two flaws to this argument. First, cultural destruction is not necessarily a corollary of the technologies that extend life and improve its quality. Western technology can be decoupled from the entailments of Western culture. Second, if I am wrong about this, I would ask, as Jesus does in the account of Saint Mark, "[W]hat shall it profit a man, if he shall gain the whole world, and lose his own soul?" For all the material progress that the West has achieved, it has paid a high price through the weakening to the breaking point of communal ties. We in the West have much to learn, and the cultures that are being destroyed in the name of progress are perhaps the best resource we have for restoring balance to our own lives. The advantage of taking a critical stance with respect to our own culture is that we become more ready to enter into a genuine dialogue with other ways of being and believing.

The Culture of the Modern West

Culture is in the last analysis a set of assumptions, often unconsciously held, about people and how they relate to one another. The assumptions of modern Western culture can be described under five headings: individualism, self interest, the privileging of "rationality," unlimited wants, and the rise of the moral and legal claims of the nation-state on the individual.

Individualism is the notion that society can and should be understood as a collection of autonomous individuals, that groups—with the exception of the nation-state—have no normative significance as groups; that all behavior, policy, and even ethical judgment should be reduced to their effects on individuals. All individuals play the game of life on equal terms, even if they start with different amounts of physical strength, in-

INSURANCE

Spending on Insurance Premiums

Region	Percent of Global Premium Market
NORTH AMERICA	**37.32**
Canada	1.91
United States	35.41
LATIN AMERICA	**1.67**
Brazil	0.51
Mexico	0.4
EUROPE	**31.93**
France	4.99
Germany	5.06
UK	9.7
ASIA	**26.46**
China	0.79
India	0.41
Japan	20.62
AFRICA	**1.03**
South Africa	0.87
OCEANIA	**1.59**
Australia	1.46

http://www.internationalinsurance.org

the experiential, elevating knowledge that can be logically deduced from what are regarded as self-evident first principles over what is learned from intuition and authority, from touch and feel. In the stronger form of this ideology, the algorithmic is not only privileged but recognized as the sole legitimate form of knowledge. Other knowledge is mere belief, becoming legitimate only when verified by algorithmic methods.

Fourth is unlimited wants. It is human nature that we always want more than we have and that there is, consequently, never enough. The possibilities of abundance are always one step beyond our reach. Despite the enormous growth in production and consumption, we are as much in thrall to the economy as our parents, grandparents, and great-grandparents. Most US families find one income inadequate for their needs, not only at the bottom of the distribution—where falling real wages have eroded the standard of living over the past 25 years—but also in the middle and upper ranges of the distribution. Economics, which encapsulates in stark form the assumptions of the modern West, is frequently defined as the study of the allocation of limited resources among unlimited wants.

Finally, the assumption of modern Western culture is that the nation-state is the pre-eminent social grouping and moral authority. Worn out by fratricidal wars of religion, early modern Europe moved firmly in the direction of making one's relationship to God a private matter—a taste or preference among many. Language, shared commitments, and a defined territory would, it was hoped, be a less divisive basis for social identity than religion had proven to be.

An Economical Society

Each of these dimensions of modern Western culture is in tension with its opposite. Organic or holistic conceptions of society exist side by side with individualism. Altruism and fairness are opposed to self interest. Experiential knowledge exists, whether we recognize it or not, alongside algorithmic knowledge. Measuring who we are by what we have has been continually resisted by the small voice within that calls us to be our better selves. The modern nation-state claims, but does not receive, unconditional loyalty.

So the sway of modern Western culture is partial and incomplete even within the geographical boundaries of the West. And a good thing too, since no society organized on the principles outlined above could last five minutes, much less the 400 years that modernity has been in the ascendant. But make no mistake—modernity is the dominant culture in the West and increasingly so throughout the world. One has only to examine the assumptions that underlie contemporary economic thought—both stated and unstated—to confirm this assessment. Economics is simply the formalization of the assumptions of modern Western culture. That both teachers and students of economics accept these assumptions uncritically speaks volumes about the extent to which they hold sway.

It is not surprising then that a culture characterized in this way is a culture in which the market is the organizing principle of social life. Note my choice of words, "the market" and "so-

tellectual capacity, or capital assets. The playing field is level even if the players are not equal. These individuals are taken as given in many important ways rather than as works in progress. For example, preferences are accepted as given and cover everything from views about the relative merits of different flavors of ice cream to views about the relative merits of prostitution, casual sex, sex among friends, and sex within committed relationships. In an excess of democratic zeal, the children of the 20th century have extended the notion of radical subjectivism to the whole domain of preferences: one set of "preferences" is as good as another.

Self-interest is the idea that individuals make choices to further their own benefit. There is no room here for duty, right, or obligation, and that is a good thing, too. Adam Smith's best remembered contribution to economics, for better or worse, is the idea of a harmony that emerges from the pursuit of self-interest. It should be noted that while individualism is a prior condition for self-interest—there is no place for self-interest without the self—the converse does not hold. Individualism does not necessarily imply self-interest.

The third assumption is that one kind of knowledge is superior to others. The modern West privileges the algorithmic over

23

cial life," not markets and economic life. Markets have been with us since time out of mind, but the market, the idea of markets as a system for organizing production and exchange, is a distinctly modern invention, which grew in tandem with the cultural assumption of the self-interested, algorithmic individual who pursues wants without limit, an individual who owes allegiance only to the nation-state.

There is no sense in trying to resolve the chicken-egg problem of which came first. Suffice it to say that we can hardly have the market without the assumptions that justify a market system—and the market system can function acceptably only when the assumptions of the modern West are widely shared. Conversely, once these assumptions are prevalent, markets appear to be a "natural" way to organize life.

Markets and Communities

If people and society were as the culture of the modern West assumes, then market and community would occupy separate ideological spaces, and would co-exist or not as people chose. However, contrary to the assumptions of individualism, the individual does not encounter society as a fully formed human being. We are constantly being shaped by our experiences, and in a society organized in terms of markets, we are formed by our experiences in the market. Markets organize not only the production and distribution of things; they also organize the production of people.

The rise of the market system is thus bound up with the loss of community. Economists do not deny this, but rather put a market friendly spin on the destruction of community: impersonal markets accomplish more efficiently what the connections of social solidarity, reciprocity, and other redistributive institutions do in the absence of markets. Take fire insurance, for example. I pay a premium of, say, US$200 per year, and if my barn burns down, the insurance company pays me US$60,000 to rebuild it. A simple market transaction replaces the more cumbersome method of gathering my neighbors for a barn-raising, as rural US communities used to do. For the economist, it is a virtue that the more efficient institution drives out the less efficient. In terms of building barns with a minimal expenditure of resources, insurance may indeed be more efficient than gathering the community each time somebody's barn burns down. But in terms of maintaining the community, insurance is woefully lacking. Barn-raisings foster mutual interdependence: I rely on my neighbors economically—as well as in other ways—and they rely on me. Markets substitute impersonal relationships mediated by goods and services for the personal relationships of reciprocity and the like.

Why does community suffer if it is not reinforced by mutual economic dependence? Does not the relaxation of economic ties rather free up energy for other ways of connecting, as the English economist Dennis Robertson once suggested early in the 20th century? In a reflective mood toward the end of his life, Sir Dennis asked, "What does the economist economize?" His answer: "[T]hat scarce resource Love, which we know, just as well as anybody else, to be the most precious thing in the world." By using the impersonal relationships of markets to do the work of fulfilling our material needs, we economize on our higher faculties of affection, our capacity for reciprocity and personal obligation—love, in Robertsonian shorthand—which can then be devoted to higher ends.

In the end, his protests to the contrary notwithstanding, Sir Dennis knew more about banking than about love. Robertson made the mistake of thinking that love, like a loaf of bread, gets used up as it is used. Not all goods are "private" goods like bread. There are also "public" or "collective" goods which are not consumed when used by one person. A lighthouse is the canonical example: my use of the light does not diminish its availability to you. Love is a *hyper* public good: it actually increases by being used and indeed may shrink to nothing if left unused for any length of time.

Economics is simply the formalization of the assumptions of modern Western culture. That both teachers and students of economics accept these assumptions uncritically speaks volumes about the extent to which they hold sway.

If love is not scarce in the way that bread is, it is not sensible to design social institutions to economize on it. On the contrary, it makes sense to design social institutions to draw out and develop the community's stock of love. It is only when we focus on barns rather than on the people raising barns that insurance appears to be a more effective way of coping with disaster than is a community-wide barn-raising. The Amish, who are descendants of 18th century immigrants to the United States, are perhaps unique in the United States for their attention to fostering community; they forbid insurance precisely because they understand that the market relationship between an individual and the insurance company undermines the mutual dependence of the individuals that forms the basis of the community. For the Amish, barn-raisings are not exercises in nostalgia, but the cement which holds the community together.

Indeed, community cannot be viewed as just another good subject to the dynamics of market supply and demand that people can choose or not as they please, according to the same market test that applies to brands of soda or flavors of ice cream. Rather, the maintenance of community must be a collective responsibility for two reasons. The first is the so-called "free rider" problem. To return to the insurance example, my decision to purchase fire insurance rather than participate in the give and take of barn raising with my neighbors has the side effect—the "externality" in economics jargon—of lessening my involvement with the community. If I am the only one to act this way, this effect may be small with no harm done. But when all of us opt for insurance and leave caring for the community to others, there will be no others to care, and the community will disintegrate. In the case of insurance, I buy insurance because it is more convenient, and—acting in isolation—I can reasonably say to myself that my action hardly undermines the community. But when we all do so, the cement of mutual obligation is weakened to the point that it no longer supports the community.

The free rider problem is well understood by economists, and the assumption that such problems are absent is part of the standard fine print in the warranty that economists provide for the market. A second, deeper, problem cannot so easily be translated into the language of economics. The market creates more subtle externalities that include effects on beliefs, values, and behaviors—a class of externalities which are ignored in the standard framework of economics in which individual "preferences" are assumed to be unchanging. An Amishman's decision to insure his barn undermines the mutual dependence of the Amish not only by making him less dependent on the community, but also by subverting the beliefs that sustain this dependence. For once interdependence is undermined, the community is no longer valued; the process of undermining interdependence is self-validating.

Thus, the existence of such externalities means that community survival cannot be left to the spontaneous initiatives of its members acting in accord with the individual maximizing model. Furthermore, this problem is magnified when the externalities involve feedback from actions to values, beliefs, and then to behavior. If a community is to survive, it must structure the interactions of its members to strengthen ways of being and knowing which support community. It will have to constrain the market when the market undermines community.

A Different Development

There are two lessons here. The first is that there should be mechanisms for local communities to decide, as the Amish routinely do, which innovations in organization and technology are compatible with the core values the community wishes to preserve. This does not mean the blind preservation of whatever has been sanctioned by time and the existing distribution of power. Nor does it mean an idyllic, conflict-free path to the future. But recognizing the value as well as the fragility of community would be a giant step forward in giving people a real opportunity to make their portions less meager and avoiding the poison.

The second lesson is for practitioners and theorists of development. What many Westerners see simply as liberating people from superstition, ignorance, and the oppression of tradition, is fostering values, behaviors, and beliefs that are highly problematic for our own culture. Only arrogance and a supreme failure of the imagination cause us to see them as universal rather than as the product of a particular history. Again, this is not to argue that "anything goes." It is instead a call for sensitivity, for entering into a dialogue that involves listening instead of dictating—not so that we can better implement our own agenda, but so that we can genuinely learn that which modernity has made us forget.

STEPHEN A. MARGLIN is Walter S. Barker Professor of Economics at Harvard University.

SELLING to the POOR

Searching for new customers eager to buy your products? Forget Tokyo's schoolgirls and Milan's fashionistas. Instead, try the world's 4 billion poor people, the largest untapped consumer market on Earth. To reach them, CEOs must shed old concepts of marketing, distribution, and research. Getting it right can both generate big profits and help end economic isolation throughout the developing world.

By Allen L. Hammond and C.K. Prahalad

When the Indian industrial and technology conglomerate ITC started building a network of Internet-connected computers called "e-Choupals" in farming villages in India's rural state of Madhya Pradesh in 2001, soy farmers were suddenly able to check fair market prices for their crops. Some farmers began tracking soy futures on the Chicago Board of Trade, and soon most of them were bypassing local auction markets and selling their crops directly to ITC for about $6 more per ton than they previously received. The same ITC network enables farmers to buy seeds, fertilizers, and other materials directly, at considerable savings, as well as to purchase formerly unavailable soil-testing services. Today, the growing e-Choupal network reaches 1.8 million farmers, and ITC is receiving demands from rural farmers for new products and services—the beginnings of consumer market power at the poorest level of Indian society.

The ITC network is one example of how access to information can increase productivity and raise incomes. It also reveals what happens when large businesses stop regarding the world's 4 billion poor people as victims and start eyeing them as consumers. For decades, corporate executives at the world's largest companies—and their counterparts running wealthy governments—have thought of poor people as powerless and desperately in need of handouts. But turning the poor into customers and consumers is a far more effective way of reducing poverty.

Why hasn't the business world caught on? The explanations are well known: Infrastructure in the developing world is often poor or nonexistent, creating the need for substantial upfront investment. Illiteracy tends to be high, requiring nontraditional marketing approaches. Tribal, racial, and religious tensions, as well as rampant crime, complicate hiring and business operations. Governments—especially local and provincial authorities—often do not function effectively or transparently. Corruption is widespread.

Yet many multinational companies already overcome such problems to serve middle-class customers in developing countries. The fundamental barriers to serving poor customers in low-income nations exist within companies and governments in rich nations, where leaders have uncritically accepted the myth that the poor have no money. In reality, low-income households collectively possess most of the buying power in many developing countries, including such emerging economies as China and India. If businesses ignore the bottom of the economic pyramid, they miss most of the market. Another myth is that the poor resist new products and services, when in truth poor consumers are rarely offered products designed for their lifestyles and circumstances, leaving them unable to interact with the global economy. Perhaps the greatest misperception of all is that selling to the poor is not profitable or, worse yet, exploitative. Selling to the world's poorest people can be very lucrative and a key source of growth for global companies, even while this interaction benefits and empowers poor consumers.

The market for goods and services among the world's poor—families with an annual household income of less than $6,000—is enormous. The 18 largest emerging and transition countries include 680 million such households, with a total annual income of $1.7 trillion—roughly equal to Germany's annual gross domestic product. Brazil's poorest citizens comprise nearly 25 million households with a total annual income of $73 billion. India has 171 million poor households with a combined $378 billion in income. China's poor residents account for 286 million households with a combined annual income of $691 billion. Surveys show that poor households spend most of their income on housing, food, healthcare, education, finance charges, communications, and consumer goods. Multinational corporations have largely failed to tap this market, even though the rewards for doing so could be substantial.

In poor countries, the distribution of households by income level is heavily skewed toward the bottom rungs of the economy. With the bulk of the population—and buying power—residing in the low-income segments of poor nations, smart companies need to start concentrating their efforts there, where demand is high and competition is sparse. Governments, too, should take note. Poor people are asking why they should not share the benefits of globalization, and there is growing awareness that traditional development solutions have not worked. The private sector can and must do better.

BUSINESS SCHOOL BASICS

Markets in the developing world can nurture global business through their sheer size, rate of growth, and consumer demands. Consider three examples: cell phones, table salt, and cosmetics.

Cellular technology was originally developed as a luxury for the rich, but today poor countries drive the explosion in wireless communications. Sub-Saharan Africa is now a leading region in percentage growth of cell phone usage, expanding 37 percent during 2003. India boasts 22 million cellular customers and is adding around 1.5 million new customers every month. By 2005, China, India, and Brazil will have a combined 500 million cell phone users, compared to 150 million in the United States. The sheer size of these markets will necessarily change the dynamics of the business—shifting to the poor the power to determine both the preferred features of cell phones and their technological makeup. The pacesetting customers will no longer be found in Tokyo and Rome, but rather in Xian and Bangalore.

The cellular industry proves that if companies wish to engage poorer markets, they must shed traditional business models developed with wealthy consumers in mind. Prepaid phone cards are now the dominant business model for the cell phone market worldwide. Such cards crush the perception that business with the poor is risky; prepaid cards eliminate phone companies' collection costs and debt, and firms are paid before they connect a call. Yet even with prepaid cards, some companies initially misjudged the nature and depth of the market. In Venezuela in 1995, for example, U.S.-based BellSouth In-

Fortune Sellers

Developing nations offer multinational corporations a vast, untapped market. But consumers, even poor ones, often associate particular brands with unsavory business practices. Increasingly, corporations seek moral and ethical legitimacy—and try to avoid charges of exploitation—when marketing their goods and services to the poor. "Customers are searching for organizations that they can trust," explains Raoul Pinnell, vice president of global brands and communications at Shell International.

Multilateral initiatives aimed at improving corporate behavior and boosting consumer acceptance, such as the United Nations Global Compact or the Caux Round Table's Principles for Business, have met with mixed results. Alternatively, some corporations that work in poor markets are undertaking innovative and transparent self-regulation projects, sometimes with assistance from governments in the developing world. In 1993, when Avon sent an army of direct-selling cosmetics merchants paddling up Amazon tributaries to sell perfumes and makeup to miners and prostitutes, Avon ladies answered only to themselves. Today, direct sellers arriving to "Avon-gelize" remote Indian villages are subject to sanctioning by an Indian judge— effectively a corporate ombudsman—if they break a code of ethics formulated by the Indian Direct Selling Association. (The code aims to weed out "fraudulent elements" among direct sellers and prevent pyramid schemes.) "It's protection for the company, the consumer, and the sales force," says David Gosling, Avon India's managing director.

Similarly, the Switzerland-based Nestlé Group, a longtime target of human rights advocates for their marketing of baby formula over mother's milk in poor countries, appointed an ombudsman in 2002 to expose any unethical promotional activities. Coca-Cola India is attempting to, in their words, regain public trust and credibility after allegations of pesticide contamination incited angry consumers in Bombay to smash thousands of Coke bottles in 2003. The company recently formed an advisory board— led by former Indian Cabinet Secretary Naresh Chandra—that will oversee Coca-Cola's practices in India. The company also appointed former chief justice of the Indian Supreme Court B.N. Kirpal to lead an advisory body called the India Environment Council, which will guide Coca-Cola India's social-responsibility practices.

ternational started selling $10 and $20 phone cards, largely aimed at the middle class. Today the company sells enormously popular $4 phone cards at more than

30,000 retail outlets, reaching even Venezuela's poorest citizens and, because of the lower unit price, reaching a far larger market. By forcing corporations to rethink costs, business models, and industry standards, poor consumers are initiating a revolution in cellular communications.

Selling to poor consumers also requires innovative research and development. In rural India, for example, only four out of 10 households use iodized table salt, even though iodized salt provides a critical and convenient nutritional supplement. Due to India's environmental conditions, much of the iodine in salt is lost during transport and storage. The remainder often disappears in the Indian cooking process. To overcome this problem, Hindustan Lever Ltd., a subsidiary of Europe's Unilever Corp., has developed a way to encapsulate iodine, protecting it from transportation, storage, and cooking, and releasing the iodine only when salted food is ingested. The new salt required Hindustan Lever to invest in two years of advanced research and development, but if its salt sells successfully, the company could sharply reduce iodine deficiency disorder, a disease that affects more than 70 million people in India and is the country's leading cause of mental retardation. The lesson: Successful product development requires a deep understanding of local circumstances, so that critical features and functionality—salt with protected iodine—can be incorporated into the product's design.

Modernizing distribution channels is also crucial for companies hoping to reach low-income markets in the developing world. "Person-to-person" cosmetic giants Amway Corp. and Avon Products, Inc. use direct-distribution strategies in India and Brazil, respectively, to sell beauty products among a wider circle of customers—increasing the corporations' reach and employing poor people as entrepreneurs. Amway, for example, has enlisted around 600,000 self-employed individual distributors in India. Hindustan Lever is mimicking the approach with a direct-distribution system for personal-care products. The company expects to sign on more than 500,000 self-employed Indian distributors within five years.

Similar transformations in business models, research and development, and product distribution are underway or imminent in healthcare, education, finance, agriculture, building materials, and other goods and services. But these changes will only lead to a true business revolution if corporate perceptions regarding the world's poor shift dramatically. Managers in multinational corporations are conditioned to think mainly of rich consumers. They are prisoners of their own logic. Poor consumers challenge virtually every preconception parroted by business schools and marketing seminars. Yet, thanks in part to role models such as Brazilian entrepreneur Samuel Klein, the number of firms doing well by doing good is growing. Klein started Casas Bahia, a successful retail chain, when he fled Europe's Holocaust and started selling inexpensive linens and blankets to poor Brazilians. Klein learned quickly that the poor are willing to pay but they are often unable to afford lump sums for purchases. Allowing customers to pay in installments was the obvious solution. What started as a one-man blanket operation has grown into a business with more than $2 billion in sales last year. Casas Bahia employs more than 22,000 people, operates over 350 stores with 10 million customers, and the company's credit system has one of the lowest default rates in Brazil.

POWER TO THE POOR

When multinational corporations attempt to penetrate new markets in the developing world, critics sometimes condemn them for preaching the gospel of consumer culture to the poor, for exploiting the poor as cheap labor, and for extracting and despoiling natural resources without fairly compensating locals. In truth, some multinationals have been guilty on all these counts. But the private sector may do more harm by ignoring poor consumers than by engaging them. After all, if the poor can't participate in global markets, they can't benefit from them either.

Poor families benefit in several ways when large companies target them as consumers. Access to new products, expanded choices, and increased purchasing power improves one's quality of life. New services and information that improve efficiency help increase productivity and raise incomes among poor citizens. Processes that are fair to the consumer and treat poor customers with respect—as when ITC uses electronic scales that give accurate weights for grain and offers a farmer a chair to sit in while the sale is completed—builds loyalty and trust in the company and in the global economic system. And the exercising of collective consumer market power forces attention to the needs of poor people.

When a retail chain in Mexico started selling chicken parts instead of whole chickens in its outlets a few years ago, sales quadrupled. Smaller unit packages—enough for a single, immediate use—enable poor consumers to buy a product that they otherwise could not afford, thus unlocking their purchasing power. The same principle applies to personal-care products. In India, Hindustan Lever, Procter & Gamble, and most of their competitors make "single-serving" versions of their products, from detergents to shampoo. More than 60 percent of the value of the shampoo market and 95 percent of all shampoo units sold in India are now single-serve. Many are designed explicitly for the poor and do not even require hot water. Because of these efforts, nearly all Indians now enjoy access to shampoo. Companies selling small unit sizes at affordable prices make money, expand markets, and generate broader access to goods and services that improve people's quality of life.

Nowhere are the benefits of access to new services more evident than in banking and the Internet. Prodem FFP, a Bolivian financial organization that targets low-

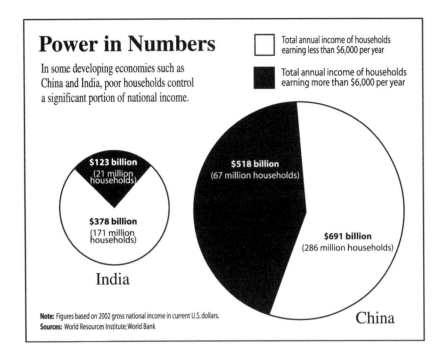

Power in Numbers

In some developing economies such as China and India, poor households control a significant portion of national income.

☐ Total annual income of households earning less than $6,000 per year

■ Total annual income of households earning more than $6,000 per year

$123 billion (21 million households)

$378 billion (171 million households)

India

$518 billion (67 million households)

$691 billion (286 million households)

China

Note: Figures based on 2002 gross national income in current U.S. dollars.
Sources: World Resources Institute; World Bank

income customers, installs automatic teller machines that recognize fingerprints, communicate via text-to-speech technology in three local dialects, and display a color-coded touch screen that illiterate customers can use. Prodem has expanded its market, and now more Bolivians have access to professional, secure banking services. On the other side of the world, in India, the wireless Internet service company n-Logue found that its customers in rural villages were slow to appreciate e-mail (many villagers do not normally communicate in writing) but quick to accept e-mail photos and video conferencing. N-Logue's customers found value in sharing a photo of a new baby with distant relatives or sending a photo of a sick cow to a government agricultural agent for quick advice. Even in traditional business sectors such as construction materials, Mexico's Cemex is expanding its market by combining a "pay as you go" system with delivery of materials and instructions as needed, enabling the poor to build better quality housing.

Beyond such benefits as higher standards of living and greater purchasing power, poor consumers find real value in dignity and choice. In part, lack of choice is what being poor is all about. In India, a young woman working as a sweeper outdoors in the hot sun recently expressed pride in being able to use a fashion product—Fair and Lovely cream, which is part sunscreen, part moisturizer, and part skin-lightener—because, she says, her hard labor will take less of a toll on her skin than it did on her parents'. She has a choice and feels empowered because of an affordable consumer product formulated for her needs. Likewise, Amul, a large Indian dairy cooperative, found an instant market in 2001 when it introduced ice cream, a luxury in tropical India, at affordable prices (2 cents per serving).

Poor people want to buy their children ice cream every bit as much as middle-class families, but before Amul targeted the poor as consumers, they lacked that option.

GLOBALIZATION'S NEW FRONTIER

In 2003, Thailand's Information and Communications Technology Minister Surapong Suebwonglee was looking for ways to extend the benefits of technology to the masses. So he challenged Thailand's computer industry to come up with a $260 personal computer and a $450 laptop. In return, Suebwonglee guaranteed a market of at least 500,000 machines. The Thai computer industry met that price. But to do so, it had to omit Microsoft's widely used (and costly) Windows and Office operating software and offer the open-source Linux operating system instead. Not wanting to be left out, Microsoft cut the price for its software to a total of $38 in Thailand, dramatically below normal retail prices. The "people's PCs" are now selling briskly (most with Linux, some with Windows) throughout Thailand. Nearly 300,000 computers were sold through early fall of 2003, with projected first-year sales of 1 million machines. In March 2004, Microsoft announced plans for a "tailored and limited" Thai-language version of its Windows XP Home software at reduced prices.

The Thai example shows that the global economy is open to both innovation and consumer market power originating in poor countries. Consumers in developing nations are increasingly willing to exercise that power, not least of all by rejecting trade or investment deals they see as unfair. Just last year in Bolivia, for example, popular discontent with the terms of foreign investment in a new pipeline to carry natural gas to global markets, including the United States, triggered protests and unrest that ultimately brought down the government of President Gonzalo Sánchez de Lozada. The president's successor, Carlos Mesa, quickly canceled the deal.

The message for the private sector is clear: Ignore poor consumers at your peril. Blocs of poor consumers increasingly have the power to reject what a multinational corporation wants to buy or sell; via their governments, they can also empower a nontraditional competitor. It may not be wise for corporations to wait for governments to smooth the path of globalization, or to depend solely on formal trade talks to make developing markets safe for their products. Businesses must learn to serve poor markets by overcoming those markets' unique constraints as

well as their own antiquated business models and misconceptions about the developing world.

Ending the economic isolation of poor populations and bringing them within the formal global economy will ensure that they also have the opportunity to benefit from globalization. That is the world's new entrepreneurial frontier.

Allen L. Hammond is vice president for innovation and director of the digital dividends project at the World Resources Institute. C.K. Prahalad is Harvey C. Fruehauf professor of business administraion at the University of Michigan Business School and author of *The Fortune at the Bottom of the Pyramid: Eradicating Poverty Through Profit* (Philadelphia: Wharton School Publishing, 2004). He is a member of the board of directors of Hindustan Lever Ltd.

THE CHALLENGE OF WORLDWIDE MIGRATION

Michael W. Doyle

In 2000, an estimated 175 million people lived outside their place of birth, more than ever before. Of these, about 158 million were deemed international migrants; approximately 16 million were recognized refugees fleeing a well-founded fear of persecution and 900,000 were asylum seekers.

They include the skilled Nigerian computer engineer working in Sweden; the agricultural worker from Guatemala working "irregularly" (without legal documentation) in the United States: the woman trafficked from Ukraine to Bosnia; and the refugee from Afghanistan now in Pakistan and about to return home.

Today, the growth of international migration and national interests both clashing and compatible call out for increased international cooperation. The international community needs to establish more widely shared norms and agreed-upon procedures in order to manage better the flow of international migrants to the benefit both of the migrants themselves and the countries of origin, transit, and destination.

MIGRATION: AN ELEMENT OF GLOBALIZATION

Globalization is a primary force that is shaping the character and impact of migration. Lower travel costs and information and communication technologies have made migration much more viable, the exchange of money and technology that is a result of migration much easier, and return or circular migration more prevalent.

Broadly speaking, the global flows of international migrants can be grouped into labor, family, and refugee categories. They may migrate voluntarily or involuntarily, and may have permanent, temporary, or no legal status.

Better opportunities abroad are the driving factor of voluntary labor migration, attracting both highly educated (e.g. medical and technical) and less-educated (farm and domestic) labor. Some migrants move with the intention of establishing permanent residence. The traditional countries of immigration admit migrants for permanent settlement and grant them the right to apply for citizenship under certain requirements. Many governments recognize the right to family reunification and often permit close family members to enter through legal channels. This has had enormous impact: Family-sponsored migrants accounted for 45 to 75 percent of all international migrants admitted to European and North American countries in 2000.

Other migrants move involuntarily. They include refugees and those with refugee-like status, including temporary protection. Refugees are forced to leave their countries owing to fear of being persecuted for reasons of race, religion, nationality, political opinion, or membership in a particular social group. In some regions, those who are outside their country of origin because of armed conflict, generalized violence, severe natural disasters, or other circumstances that have seriously disturbed the public order are also accorded refugee-like status.

The situations of documented and undocumented (irregular or "illegal") migrants greatly differ, with the former having a wide range of legal and social protections that the latter lack. An increasingly important dimension of undocumented migration is smuggling and trafficking in human beings. Smuggling involves the facilitation of illegal border crossing or residence in another country with the complicity of the migrant. Trafficking is the coercive, exploitative, and non-consensual movement of persons for profit. It does not necessarily involve a cross-border movement. Although hard to quantify, undocumented migration has risen significantly in the last 10 years, with migrant smuggling and trafficking becoming one of the most profitable branches of organized crime.

Over the longer run, the so-called "Birth Rate Crisis" will make international migration an even more significant factor in globalization. The number of persons aged 60 and above was 600 million in 2000, triple the number in 1950, but less than one-third the projected 2 billion of 2050. These changes will have profound consequences

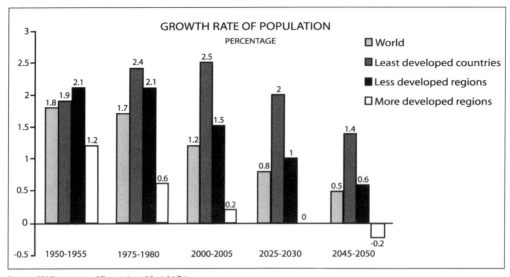

Source: UN Department of Economic and Social Affairs

and far-reaching implications for pension schemes, health-care systems, education programs, and housing plans, as well as for the economic vitality and growth of a country. Governments will have to radically overhaul migration policies, as well as other established policies and programs, including those on retirement ages, labor-force participation, levels of contributions of workers to retirement and health-care schemes, and benefits for the elderly.

The potential support ratio (the number of persons aged 15 to 64 for each person 65 or older), which measures social security and other potential burdens, decreased from 12 to nine from 1950 to 2000 and will fall to four by 2050. This will particularly impact the health-care sector, where the number of persons older than 85 is also rising steeply. One implication is that the demand for immigrant labor is very likely to rise, particularly in Western Europe and Japan.[1] (U.S., Australian, and Canadian demand, which face similar trends, are already being met so far by legal and illegal immigration.)

NATIONAL INTERESTS

National interests vary widely by type of country. For countries of origin, international migration can reduce unemployment, contribute to an increase in real wages, provide significant remittance flows (according to estimates, from $60 to $100 billion per annum worldwide, or double official development assistance), and raise standards of living. But it can also induce losses of highly skilled personnel. For countries of destination, international migration tends to have a mixed impact on the economy, particularly on the employment and salaries of non-migrants, and on social transfers. Socially, migrants contribute to cultural diversity in countries of destination, but they also can impose significant costs that include difficulties in integration that lead to welfare

dependency, and the observance of cultural practices that some native inhabitants can perceive as threats to their established way of life.

A particularly disturbing trend in recent years is an asylum crisis that is emerging in an increasing number of developed democracies in North America, Europe, and Australasia. Although the vast majority of the world's refugees are in and supported by the developing countries, in the developed countries there is an emerging loss of confidence in the asylum process. Developed countries have spent up to $ 10 billion per year on asylum, and almost two-thirds of asylum seekers have failed to qualify for asylum. Of those who fail, only a fifth return to their countries of origin. This has led to tighter border controls and, ironically, to increased smuggling and trafficking. Rather than addressing the problem, some politicians in some countries of destination have seized upon asylum as a wedge issue, dividing citizens and vilifying asylum seekers, who are portrayed as threats to cultural integrity, as rivals for welfare provisions, and as criminals. Genuine solutions are needed to ensure that asylum systems determine status quickly and fairly.

INTERNATIONAL COOPERATION

Migration has thus become a key global concern for a growing number of countries of origin, transit, and destination. This is because it touches a wide range of issues, including labor shortages, unemployment, brain drain and brain gain, worker remittances, human rights, refugee/asylum crises, social integration, xenophobia, illegal migration, human trafficking, and national security. And while migration has long been a sensitive matter of national sovereignty, it is beginning to encompass regional and international dimensions.

Increasingly, States are looking for regional and international mechanisms through which to discuss migration issues and trends. Many States want to coordinate their actions better and to develop regional and international guidelines for what is recognized as a global phenomenon with broad economic, social, and security ramifications. And many migrants are looking for someone, or some institution, to advocate for their rights and address their needs.

Unfortunately, where there are laws, organizations, and advocates for migrants and on migration, they tend to be incomplete or insufficient, and to compound the challenges of internationalization. To ensure that migration is a win-win story for States and migrants, stronger international cooperation is warranted.

Such cooperation must ensure that the basic human rights of all people, including migrants, are protected, that the burdens and responsibilities of providing assistance for refugees are shared fairly, and that the positive potential of international migration for migrants, and for sending, transit, and receiving countries, is fully realized. All countries should strive for migration regimes that are transparent, just, and responsive to both the realities of the labor market and rights to family unity.

A first step toward closing the normative gaps can be achieved by disseminating widely the international norms that already exist on migration, such as the Cairo Programme of Action and provisions of human-rights treaties, and by campaigning for broader ratification of the Migrant Workers Convention.

A longer-term, incremental strategy should be focused on issue-centered multilateral agreements. It could begin with issue-specific international conferences designed to build common understanding and reach agreement. Led by groups of Member States, these conferences could address issues such as remittances, citizenship, family reunification, and asylum. This option could replicate the World Trade Organization model by proceeding through successive rounds of negotiation toward issue-specific multilateral agreements.

International migration is lightly institutionalized within the United Nations system, with no agency working systematically across the whole spectrum of the roles listed above. No organization has the broad mandate that would allow the international community better to meet the challenges of internationalization by coordinating action, developing preventive strategies, and fostering constructive solutions. As the absence of an authoritative United Nations "voice" on migration becomes more keenly felt, the question arises as to how the Organization might develop links to non-UN organizations such as the

International Organization of Migration and foster additional forms of interagency cooperation, in order to effectively fulfill a role in migration governance and contribute to the migration debate.

The United Nations also needs to influence the migration debate with rigorous and thoughtful advocacy, and by facilitating a series of informal meetings that include civil society, labor, and the private sector, to develop a deeper and wider understanding of the new challenges and opportunities that migration poses.

Migration policy is sufficiently important and complicated that it would benefit from a comprehensive effort to explore the issue in more depth and to gauge the depth of international support for substantial initiatives. Similar challenges have been addressed by global commissions, such as the Brundtland Commission on health. A commission could help mobilize attention, assemble expertise, conduct research, and identify areas of emerging consensus and choice for the international community. Above all, such a group should focus on public education to mobilize the political attention and foster the international understanding that international migration so very much requires. Fortunately, under the leadership of former minister Jan Karlsson (of Sweden) and Dr. Mamphela Ramphele (a managing director of the World Bank, from South Africa), such a commission is now being formed.

But the most important area for international action is political leadership. Politicians can choose to embrace the potential that migrants and refugees represent, or use them as political scapegoats. To avoid the latter, they should begin by "de-mythologizing" migration, addressing negative myths and fears, and informing their publics of the benefits that well-managed migration can produce. Political leaders must share best practices on how to integrate migrants into their societies as full members, with the privileges as well as the obligations that entails Immigrants and refugees should not be portrayed simply as a burden, but as people who want a safer, more prosperous future for their children, and who are willing to work for it.

---

1. The numbers of migrants needed to offset labor shortages are comparable to those in recent past experience for some developed countries, but much higher for others. The levels of migration needed to offset population aging and maintain current ratios of working to non-working residents are extremely high, and entail vastly more immigration than has occurred in the past.

An earlier version of this article appeared as "Chez soi a 1'etranger" in *Le Suisse et le Monde* 3 (2003), 10–11, published by the Federal Department of Foreign Affairs of Switzerland.

UNIT 2

Political Economy and the Developing World

Unit Selections

Key Points to Consider

- What are some misconceptions about international trade?
- What caused the collapse of the 2003 Cancun trade talks? What is the effect of this collapse likely to be?
- Has NAFTA had positive or negative results?
- What have been the effects of neoliberalism?
- Are the IMF and World Bank unaccountable? How does the IMF answer its critics?
- What industrialized countries are the most committed to development in poor countries?
- Why should aid to Africa be increased?

Student Website

www.mhcls.com/online

Internet References

Further information regarding these websites may be found in this book's preface or online.

Center for Third World Organizing
http://www.ctwo.org/

ENTERWeb
http://www.enterweb.org

International Monetary Fund (IMF)
http://www.imf.org

TWN (Third World Network)
http://www.twnside.org.sg/

U.S. Agency for International Development (USAID)
http://www.info.usaid.gov

The World Bank
http://www.worldbank.org

Economic issues are among the developing world's most pressing concerns. Economic growth and stability are essential to progress on the variety of problems confronting developing countries. The developing world's position in the international economic system contributes to the difficulty of achieving consistent economic growth. From their incorporation into the international economic system during colonialism to the present, the majority of developing countries have been primarily suppliers of raw materials, agricultural products, and inexpensive labor. Dependence on commodity exports has meant that developing countries have had to deal with fluctuating, and frequently declining, prices for their exports. At the same time, prices for imports have remained constant or have increased. At best, this decline in the terms of trade has made development planning difficult; at worst, it has led to economic stagnation and decline.

With some exceptions, developing nations have had limited success in breaking out of this dilemma by diversifying their economies. Efforts at industrialization and export of light-manufactured goods have led to competition with less efficient industries in the industrialized world. The response of industrialized countries has often been protectionism and demands for trade reciprocity, which can overwhelm markets in developing countries. Although the World Trade Organization (WTO) was established to standardize trade regulations and increase international trade, critics charge that the WTO continues to disadvantage developing countries and remains dominated by the wealthy industrial countries. The developing world also asserts that they are often shut out of trade negotiations, must accept deals dictated by the wealthy countries, and that they lack sufficient resources to effectively participate in the wide range of forums and negotiations that take place around the world. Moreover, developing countries charge that the industrialized countries are selective in their efforts to dismantle trade barriers and emphasize trade issues that reflect only their interests. Delegates from poor countries walked out of the 2003 WTO ministerial meeting in Cancún, Mexico to protest rich countries' reluctance to eliminate agricultural subsidies and their efforts to dominate the agenda.

A successful conclusion to the Doha round rests on the willingness of both industrialized and developing countries to make concessions but prospects for this do not look promising. The economic situation in the developing world, however, is not entirely attributable to colonial legacy and protectionism on the part of industrialized countries. Developing countries have sometimes constructed their own trade barriers and have relied on preferential trade relationships. Industrialization schemes involving heavy government direction were often ill-conceived or resulted in corruption and mismanagement. Industrialized countries frequently point to these inefficiencies in calling for market-oriented reforms, but the emphasis on privatization does not adequately recognize the role of the state in developing countries' economies and privatization may result in foreign control of important sectors of the economy as well as a loss of jobs.

Debt has further compounded economic problems for many developing countries. During the 1970s, developing countries' prior economic performance and the availability of petro dollars encouraged extensive commercial lending. Developing countries sought these loans to fill the gap between revenues from exports and foreign aid, and development expenditures. The second oil price hike in the late 1970s, declining export earnings, and worldwide recession in the early 1980s left many developing countries unable to meet their debt obligations. The commercial banks weathered the crisis, and some actually showed a profit. Commercial lending declined in the aftermath of the debt crisis and international financial institutions became the lenders of last resort for many developing countries. Access to World Bank and International Monetary Fund financing became conditional on the adoption of structural adjustment programs that involved steps such as reduced public expenditures, devaluation of currencies, and export promotion, all geared to debt reduction. The consequences of these programs have been painful for developing countries. Declining public services, higher prices, and greater reliance on the exploitation of resources have resulted.

The poorest countries continue to struggle with heavy debt burdens, and the IMF and World Bank have come under increasing criticism for their programs in the developing world. Though these institutions have made some efforts to shift the emphasis to poverty reduction, some critics charge that the reforms are superficial and that the international financial institutions lack accountability. There is renewed interest in eliminating the debt of the world's poorest countries and studies suggest that the benefits of debt reduction could be substantial—especially if repayments are redirected toward reducing poverty.

Globalization has complicated international economic circumstances and views differ regarding benefits and costs of this trend for the developing world. Advocates claim that closer economic integration, especially trade and financial liberalization, increases economic prosperity in developing countries and encourages good governance, transparency, and accountability. Critics respond that globalization's requirements, as determined by the powerful nations and reinforced through the international financial institutions, impose difficult and perhaps counterproductive policies on struggling economies. They also charge that globalization undermines workers' rights and encourages environmental degradation. Moreover, most of the benefits of globalization have gone to those countries that were already growing—leaving the poorest even further behind.

In part, because poverty in the developing world contributes to the despair and resentment that leads some to terrorism, there has been increased attention focused on foreign aid. Although the United States has recently committed to increasing foreign aid, a recently developed Commitment to Development Index shows that the smaller industrialized countries actually provide more aid to the developing world—suggesting that the G-7 countries could do more to help the poor. While aid has often been criticized, it does produce benefits. Those benefits, however, could be enhanced by a more effective aid bureaucracy and enhancing local capacity-building. Microfinancing continues to be a promising strategy to alleviate poverty. Some of this financing goes toward the purchase of cell phones that evidence shows have a major economic impact and helps bridge the digital divide. In any event, more effective ways to promote economic growth and stability are essential to progress in the developing world.

INTERNATIONAL TRADE

Why have disagreements between rich and poor nations stalled the global trading system? Because vapid debates over "fair trade" obscure some inconvenient facts: First, notwithstanding their demands for equity, poor countries are more protectionist than advanced economies. Second, if rich nations cut their self-defeating agricultural subsidies, their own publics would benefit, but consumers in many poor countries would not. Finally, despite criticisms to the contrary, the WTO can help promote economic development in low-income countries—but only if rich nations let the global body do its job.

By Arvind Panagariya

"Economies That Are Open to Trade Grow Faster"

True. In low-income countries, openness to international trade is indispensable for rapid economic growth. Indeed, few developing nations have grown rapidly over time without simultaneous increases in both exports and imports, and virtually all developing countries that have grown rapidly have done so under open trade policies or declining trade protection. India and China are the best recent examples of countries that started with relatively closed trade policy regimes in the 1980s but subsequently achieved accelerating growth while opening up their economies. From the mid-1950s through the mid-1970s, industrial countries also enjoyed rapid growth while dismantling their high post-World War II trade barriers and embracing new technologies. Japan offers the most dramatic example, but countries such as Denmark, France, Greece, Italy, the Netherlands, Norway, and Portugal exhibited similar patterns.

Openness to trade promotes growth in a variety of ways. Entrepreneurs are forced to become increasingly efficient since they must compete against the best in the world to survive. Openness also affords access to the best technology and allows countries to specialize in what they do best rather than produce everything on their own. The fall of the Soviet Union was in no small measure due to its failure to access cutting-edge technologies, compete against world-class producers, and specialize in production. Even as large an economy as the United States today specializes heavily in services, which account for 80 percent of total U.S. output.

Of course, openness to trade is not by itself sufficient to promote growth—macroeconomic and political stability and other policies are needed as well—so some countries have opened up their markets and still not seen commensurate increases in economic growth. That has been particularly true of African countries such as the Ivory Coast during the 1980s and 1990s. But such instances hardly disprove the benefits of openness. Economists do not understand the process of growth well enough to predict precisely when the opportunity will knock on a country's door. But when it does knock, an open economy is more likely to seize it, whereas a closed one will miss it. Even globalization skeptics such as economists Dani Rodrik and Joseph Stiglitz recognize this point; neither chooses trade protection over freer trade.

"Rich Countries Are More Protectionist Than Poor Ones"

Not even close. On average, poor countries have higher tariff barriers than high-income countries. For instance, rich nations' tariffs on industrial products average about 3 percent, compared to 13 percent for poor countries. Even in the textiles and clothing sectors, tariffs in developing nations (21 percent) are more than double those in rich countries (8 percent, on average). And while textiles and clothing are subject to import quotas in rich economies, such restrictions are due to be dismantled entirely by January 1, 2005, under existing World Trade Organization (WTO) agreements.

Of course, not all poor countries are equally protectionist; some are even more open to trade than rich nations. For many years now, Singapore and Hong Kong have been textbook cases of free-trading nations. Likewise, middle-income economies such as South Korea and Taiwan are not significantly more protectionist than developed countries. But overall, the countries that stand to benefit most from greater competition and openness are those nations that display the highest protection, including most countries in South Asia and some in Africa.

The highest tariffs—or "tariff peaks"—in rich countries apply with particular strength to labor-intensive products exported by developing countries. In Canada, the United States, the European Union (EU), and Japan, product categories with especially high tariff rates include textiles and clothing as well as leather, rubber, footwear, and travel goods. But developing countries themselves are often quite zealous in protecting their markets from goods exported by other poor nations. Labor-intensive products such as textiles, clothing, leather, and footwear, which developing countries also export to each other, attract high duties in countries such as Brazil, Mexico, China, India, Malaysia, and Thailand.

Traditionally, rich economies such as the United States and the EU have been quick to engage in antidumping initiatives—erecting trade barriers against countries that allegedly export goods (or "dump" them) at a price below their own cost of production, however difficult it may be to quantify such a charge. But developing countries have been learning the same tricks and initiating antidumping measures of their own, and now the number of such actions has converged between advanced and poor economies. For example, according to the "WTO Annual Report 2003," India now ranks first in the world in initiating new antidumping actions, and third (behind the United States and the EU) in the number of such actions currently in force.

"Freer Trade Increases Poverty in the Third World"

Not true. Historically, countries that have achieved large reductions in poverty are generally those that have experienced rapid economic growth spurred in significant measure by openness to international trade. Newly industrialized economies such as Hong Kong, Singapore, South Korea, and Taiwan have all been open to trade during the past four decades and have been entirely free of poverty, according to the dollar-a-day poverty line, for more than a decade. By contrast, during the 1960s and 1970s, India remained closed to trade, grew approximately 1 percent annually (in per capita terms), and experienced no reduction in poverty during that period.

Trade helps produce rapid growth, and rapid growth helps the poor through three channels. First, it leads to what Columbia University economist Jagdish Bhagwati calls the active "pull-up" rather than the passive "trickle-down" effect—sustained growth rapidly absorbs the poor into gainful employment. Second, rapidly growing economies can generate vast fiscal resources that can be used for targeted anti-poverty programs. And finally, growth that helps raise incomes of poor families improves their ability to access public services such as education and health.

The current impression that the freeing of trade has failed the world's poor is partially rooted in disputable "official" World Bank poverty figures. The bank reports that though the proportion of the poor in developing countries declined from 28.3 percent in 1987 to 23.2 percent in 1999, increased population has left the absolute number of poor unchanged at 1.2 billion. And since that period also witnessed further freeing of trade, some conclude that trade has failed the poor. Yet, independent research by economists Surjit Bhalla in New Delhi and Xavier Sala-i-Martin at Columbia University has persuasively shown that the absolute number of poor declined during 1987–99 by at least 50 million and possibly by much more.

"Agricultural Protectionism in Rich Nations Worsens Global Poverty"

Not necessarily. If developed countries eliminate all forms of agricultural protection, including subsidies to domestic producers and quotas on foreign imports, their agricultural production will decline and the worldwide price of agricultural products will increase. Therefore, poor countries that are efficient agricultural producers will benefit from higher prices and access to new export markets. But consider the flip side: Poor countries that import agricultural products will suffer from higher prices. In 1999, as many as 45 of the 49 least developed countries imported more food than they exported. In 2001, for example, Senegal spent as much as $450 million on food imports, equivalent to about 10 percent of its gross domestic product and one third of its annual export earnings. Certainly, if agricultural trade is liberalized and prices rise, some poor countries will become net agricultural exporters, but many will not.

Some may argue that even if the poor countries pay higher prices for agricultural imports, their poor farmers will still benefit from those increased prices. But, in fact, high domestic prices do not require high world prices. Even under current world trading rules, the least developed countries can offer higher than world prices to their own farmers. In India, for example, the government buys food grains from farmers at prices higher than (and unrelated to) world agricultural prices.

Ironically, the major beneficiaries of widespread agricultural liberalization would be rich countries themselves, which bear the bulk of the cost of the subsidies and protection, and their domestic consumers. Other potential beneficiaries include nations such as those belonging to the Cairns Group—a coalition of 17 agriculture-exporting countries (9 of them from Latin America but also including advanced economies such as Canada and Australia) that enjoy efficient agricultural sectors and lobby for more open trade in agriculture.

Ultimately, even if some poor countries did suffer from more open agricultural trade, the case for liberalizing global agricultural markets remains unimpeachable. The current trading system in agriculture grossly distorts prices and production patterns and results in an inefficient global agricultural market.

"Poor Countries Should Not Open Their Markets If Rich Countries Maintain High Trade Barriers"

Big mistake. As the late British economist Joan Robinson once remarked, "if your trading its partner throws rocks into his harbor, that is no reason to throw rocks into your own." Responding to protectionism with more protectionism may seem "fair," but it is downright silly. Many Western advocacy organizations and religious groups that make this argument fail to understand that such talk hardly helps poor nations. It is hard enough for leaders in these countries to convince domestic producers that opening national markets is a worthy objective; loose talk of "hypocrisy" and "unfairness" only makes it harder. Even people who should know better fall into this trap. "It is surely hypocritical of rich countries to encourage poor nations to liberalize trade," former World Bank chief economist Nicholas Stern reportedly stated in a March 2001 speech in New Delhi, "whilst at the same time succumbing to powerful groups in their own countries that seek to perpetuate narrow self-interest."

Certainly, trade protectionism by rich nations merits opposition. But whether or not rich nations lower their barriers, poor countries should unilaterally dismantle their own protectionist policies in order to increase trade and stimulate economic growth. Trade barriers are often porous rather than absolute, so that countries with outward-oriented policies often succeed in expanding exports even when markets in partner nations are not fully open. Trade-oriented East Asian economies such as Hong Kong, Singapore, South Korea, and Taiwan have registered excellent export performance since the early 1960s. By contrast, relatively protectionist countries such as India, China, Argentina, and Egypt have hurt their own export growth and, as a result, stifled their overall economic performance in those years. Yet all these countries faced virtually the same trade protectionism abroad. Economic history since the end of World War II confirms that export pessimism is self-fulfilling, whereas nations that adopt export-oriented trade policies manage to exploit international markets despite foreign protectionism.

"There Is No 'Development' in the Doha Development Agenda"

False. Judging by the anger many poor nations displayed at the recent WTO talks in Cancún, it would seem that the current round of WTO trade negotiations—ambitiously dubbed the "development round" when the talks were launched in Doha, Qatar, in late 2001—have nothing to offer the cause of development. But such a conclusion would be mistaken. Insofar as the WTO negotiations aim to liberalize trade in nations both rich and poor, development cannot and will not be missing from the agenda.

For more than four decades, developing countries have demanded that rich economies remove their tariff peaks, which apply in particular to labor-intensive goods (such as textiles, apparel, and footwear) from developing countries. The Doha declaration explicitly addressed this objective. The declaration also addressed the substantial relaxation of agricultural protection in rich nations, including the removal of farm subsidies, which developing nations consider crucial. Brinkmanship by both rich and poor countries produced the failure in Cancún, but the negotiations are far from buried. When they eventually conclude, development concerns will be central to the agreement.

However, even well-intentioned advocates can go too far in linking trade policy with development. Former WTO Director-General Michael Moore has argued that investment and competition policy, transparency in government procurement, and trade facilitation (i.e., less red tape when goods enter a country and adequate information on import and export regulations) are also development issues. The EU has placed these issues on the Doha agenda, even though a large number of developing countries oppose their inclusion.

The expansion of the WTO into these areas contributed in no small measure to the breakdown of talks in Cancún. Agreement in these areas would require developing countries to adopt existing developed-country practices and regulations; this action would therefore impose "asymmetric" obligations on developing countries. Many poor countries lack even the resources necessary to implement these obligations. Finally, differences in local conditions require local solutions rather than an externally imposed and globally uniform regime in these areas. "One size fits all" is the wrong answer.

"The World Trade Organization Harms Poor Countries"

No. Contrary to popular belief among many Western nongovernmental organizations and politicians in developing countries, the WTO is the best friend available to exporters in poor nations. The General Agreement on Tariffs and Trade (GATT), signed in 1947 and incorporated into the WTO at the latter's inception in 1995, substantially opened markets in rich countries during the first 40 years of the GATT's existence. Under its "most favored nation" provision, GATT required that such markets be open to all GATT members, including developing countries. Therefore, even without undertaking any trade liberalization of their own, developing nations became the beneficiaries of the market opening in the developed world.

The GATT's Uruguay Round of trade negotiations, which began in 1986 and culminated in the establishment of the WTO, marked the first time rich nations insisted that developing countries fully participate in the negotiations. Developing countries felt shortchanged in this round on three counts: Their expectations of opening agricultural markets in rich countries were not realized; developing countries committed themselves to cutting industrial tariffs more deeply than developed economies; and developed countries successfully enacted a global intellectual property rights regime that undermined poor countries' access to cheap medicines.

Although the Uruguay Round benefited developed countries more than developing ones, poor nations still gained. First, developing countries liberalized more because they had higher trade barriers to begin with (and remember, in economic terms, greater liberalization is a benefit, not a cost). Second, after years of complaining, developing countries convinced developed nations to commit to dismantling quotas on imports of textiles and

clothing. Third, while the Uruguay Round did not enhance developing countries' access to global agricultural markets, it opened the way for future liberalization in this important arena.

Despite the dominance of developed countries and skewed distribution of the bargaining power within the WTO, the global body offers low- and middle-income countries a rules-based forum in which to defend their trading interests and rights. For example, the strength of the WTO has helped developing nations deflect pressures from rich nations to link further trade opening to the creation of stronger labor standards in poor nations. Without the WTO, developed countries simply could have resorted to unilateral trade sanctions to enforce their desired standards. Moreover, at the September 2003 trade talks in Cancún, this rules-based bargaining allowed developing countries to delay negotiations on investment and competition policy.

"Free Trade Is Bad For the Environment"

No. Certainly, trade forces can hurt the global environment. For instance, the rapid expansion of coastal shrimp farming in several countries in Asia and Latin America in the 1980s, driven principally by the demand for exports, led to the contamination of water supplies and destruction of surrounding mangrove forests. But trade opening can bring

Want to Know More?

For a discussion of trade policy issues, visit the Web site of the **World Trade Organization** (WTO). In particular, consult the **"World Trade Report 2003"** (Geneva: World Trade Organization, 2003). For insights on the debates over free trade, see Jagdish Bhagwati's *The Wind of the Hundred Days: How Washington Mismanaged Globalization* (Cambridge: MIT Press, 2001). Dani Rodrik argues that the free trade agenda misleads poor countries in **"Trading in Illusions"** (FOREIGN POLICY, March/April 2001). On the politics behind U.S. trade policy, consult Susan Ariel Aaronson's *Taking Trade to the Streets: The Lost History of Public Efforts to Shape Globalization* (Ann Arbor: University of Michigan Press, 2001) and I.M. Destler's *American Trade Politics, 3rd edition* (Washington: Institute for International Economics, 1995).

Bruce Stokes contends that the influence of anti-trade forces has been exaggerated in **"The Protectionist Myth"** (FOREIGN POLICY, Winter 1999–2000). On the relationship between trade and the environment, see Scott Vaughan's **"Trade Preferences and Environmental Goods"** (Washington: Carnegie Endowment Trade, Equity, and Development Series, No. 5, February 2003). The British-based group Oxfam assails rich countries for their trade barriers against poor nations in **"Rigged Rules and Double Standards: Trade, Globalisation, and the Fight Against Poverty"** (Oxford: Oxfam, 2002).

This article builds upon recent work by Arvind Panagariya, including **"Developing Countries at Doha: A Political Economy Analysis"** (*World Economy*, Vol. 25, No. 9, September 2002) and **"Wanted: Jubilee 2010, Dismantling Protection"** (*OECD Observer*, No. 231/232, May 2002), coauthored with Bhagwati.

For links to relevant Web sites, access to the *FP* Archive, and a comprehensive index of related FOREIGN POLICY articles, go to www.foreignpolicy.com.

environmental benefits as well. For example, the agricultural liberalization proposed in the two's Doha negotiations would not only bring economic and efficiency benefits by shifting production from high-cost to low-cost producers, but it would also yield environmental benefits by replacing Europe's pesticide-intensive agriculture with natural manure-intensive agriculture in developing countries.

Activists who decry the environmental impact of trade should realize that trade protectionism often brings environmental costs as well. During the 1980s, the United States imposed quotas on Japanese small-car imports; the policy not only hurt U.S. consumers but also harmed the environment by reducing access to lower-pollution vehicles. More broadly, closed-door policies in pre-1989 Eastern Europe were accompanied by an extremely poor environmental record.

When trade produces adverse environmental effects, the solution is not to ban or restrict trade. Instead, governments should adopt appropriate environmental policies to achieve environmental objectives and allow trade policy to target economic objectives. In the shrimp farming case, shrimp producers should be taxed for the pollution they create but then left to trade freely. Such a policy normally will reduce exports and economic output, but that result would be offset by reduced pollution. Reliance on a single instrument (trade policy) to target both economic and environmental objectives is like trying to kill two birds with one stone—a strategy successful hunters would not recommend. Just as governments should not subsidize trade to help the environment, neither should they restrict it to avoid harming the environment.

Arvind Panagariya is professor of economics and codirector of the Center for International Economics at the University of Maryland, College Park. He is co-author of Lectures on International Trade *(Cambridge: MIT Press, 1998) with Jagdish Bhagwati and T.N. Srinivasan.*

Trade Secrets

*The real message of the collapse of trade talks in Cancún:
Business as usual is over for the WTO.*

By Lori Wallach

The World Trade Organization's (WTO) September 2003 meeting in Cancún, Mexico, had barely collapsed before major corporations and their government and media mouthpieces launched into damage control: Blame developing nations—those countries making up the majority of WTO members, and that U.S. Trade Representative Robert Zoellick dismissed as "won't do" countries—for defending their interests. Blame Mexico's Foreign Minister Luis Ernesto Derbez, who hosted the meeting. Or blame advocacy groups and nongovernmental organizations.

This blame game is a wasteful distraction from the reality that, after nine years of existence, business as usual at the WTO is over.

To overcome opposition when the WTO was established in 1995, promoters promised benefits eerily similar to those trotted out before Cancún: billions of dollars in global economic growth and the reduction of poverty in poor nations. Not only has the WTO failed to deliver on such promises, but numerous countries are suffering economic, environmental, and social harm after implementing the global body's mandates. This harm highlights the WTO's key contradiction: Shouldn't those living with the results determine the policies versus having them imposed by the WTO?

Not according to the old powers that be. In the run-up to Cancún, a bloc of powerful economies—the European Union, Japan, the United States, and Canada—plotted with the allegedly neutral WTO secretariat to set the agenda in advance. Their plan was not to negotiate, but to dictate four more WTO agreements that have little to do with trade and that require signatories to rewrite their domestic laws to conform even more to a "Washington Consensus" set of one-size-fits-all policies. Developing countries were expected to provide more privileges for foreign investors, subjugate their government procurement policies to WTO disciplines, prioritize facilitation of imports over other domestic policy goals, and adopt uniform "competition" policies that enable mega-conglomerates to further consolidate

markets. The proposal would have undermined economic development and made the WTO even more untenable for scores of poor nations still unable to digest their Uruguay Round commitments, such as requirements to create new corporate patent rights over medicines and seeds.

Unfortunately, the genuine demands of the developing world were not on the agenda. The so-called G-21, a group led by Brazil, China, South Africa, and India—representing half the world's population and two thirds of its farmers—pressed hard to cut rich countries' farm subsidies. Many developing countries also presented a unified position on nonagricultural market access and dismissed an "Implementation Agenda" of 105 specific changes to current WTO rules.

The Cancún agenda shut out the possibility of real transformation of the existing terms of globalization.

Yet, wealthy countries were shocked (yes, shocked) when Cancún broke down, even though more than 80 developing countries had already rejected the rich nations' agenda. But Cancún followed 18 months of deadlocked WTO talks in Geneva, which began in Doha in 2001 after a similar agenda had fallen apart in Seattle in 1999. Clearly, Zoellick's triumphant remark in Doha that the "stain of Seattle" was erased proved laughable given the outcome of the Doha WTO meeting. Rich countries called it the "Doha Development Agenda" and poor countries called it the "Everything but Development Round." By any name, it papered over gaping disagreements between rich and poor.

On the road to Cancún, the small bloc seeking WTO expansion merrily lectured the developing countries about the WTO's great success to date and imperiously declared that more of the same was in developing countries' interests. But poor nations have lived with nine years of WTO policy, along with 20-plus years of "field testing" the WTO model through International Monetary Fund structural adjustment programs. Today, the least developed countries' share of world trade is more than 45 percent less than its share before the establishment of the WTO. Excluding China, the percentage of people living on $1 per day worldwide has increased. WTO agriculture rules have led to record low prices paid to farmers and increased food dumping, destroying livelihoods and undermining food security for millions. Countries that most faithfully adopted the "neoliberal" policy package (trade, finance, and investment liberalization; privatization; deregulation; and new property protections) have been the hardest hit. Few developing-country officials believe studies touted in the United States claiming that countries most open to the global economy grow the fastest. They need only look at Argentina's collapse and China's remarkable growth. Ironically, the countries exalted as the most economically open—such as China, Vietnam, and Malaysia—flouted major elements of the model by having no convertible currency, tightly regulating foreign investors, limiting imports, and planning industrial policies. These countries engaged the global economy on their own terms, using trade policy strategically when they deemed it useful.

Given the WTO's record in rich countries, developing countries also question what they can expect in the future. Under the current trading system, the U.S. trade deficit exploded to $435 billion in 2002, equal to 5.2 percent of U.S. gross domestic product. Nearly 3 million U.S. manufacturing jobs (one in six) have been lost over 10 years under the North American Free Trade Agreement and the WTO, with cascading effects on state and local tax revenues. U.S. real median wages remain below 1972 levels. This record does not even touch on the string of domestic environmental, health, and food safety policies under assault by the WTO. U.S. voters' dismay over this record has forced trade to the top of the presidential primary debate in the United States.

Trade can be beneficial, but under what rules? The Cancún agenda shut out the possibility of real transformation of the existing terms of globalization. The true loss in Cancún was not that a bad agenda was rejected, but that the real issues were never even discussed. No one is calling for an end to trade or trade rules. But what will those rules be, and who will write them? Either those desperately defending the status quo will come to realize that their failed project is over and that change is inevitable, or their ideological intransigence will cause autarchy following the current system's inevitable implosion.

Lori Wallach is director of Public Citizen's Global Trade Watch in Washington, D.C.

Why Prospects for Trade Talks Are Not Bright

Aaditya Mattoo and Arvind Subramanian

NOTWITHSTANDING some recent fresh impetus, the Doha Round has been hobbled by a frustrating "now-it's-on-now-it's-off" quality. If only it could be brought to conclusion, the world would be so much better off, especially if one is to believe the estimates of the gains from trade liberalization that the Round would deliver—running as high as $500 billion. But a meaningful conclusion to the Doha Round will not be easy to achieve. Aggressive interests in trade liberalization are fewer now than in the past and industrial countries' defensiveness is greater.

Substantial opportunities for liberalization undoubtedly exist in both industrial and developing countries. Moreover, the World Trade Organization (WTO) system and associated trade negotiation rounds provide an institutional framework for making these opportunities feasible. This framework relies on reciprocal market opening: the domestic political pain of liberalizing imports is countervailed by the domestic political benefits of providing greater market opportunities for exporting interests. But right now, this framework is in trouble.

Market opening bargains between industrial and developing countries are increasingly difficult to strike. First, exporting interests in industrial countries have declining enthusiasm for the multilateral trade system as a way to achieve access to desired markets. Second, industrial countries are finding it difficult to overcome defensive interests within their own borders to deliver greater access to developing countries. Ironically, this disengagement coincides with an apparent effort by (at least) the larger developing countries to negotiate seriously.

Private sector disinterest

Historically, multilateral trade liberalization has been driven by corporate interests, notably in the United States and Europe, seeking access to foreign markets. Early rounds of trade liberalization under the WTO's predecessor, the General Agreement on Tariffs and Trade (GATT), were driven by U.S. private sector interests threatened by trade diversion consequences of

the then-European Economic Community's formation and subsequent enlargement. Looking to boost sales and profitability during the difficult economic times of the 1980s, U.S. and European services sectors—especially those with intellectual property interests at stake—provided much of the impetus for the Uruguay Round (1986–94).

> "Industrial countries are finding it difficult to overcome defensive interests within their own borders to deliver greater access to developing countries."

The WTO would seem to be the best vehicle for advancing the current interests of the industrial countries' private sectors that are seeking the opening of markets in developing countries for manufactured goods, particularly the larger developing countries where barriers remain high. But the Doha Round, launched in 2001 to address developing countries' discontent with globalization and the multilateral trading system, has always been plagued by a private sector interest deficit. Attempts to make the Doha Round a "development" round have obscured the fundamental problem of the relative absence of the industrial countries' corporate sectors from the negotiations. This absence results from an interesting combination of trade liberalization success unilaterally and defensiveness at the multilateral level.

Many developing countries, having adopted the policy prescriptions of the "Washington Consensus," have been unilaterally dismantling their trade barriers, typically at the urging of the World Bank and the IMF. Since the early 1980s, quantitative restrictions have been eliminated and tariff barriers have been lowered considerably. With this happening outside the

WTO framework, industrial countries do not have to negotiate within the WTO to secure access to new markets that their firms are obtaining without cost. The WTO process is thus a "victim" of the success of the World Bank and IMF.

Just as countries continue to lower trade and investment barriers unilaterally, they increasingly do so in the context of regional trade agreements. Weakly constrained by WTO law, regional integration may, however, be lessening scope for—and private sector interest in—pursuing bargains at the multilateral level. The systemic effects of regional agreements for multilateral bargaining may in fact be perverse: countries in a regional arrangement may actually want less broad-based liberalization in the WTO because their preferential access to each other's main export markets would likely be eroded. For example, some recent simulations show Mexico as a loser from a successful Doha Round.

Services is an area where opportunities for enhancing national and global welfare have only begun to be tapped. Despite significant unilateral liberalization, most countries have so far been wary of engaging in multilateral talks. One reason is because it is difficult to make the deep legislative and regulatory changes needed to open services markets in the context of international trade negotiations. More importantly, scope for reciprocity within service sectors has been drastically curtailed by industrial countries' unwillingness to consider greater openness where developing countries have a comparative advantage—notably, in the supply of services through the movement of persons.

The generally weak and unfinished framework of rules governing trade in services provides evidence of the overall poor results—as well as the minimal level of liberalization commitments undertaken by countries. The wedge between the reality of present-day regulatory regimes and the level of bound commitments often remains large. As a result, the private sector has concluded that the multilateral system is ineffective —and decidedly slow—in delivering real opening of services markets worldwide.

This perception is strengthened by the increasing disconnect between the accelerating product cycle firms face in global markets and the lengthening negotiating cycle with which governments must contend. The blistering pace of technological progress has had profound implications for key sectors such as telecommunications, transport, and finance. But if the multilateral response remains slow and ponderous, private sector enthusiasm will wane further. Nongovernmental routes to securing market access and standard setting—as well as the call of regional intergovernmental sirens—will prove more attractive.

Even in the area of intellectual property, Northern corporate interests are not looking to the Doha Round. Many of their goals were accomplished in the Uruguay Round. They have been pursuing their remaining interests in raising standards of intellectual property protection through the regional route—an effort that has achieved some success. For example, under regional agreements negotiated by the United States with Jordan, Morocco, and Vietnam, these countries have had to go beyond the WTO's Trade-Related Aspects of Intellectual Property Rights (TRIPS) agreement in providing protection for pharmaceuticals and testing data used in obtaining regulatory approval for pharmaceuticals. Overall, this has increased the monopoly power of the patent holder and limits the ability of generic producers to compete.

Growing Northern defensiveness

Since the Uruguay Round, it has become increasingly clear that striking any bargain in the multilateral arena would require meeting the demands of the larger developing countries for market opening in the North. With the emergence of large developing countries as important global traders, the WTO system must increasingly accommodate their market access priorities. What are these priorities and can the North deliver?

Given the South's comparative advantage, it seeks market access in four areas: agriculture, textiles, labor mobility, and cross-border supply of services. The political problems of opening vary across these areas. In agriculture and textiles—two of the traditionally most protected sectors in the United States, European Union (EU), Japan, and Canada—the political difficulties are well known. Farming interests in France and the clothing industry in the United States have been formidably effective over the years in resisting liberalization. But for different reasons, they are experiencing (or will soon experience) a bout of wrenching disruption.

Enlargement of the EU's boundaries to include new countries in Eastern Europe and beyond—and the consequent budgetary pressures—make it necessary to reduce subsidies. The dismantling of clothing quotas in the United States under the Uruguay Round is now going to expose domestic firms to greater competition. Both developments are, in a sense, predetermined. The EU may wish to present its subsidy cuts as a potential concession and seek concessions in return from its trading partners. In a world where budgetary difficulties are common knowledge and trade negotiators are far from naive, however, real return "payment" will come only if the EU goes further in its reform than it is obliged to. At a time of such change, the appetite for further liberalization in agriculture in the EU and textiles in the United States, as developing countries are seeking, is not likely to be great. How much can industrial countries commit to in the Doha Round?

In negotiations on services, labor mobility has always been a difficult issue, but now even the openness of cross-border trade in services seems uncertain. Developing countries such as Egypt, India, the Philippines, and Sri Lanka have a profound stake in ensuring the mobility of skilled personnel—and most developing countries in ensuring the mobility of the unskilled. Notwithstanding the large mutual gains that would be derived from allowing greater labor mobility, immigration policy has yielded only grudging concessions so far. And with traditional political difficulties compounded by a new fear of terrorism, greater openness seems elusive.

Industrial countries account for over three-fourths of all cross-border trade in services. But Brazil, Costa Rica, India, and Israel are among the 20 developing countries whose exports of business services have grown by more than 15 percent annually in the last decade. This growth and the outsourcing of service jobs have provoked deep concerns in many industrial countries—obscuring their own comparative advantage in services. A fuller reckoning of the forgone benefits by the United States and other countries might still occur in the future, leading to a more enlightened strategy. But for the moment, industrial countries, far from seeking greater openness abroad, seem reluctant to lock in current openness of cross-border trade.

In contrast, developing countries are showing a new willingness to engage. Some of the larger developing countries seem increasingly willing to contemplate serious liberalization of hitherto protected manufacturing and services subsectors in the context of the WTO, provided their industrial country partners are willing to reciprocate. For example, Latin American countries are willing to open their financial and telecommunications sectors further in return for meaningful concessions by the United States and the EU on agriculture. Similarly, if labor mobility were seriously on the agenda, countries such as India and the Philippines could well display an openness to reducing trade barriers.

Are we being woolly-headed about this? Developing countries such as Brazil and India have been notorious procrastinators in the multilateral system. And these countries may continue to find it difficult to acquiesce to multilateral liberalization of services, which poses challenges of sequencing regulatory changes and other domestic reforms. But what is different today—and this leads us to feel more optimistic—is both a conviction in these countries about the need for their own reforms and a recognition that these domestic reforms will be easier politically if they can be coupled with market opening abroad. These developing countries are not only interested in greater access for their agriculture and labor-based services exports for its own sake but, crucially, also as a means of furthering domestic reforms.

Hazarding a prediction

If our analysis here is correct, the prospects for a meaningful Doha Round may not be too bright. We fear a scenario in which a limited set of concessions is agreed to, based largely on what has already been done—subsidy reduction in agriculture in the EU and locking in ("binding") the already undertaken services reform in developing countries—and this package is trumpeted as a successful Doha Round.

Are we being overly cynical? Ten years ago, a "successful" Uruguay Round was concluded, leading to estimates of large global welfare gains. But liberalization assumptions built into the models were disconnected from what the Round actually achieved. The models assumed substantial liberalization in agriculture and manufacturing by developed and developing countries. For many developing countries, however, "liberalization" attributed to the Round was notional, even illusory. Very little incremental liberalization took place: in both agriculture and manufacturing, developing countries agreed to bind tariffs at levels that often were higher than prevailing levels.

For industrial countries, meaningful liberalization took the form of quota dismantling, but apart from that very little was achieved. In agriculture, countries set tariffs at very high levels to offset the elimination of quotas ("dirty tariffication"). Cuts in tariffs were rendered notional by an arcane process of choosing a base year well before unilateral reductions had been made. Furthermore, the model estimates conveniently ignored the impact of the intellectual property agreement, which would have reduced welfare gains, especially for developing countries.

We are not saying that there was no liberalization during the 1990s. Nor are we claiming that there is no value in locking in reforms that have already occurred. What we are saying is that the benefits of the Round were exaggerated and its costs were underplayed. Cutting through all the hype, the Uruguay Round was all about industrial countries eliminating clothing quotas in return for which developing countries increased their intellectual property protection. The rest did not amount to much. While framework agreements in services and tariffication in agriculture set the stage for future liberalization, much greater claims were made on their behalf. Yesterday's future has arrived and we will have to see how much these framework agreements deliver in terms of actual liberalization.

> ## "There will be overwhelming pressure to again create the illusion of a successful negotiation with a delusory development dimension."

We highlight these concerns to warn that there will be overwhelming pressure to again create the illusion of a successful negotiation with a delusory development dimension. A meaningful development round faces forbidding challenges. In the industrial countries these include cutting textile tariffs at a time when industry is coming to terms with eliminating quotas; cutting agricultural tariffs when farmers are struggling to accept cuts in production subsidies; securing greater labor mobility when terrorism has cast its shadow over immigration policy; and preserving openness of cross-border trade in services when the boom in outsourcing is creating deep anxieties about job security. In developing countries, the challenge is to use the WTO to undertake meaningful liberalization of goods and services in a way that these countries have not yet done.

Now is the time to muster the political will to overcome these challenges. If we cannot, then when the dust settles on the Doha Round, let us be honest in how we assess it—using a clear benchmark of how much additional liberalization it truly caused.

Aaditya Mattoo is the Lead Economist in the World Bank's Development Economics and Research Group. Arvind Subramanian is a Division Chief in the IMF's Research Department.

Why Developing Countries Need a Stronger Voice

Running faster at the IMF, but where's the progress?

Cyrus Rustomjee

"TAKE my hand," said the Red Queen from Lewis Carroll's *Alice Through the Looking Glass,* "and I will teach you something." And Alice took the Red Queen's hand. "Run," said the Red Queen, and Alice ran. "Faster, faster," said the Red Queen, and Alice ran faster.

And so began the IMF's extraordinary foray, in December 1999, into the world of poverty reduction and growth in low-income countries, a world of high-flown objectives and obscure acronyms. I recall vividly the three critical Executive Board decisions, made in quick succession, to launch the Poverty Reduction and Growth Facility (PRGF), the Poverty Reduction Strategy Paper (PRSP) process, and the qualification linking the enhanced Heavily Indebted Poor Countries (HIPC) Initiative to both the PRGF and the PRSP.

The initiatives were designed to make poverty reduction and economic growth the clear, central objectives of the IMF's work in low-income countries. The traditional focus on short-term stabilization measures, driven particularly by the IMF's focus on financial programming, was intended to give way to a much more nuanced, longer-term approach to economic development that acknowledged institutional, political, economic, and historical limitations in low-income countries and that balanced traditional short-term objectives of macroeconomic stabilization with longer-term objectives of poverty reduction and economic growth.

But, to those unacquainted with the IMF's previous work in low-income countries through the Enhanced Structural Adjustment Facility, it was as if low-income countries were being lured into a vast dungeon, a Shelob's lair, with the Executive Board's decisions representing the final closing of the web around low-income countries' sovereign decision making.

The Executive Board's decisions were not easily achieved. Some constituencies on the Board, such as my own, accepted the proposals very reluctantly. We argued that the success of the three-pronged initiative would hinge on several factors that had not been thought through properly and that, if not adequately addressed, would precipitate the failure of the IMF's venture into an area in which it had little substantive experience. Among these factors were the need to redesign the IMF's approach to stabilization; the need to provide adequate debt relief resources; the need for parallel multilateral trade initiatives that would alter the extraordinary terms of trade disadvantage suffered by low-income countries; the need to rapidly increase resources for human resource capacity building and development, including the IMF's technical assistance in low-income countries; and the need to improve the voice of developing countries in decision making.

Five years later, there have been some successes and a series of important limitations and failures, particularly in sub-Saharan Africa. As the IMF marks its 60th year, several major factors, identified in 1999, can be seen as key reasons for the lack of success of the PRGF in the sub-Saharan African countries:

• the region's wholly inadequate level of voice and representation in the IMF;

• a program design that continues to focus predominantly on the traditional stabilization objectives of the IMF; and

• a series of specific challenges that have been left unaddressed since the launch of the PRGF, PRSPs, and the enhanced HIPC Initiative.

If these factors remain unaddressed , these poverty initiatives will fail and poverty will deepen.

Contrasting outcomes

A number of low-income countries outside sub-Saharan Africa are registering modestly improved GDP growth and are making some headway in reducing poverty. But for many sub-Saharan African countries, GDP growth remains insufficient to establish the momentum they need to exit from profound poverty. Worse still, in many cases, poverty is deepening. This would not be particularly problematic for the integrity of the IMF's overall effort in low-income countries but for the fact that by far the largest proportion of PRGF, PRSP, and HIPC cases are in sub-Saharan Africa, suggesting that, at their center of gravity, the initiatives are failing.

> "When an issue is in dispute among the various shareholders, there is no reasonable possibility of forging a consensus that enables the African members' views to prevail."

The sub-Saharan African cases can be split into three categories. The first is a tiny handful of countries that are succeeding in addressing poverty and are registering strong economic growth—for example, Mozambique and Tanzania. The second—comprising the bulk of countries that have entered into PRGF arrangements, many of which have been recipients of HIPC debt relief—are registering no discernible change in growth rates; in several cases, countries are slipping further into poverty. The third category comprises a growing list of failed states that are exhibiting protracted political, economic, and social instability and on which no international multilateral efforts to date have had any significant effect. In addition, in several of the countries that have passed the completion point for HIPC Initiative resources, the combination of adverse terms of trade and other external shocks, coupled with the inadequacy of HIPC resources, has resulted in their slipping back into unsustainable external debt and an adverse balance of payments position.

Too small a voice

The need for improved representation of low-income countries, particularly in sub-Saharan Africa, was a strong theme during my term as Executive Director. This was not a crude attempt to snatch votes and voice but an appeal to reason. Developing countries now account for

by far the largest client base of the institution and are the focus of the significant majority of the IMF's policies, the entirety of its financing, almost all of its technical assistance, and a large part of IMF surveillance activity. They correctly assert that they should exert a proportionately larger influence over decision making in the institution. But voting strength is extraordinarily skewed in favor of creditors, which command 71 percent of the voting strength in the IMF Executive Board. Extreme creditor domination has resulted in poor decision making, particularly when decisions affect low-income countries, whose representatives on the Board are unable to influence decisions effectively.

Sub-Saharan Africa illustrates the case. Approximately one-fourth of the IMF's member countries are in sub-Saharan Africa. Yet these countries share a combined voting power of 4.4 percent. Forty-four of them are accorded only two Executive Board seats. They represent the majority of PRGF cases, the majority of PRSP cases, and the majority of enhanced HIPC Initiative cases. They also constitute all of the IMF's protracted arrears cases and almost all of the IMF member countries that receive post-conflict emergency assistance.

Much has been made of the consensus style of decision making in the IMF's Executive Board. During my term on the Board, I witnessed its merits, which were real and effective in many instances. But as a Board member representing a large group of primarily small countries, I also experienced the trap that this model represents. For, in the case of sub-Saharan Africa, when an issue is in dispute among the various shareholders, there is no reasonable possibility of forging a consensus that enables the African members' views to prevail. Instead, consensus means joining a creditor-dominated perspective only to see the specific points of objection that were raised slip away.

Consequences of unbalanced representation

Unbalanced representation is a problem for a number of reasons. First, it breeds inefficiency. A system of decision making that makes it almost impossible for one segment of the membership to achieve an outcome on any matter at all that, on balance, favors its interests above those of the rest of the membership is alienating to that group. In practice, because decisions are always seen to be arrived at based, at best, on a consensus centered on the interests of creditor members, there has been a growing sense of loss of ownership by the developing countries of policies agreed to by the Board. This can be and, indeed, has been costly to the institution and to individual members.

Adequate evidence exists to demonstrate how significantly the imbalance in representation arrangements has affected the efficiency of decision making in the long term. It has been illustrated most vividly during periods of crisis. For example, during the Asian crisis in 1997–98, a host of commentators, including several Asian member countries, argued that IMF program design was inappropriate and failed to take account of the specific circumstances of member countries. The quality of decision making and, in turn, program design and content would have been far improved and the prospects of success strengthened had the recipient members had a stronger influence. Similar arguments and criticisms have been made not only about many other individual country programs but also about IMF policies in general.

One example is the policy conditions—known as conditionality—attached to lending. IMF conditionality has been the subject of extensive criticism and has affected all developing countries with IMF programs. Despite clear and mounting evidence over many years that program conditionality had become excessive, in many cases irrelevant and in many cases counterproductive, decisions approved by the Executive Board continued, over several years, to favor excessive conditionality, despite repeated and well-argued objections by the debtor countries. Developing countries argue that the lack of votes to carry their view resulted in substantive failure of the IMF's conditionality policy, caused unnecessary reputational damage to the institution, and contributed to program failure in many cases. Fortunately, a fundamental change in conditionality policy was finally agreed after an extensive consultative process, though only after many years of growing policy failure. Inappropriate conditionality was not a costless exercise. In many instances, it resulted in excessively contractionary policies, stunted growth, and postponed poverty reduction. In the process, it alienated even the most committed policymakers.

Similar problems existed with PRSP and HIPC decision making when concerns raised by developing countries were overridden because of the overwhelmingly superior voting power of the creditor group. The consequence was that in almost all instances, both the PRSP and the PRGF processes encountered precisely the challenges and difficulties that developing countries, particularly the PRGF members, had expected. Some of these were subsequently corrected during important reviews of both the PRGF and the PRSP processes in 2002, based again on evidence in the field of mounting and valid objections to several aspects of the process. Yet the objections had been raised three years earlier.

What should be changed?

So, in the context of low-income countries, what should the IMF do differently? Several decisions can be made that can improve the prospects of success.

First, *the orientation of the PRGF must be shifted away from the overriding preoccupation with demand compression and excessively short term macroeconomic stabilization objectives to more balanced programs incorporating strategies for growth.* This will necessitate

far more PRGF financial resources, a reduction in both the force and the number of quantitative and structural conditions, and improved program design, with a more decisive shift away from the financial programming model to a broader growth programming orientation that makes possible the shifting of certain responsibilities to the World Bank. Fortunately, there is a solid internal literature on a mixed approach to financial and growth programming, although the underlying logic has not been accepted within the IMF. The evidence of growing failure of both the growth and the poverty reduction elements of the PRGF in sub-Saharan African should prompt a re-reading of this literature, to find ways to reorient the PRGF or, alternately, to introduce a new facility for low-income countries. Of the two approaches, having witnessed the significant diversion of human resources in low-income countries between 1999 and 2003 to adjust to the PRSP and the PRGF, I believe a gradual reorientation of the PRGF, over three to five years, would probably be preferred over a sudden adjustment to a new IMF approach to financing low-income countries.

Second, *the definitions established for debt sustainability, even in the enhanced HIPC Initiative, need revision.* They were extraordinarily overoptimistic when established in 1999. They might have proved sufficient in an environment of stable exchange rates and steady growth in global trade and financial flows but have certainly been insufficient to afford low-income countries a cushion of sustainability in the face of acute, repetitive, and extraordinarily destabilizing external shocks. In practice, this points to the need for a new initiative to further augment the financial resources available for debt relief.

Third, *representation arrangements need to be altered in a manner that strengthens the debtors' decision-making capacity.* The current margin of voting share in favor of creditors beyond that required to ensure a simple majority strikes at the foundation of the principles of collaboration and consensus decision making upon which the IMF operates. It weakens the institution, reduces operational efficiency, gnaws at the institution's legitimacy, erodes ownership of programs and policies by the collective membership, offers no tangible benefit to the collective membership, and has bred understandable resentment in the debtor group. Various potential options to achieve a better balance exist. All require political consensus within the membership, and some would make it possible to preserve some of the factors to which creditors attach importance, including the principle of a perpetual creditor majority, U.S. and European veto power, and relative ranking of creditors.

Fourth, *global financial resources allocated to human resource development, capacity building, and technical assistance need to be significantly increased.* One of the clearest lessons of the experience with the PRSP and the PRGF has been the absence of adequate human resources to sustain reforms. The IMF needs to redouble its own efforts, in conjunction with other international agencies, to establish a neutral, effective international agency to address this crucial foundation for sustainable development.

* * * * *

When the journey was over, the Red Queen asked, "Well, Alice, what did you learn from that experience?" "I'm not sure," said Alice. "But I did notice that the faster I ran, the faster I seemed to stay in the same place." The PRGF/PRSP/HIPC complex has helped some countries. But

most remain in the same place, and some, particularly in sub-Saharan Africa, are slipping back. With the IMF at 60, it is time for the institution to take a new road in low-income countries. It is time for the Red Queen to take Alice's hand and learn.

Cyrus Rustomjee, a former Executive Director at the IMF representing about half of the sub-Saharan African countries, is Managing Director of the Centre for Economic Training in Africa, based in Durban, South Africa.

Why Should Small Developing Countries Engage in the Global Trading System?

3 Three points of view on a hot topic in the Doha Round

Over the past few years, developing countries, especially the smallest and most vulnerable, have increasingly worried about greater participation in the global trading system—fearing they may get swamped by products from rich countries or lose out to cheaper rivals. To gain some insight into these fears, which are a major factor holding up progress in the Doha global trade talks, *F&D* turned to three experts on the topic. Rubens Ricupero, former Secretary-General of the United Nations Conference on Trade and Development (UNCTAD), stresses that practical initiatives are needed to exorcize the fears of developing countries about the impact of liberalization. Faizel Ismail, head of South Africa's delegation to the World Trade Organization (WTO), underlines the need for special consideration for the smallest members. And Sok Siphana, Cambodia's Secretary of State for Commerce, emphasizes that openness, not isolation, is the key, noting that Cambodia has managed to make WTO rules work in its favor.

1 Overcoming Fear First

Rubens Ricupero

Our question could be rephrased as: Why are small developing countries afraid of multilateral trading negotiations? To emphasize fear as the central element of the problem does not mean we are adopting a negative approach; rather, it is simply a recognition of the truth of the matter. Fear is, in fact, the basic explanation of why small developing countries are reluctant to engage in the multilateral trading system. Fear is not always irrational or unexplainable. When it results from a real danger, there is no use pretending that it will go away under the influence of academic arguments about the theoretical gains from trade openness.

The specific causes of fear must be addressed and exorcized. In the case of trade negotiations, they can be reduced to three generic types:

• the fear of not understanding the issues at stake or of not possessing the negotiating skills and resources to effectively take part in the difficult game of negotiations;

• the fear of very tangible negative consequences of negotiations: the loss of jobs, of preference margins, of food security, the deterioration in terms of trade for net food importers, and the extra cost of implementing complex enforcement systems like those required for intellectual property rights; and

• the fear of not being competitive in quality, price, and range of products.

Lack of skills

The first variety of fear can be dealt with only through a systematic drive to train skilled negotiators and boost analysis of the pros and cons for developing countries of adopting certain positions. Boosting negotiating capacity has to go much beyond the traditional concept of technical cooperation to explain the content of trade agreements and to provide technical advice in particular cases. As we envisaged in UNCTAD, when we launched the "Positive Agenda for Trade Negotiations" in 1996, the ultimate goal has to be more ambitious. The aim should be to build up a country's capacity to formulate its own negotiation strategy based on its potential competitive advantages

51

and capacity to supply certain products. Several agencies are involved in this kind of work. Among the best results, at least in terms of conceptual planning, have been those from the interagency Integrated Framework for Trade-Related Technical Assistance to Least Developed Countries. Difficulty in financing the implementation of projects has been a stumbling block, however. There may be a case, therefore, for making trade-oriented technical cooperation an integral and enforceable commitment for future negotiations.

Uncertain outcome

The fear of specific losses has to be faced, case by case, with concrete and practical initiatives, such as the welcome decision of the IMF to provide financial support to countries severely affected by the erosion of preferential margins. There are other proposals of the same nature, like the joint work of UNCTAD and the UN's Food and Agriculture Organization that deals with the problems of net food importing countries. Whenever an agreement under negotiation will result in a substantial increase in administrative costs in developing countries—as in the case of the WTO's intellectual property rights agreement—this downside has to be taken into account in the overall balance of costs and benefits. In addition, implementation and transitional periods will have to reflect the availability of financial and human resources in the country under consideration.

In this second category of concerns, the fear about job losses is perhaps the most serious because of its potential social destabilizing effect. This problem should not be underestimated or minimized as there are many indications of the short- or medium-term dislocations in employment in African and Latin American economies that have undergone rapid trade liberalization without complementary social safety nets. An example is that of the United States, which, since the days of President John F. Kennedy more than 40 years ago, has always had a Trade Adjustment Act to serve as a complement to major trade negotiations. The most recent version, adopted in 2002 together with the Trade Promotion Authority, earmarked millions of dollars to spend on retraining, education, health insurance, and pension benefits. If the most competitive economy in the world considers adjustment an indispensable tool of trade liberalization, would it not make sense for the international community to envisage a similar multilateral aid program for the countries that lack the resources internally?

Problems competing

Finally, there is the fear arising from supply-side constraints. Many small developing countries still rely on one, two, or three commodities for the bulk of their exports—such as coffee, cocoa, cotton, sugar, palm oil, and petroleum. It is very hard to convince these countries that they have something to gain from negotiations that have

little or nothing to do with their main concerns: oversupply, excessive price volatility, and increasing erosion in the percentage of the final price that goes to the producer. Countries in that situation need first and foremost to diversify and enhance their productive sector through investment, technology, and managerial skills. Successful examples are Cambodia, Lesotho, and Mauritius. Once countries improve their production structure, their interest in trade negotiations is a natural, logical, and spontaneous consequence.

Many, perhaps all, of these fears could be addressed if, in the case of small and vulnerable developing countries, major trade negotiations were supported by social impact studies of the likely outcome. Nowadays, nobody dreams of undertaking an infrastructure construction project without evaluating its possible environmental impact. Why shouldn't we adopt the same kind of approach for the social evaluation of trade liberalization? If conducted in a balanced and responsible way, this new approach could become a powerful instrument to dispel the fear of negotiations, build confidence in the trading system, and increase the integration of small developing economies into the system.

Rubens Ricupero, a former Finance Minister of Brazil, was UNCTAD's fifth secretary-general from 1995 to 2004.

2 Help Needed to Liberalize

Faizel Ismail

Ever since the Doha trade talks were launched, WTO members have been debating how to tackle a plethora of complex trade and development issues raised by developing countries. The issues have been made more complex by the changing patterns of global integration of developing countries since the conclusion of the Uruguay Round. While some developing countries have begun to successfully integrate into the world economy and significantly improve their share of world markets, others—especially the smaller economies—have been increasingly marginalized, with their share of world markets declining.

This changing global context has prompted some developing countries to request special consideration in light of their development needs. But so far, no significant progress has been made on how to handle what is known in trade circles as special and differential treatment.

A new approach

If the Doha talks are to advance in this area, a new conceptual approach will be needed. I would like to suggest the following four-pronged strategy.

(1) Financial help. Several studies indicate that while a successful Doha trade round could lift at least 140 million

people out of poverty—and integrate developing countries into the world economy—the trade reforms would need to be accompanied by complementary actions in low-income countries to support adjustment and a strong supply response. If developing countries are to take these steps, however, they will need significant additional financial assistance from developed countries. The history of European economic development provides some valuable insights into past trade integration efforts. The Marshall Plan after World War II was partly initiated to "neutralize the forces moving Western Europe permanently away from multilateral trade (Foreman-Peck, 1983)." European integration itself was facilitated by economic assistance provided to weaker countries and regions (Tsoukalis, 2003).

Where would the needed money come from? One source could be recycled funds from Doha Round gainers—which studies show include both developed and developing countries, with developed countries being the major beneficiaries. Gains would accrue to both consumers and producers and would boost tax revenues for many governments. A small share of these funds could be set aside to help low-income countries meet adjustment costs and build needed capacity.

(2) Capacity building. Many developing countries have argued that their access to developed country markets is further denied by high costs associated with meeting health and technical standards. The WTO has recognized the need to address the capacity constraints of developing countries and committed the WTO to provide enhanced technical and capacity building assistance to increase their effective participation in the negotiations, to meet WTO rules and health and technical standards, and to enable them to adjust and diversify their economies.

(3) Vulnerabilities. Developing countries stand to gain considerably from ambitious multilateral reductions in tariffs and agricultural subsidies. But it is important to recognize that some of them will also be seriously hurt by the decline in the value of their preferential access to developed country markets as a result of global liberalization. Policymakers will have to tackle the anticipated revenue loss, higher cost of food imports, and need for adjustment and diversification, but they should do it in a way that does not increase the existing distortions in world markets and perpetuate existing levels of protection for developed countries.

(4) Policy space. In the agriculture negotiations on market access, developing countries have argued for increased protection and space to adjust for special products, based on the criteria of food security, livelihood security, and rural development. A similar case could be made for industrial products based on development needs. Any granting of "policy space" should go hand in hand with developing countries increasing the binding (or fixing) of tariffs so as to benefit from the discipline of a rules-based system and not face further marginalization.

Greater flexibility in WTO rules is also required for many developing countries that may need more discretion to use some trade policy instruments to enhance economic development. How could WTO flexibility be monitored? A multilateral monitoring mechanism could be set up to identify where WTO disciplines would be inappropriate and would need to be made more flexible in response to development needs (see article by Bernard Hoekman in this issue). The result would be to create a more development-friendly WTO for these countries.

In sum

Adopting this four-pronged approach to the challenges faced by developing countries in the global trading system would go a long way toward making WTO rules more compatible with the development objectives of small developing countries and bringing the Doha Round to a successful conclusion.

Faizel Ismail joined the new democratic government of South Africa in 1994 and led its trade negotiations with the European Union, the Southern African Development Community, the Southern African Customs Union, a number of bilateral trading partners, and the WTO. He joined South Africa's Mission to Geneva in May 2002.

References:

Foreman-Peck, James, 1983, "A History of the World Economy" (Totowa, New Jersey: Barnes & Noble).
Tsoukalis, Loukas, 2003, "What Kind of Europe?" (Oxford: Oxford University Press).

3 Cambodia—No Turning Back
Sok Siphana

Cambodia's accession to the WTO in October 2004 marked its reintegration into the global trading system. Cambodia restored relations with the IMF and World Bank more than a decade ago and joined the Association of South East Asian Nations in 1999. Joining the WTO has marked the final step in bringing Cambodia back into the major regional and international organizations that govern international economic relations.

Like its trading partners, Cambodia views its participation in the multilateral trading system as a means of integrating into the global economy and maximizing the benefits from international trade. But skeptics have questioned whether Cambodia will really benefit from WTO membership. They argue that the 30 least developed countries in the WTO have been unable to secure trade opportunities commensurate with their development needs. Given the length and difficulty of the accession process, they have suggested that it is surprising so many members even want to become part of the WTO. However, in the case of a poor country like Cambodia, accession is seen as a necessary means to achieve economic growth. In the words of the Cambodian chief negotiator Prasidh Cham: "In a time of harsh and fierce global competition, the survival of our country depends on our ability to capture the right opportunities at the right time. We believe entry into the WTO is such a case."

WTO accession provides the 12 million Cambodian people with secure, predictable, and nondiscriminatory access to the markets of 147 trading partners. Joining the WTO has shifted the process of trade reform in Cambodia away from an incremental approach to one incorporating quite detailed rules for trade policy. It is shaping, in very practical terms, the way business is done in Cambodia.

During the negotiations to join the WTO, Cambodia was able to achieve all of its main objectives: First and foremost, accession enables Cambodia to protect its garment industry— its main export industry—from having quotas imposed on exports in 2005 and beyond. In the process, of course, Cambodian employment has also been protected. Cambodia has avoided any obligation to lower applied tariffs; it has retained the ability to offer exporters duty-free access to imported raw materials and other inputs.

WTO membership can also help Cambodia improve the investment climate by ensuring unimpeded access to foreign markets. We are improving the legal framework for businesses and enhancing the business services infrastructure. The country retains ample flexibility to develop appropriate policies to support the agricultural sector and will benefit fully from the Doha Declaration on Trade-Related Aspects of Intellectual Property Rights and Public Health. Cambodia has recently passed and promulgated a law excluding pharmaceutical patenting altogether until the 2016 deadline agreed in the Doha Declaration. This means the Cambodian government and the public will not have to pay higher prices for drugs. Cambodia has thus succeeded in harnessing WTO accession in a way that advances and reinforces its outward-looking development strategy.

The full significance of this accomplishment, however, will depend on what happens next. The reality of WTO accession is that it is no guarantee of success in world trade. It is a tool to be used by the government, the private sector, and other stakeholders that provides opportunities and safeguards. That is the contract: living up to the obligations and taking advantage of the opportunities. If joining the WTO opens up a new range of possibilities for Cambodia, it does not guarantee that these possibilities will be automatically translated into advantages. In order to transform potential into reality, much hard work will need to be done in the months and years ahead.

While WTO accession will afford Cambodia access to member country markets on a most-favored-nation basis, it will also intensify competition from foreign goods and enterprises in both domestic and international markets. Because of this, Cambodia's investment climate will need to converge toward international norms by removing impediments that put its exporters at a disadvantage relative to other exporters competing for similar markets.

The message is clear: Cambodia has chosen not to protect, but to compete. Cambodia cannot be regarded as a "free rider" (taking the benefits of WTO membership without making its own contribution) and thus devoid of individual negotiating weight.

Making the most of the many exemptions (classified as special and differential treatment) and the generous technical assistance it has obtained under the WTO-led Integrated Framework, and stimulated by proactive pro-integration policies, Cambodia is poised to rise to the challenge of post-WTO integration.

The more open and competitive the Cambodian economy becomes, the more exporters and importers and all Cambodian citizens, as well as foreign investors, will benefit from the legal security of the rules-based trading system. After all, Cambodia has tasted two decades of economic embargo and isolation, and it has also experienced a decade of integration and prosperity.

If anyone has doubts about openness, try isolation. Cambodia did, and we sure do not want to go back. With our accession to the WTO we have turned a page of history. It is time to open a new page and look forward to better days ahead.

Sok Siphana is Cambodia's Secretary of State for Commerce. He led his country's negotiations to join the WTO.

Without Consent: Global Capital Mobility and Democracy

by Jeff Faux

Shortly after he became the first general secretary of the World Trade Organization, Renato Ruggiero observed, "We are no longer writing the rules of interaction among separate national economies. We are writing the constitution of a single global economy."

The word constitution—with its implication of world government—shocked some international trade officials. Like a reference to sex at a Victorian dinner table by an otherwise respectable gentleman, it was resolutely ignored by the business press and the policy academics, whose public commentary acts as a Greek chorus for what George Soros so aptly named "free-market fundamentalistism." The WTO, sings the chorus, is *not* a constitution. Its purpose is "free trade," an arrangement that presumably requires less, not more, government.

Yet Ruggiero was simply acknowledging the obvious. Markets are not found in a state of nature. They are human creations, defined by enforceable rules. Even the most primitive markets require rules for what constitutes private property, valid contracts, weights and measures, and so on. And they always reflect a social contract.

In modern, civilized economies, rules are enforced by public institutions—legislatures, courts, regulatory agencies, central banks. The social contract includes protection of labor, the environment, and public health from the brutalities of unconstrained capitalism.

The precise content of a market's rules has major consequences for who gets to be rich and who gets to be poor. Therefore, all markets have a *politics*. Political science, as a famous American scholar once observed, is the study of "who gets what."

When markets expand their boundaries, so must the rules. In our own history, advances in technology, business organization, and westward migration expanded the U.S. economy from a series of regional markets, regulated by state governments, to a continental economy regulated primarily by the federal government. Note that the federal government did not just impose rules on trade *among* the states, but market rules *within* the states as well. Because we had a Constitution guaranteeing some form of democracy and a Bill of Rights, the new rules were subject to public debate. Political parties evolved around class-based conflicts over land settlement, the gold standard, anti-trust, child labor, social security, environmental protections, and so forth.

Today, technology, business organization, and migration are relentlessly expanding markets beyond the capacity of individual nation-states to regulate them. Because business must have rules, a constitution for the global market is being written—at the World Trade Organization, the International Monetary Fund, and the World Bank. Befitting a world dominated by one superpower, the U.S. Treasury and the Pentagon play leadership roles. Because there is no prior framework of democracy or accountability, the new constitution is being written piecemeal, in secret, and publicly unacknowledged, except for an occasional slip of the tongue, as in the case of Ruggiero.

Who Decides?

But if all rule-setting generates politics, what are the politics of the setting of the new rules for the global economy? Who gets to decide "who gets what?"

To the typical reader of the world's major newspapers or watcher of the nightly news, the rules for a borderless economy seem to be set by a sort of parliament of nations, where finance ministers at the IMF, trade ministers at the WTO, and economic ministers at the World Bank pursue their national interests. Interestingly enough, the new constitution is not being written at the United Nations, which is presumably our principal world legislature.

This notion of "national" interests dominates the language of globalization. Thus, the reports from the recent WTO meetings in Cancun speak of *U.S. interests* vs. *Brazil's interests* vs. *South Africa's interests*, and so on.... The implication of this language is that when George W. Bush or Lula or Thabo Mbeki turns his gaze to foreign economic affairs, the domestic conflict over "who gets what" stops at the border.

National interests are then aggregated into international blocs. Global economic politics is presented as a conflict between rich countries and poor countries, the North and the South, the producers of raw materials and the producers of software.

Yet, as the late Michael Harrington once remarked, there are poor people in rich countries and rich people in poor countries. And just as politics in an expanding American economy developed around class and other interests across state lines, a similar process is going on in the current globalizing economy.

The individuals who negotiate trade and investment agreements and who sit on the boards of the IMF, the World Bank, and international financial agencies formally represent different national interests. But they increasingly act as agents for an international class interest as well. Globalization has created a global elite—people with mutual economic interests regardless of nationality. They include the leaders of multinational corporations and their financiers, their political partners, and their clients and retainers among the punditry, the military, the international bureaucracies, and the academy.

After a speech I gave a few years ago at the Council on Foreign Relations in New York, a retired State Department official bluntly underlined the fundamental reality. "What you don't understand," he said, "is that when we negotiate economic agreements with these poorer countries, we are negotiating with people from the same class. That is, people whose interests are like ours."

I call this global governing class the Party of Davos, after the Swiss site of one of the annual conferences of the global elite. As Adam Smith reminded us, "People of the same trade seldom meet together, even for merriment and diversion, but the conversation ends in a conspiracy against the public." We should expect no less when people from different countries with the same interests meet at the global economy's watering holes for merriment, diversion … and conspiracy. It would be odd if it were otherwise. So it should be no surprise that the rules of the global market written by the Party of Davos protect and promote the positions of its membership—those who control large amounts of capital. The rules thus encourage trade deregulation, privatization, weakening of unions, financial market liberalization, and a general shredding of the social contract.

This is not to say that the world's governing class is always of exactly one mind, or that nationality plays no role in the pursuit of self-interest. Bankers in Miami see the world differently than bankers in Portland, Oregon. Those in London have a different perspective from those in Singapore. But when it comes to protecting the generic rights of capital, the elites of Miami, Portland, London, and Singapore are united.

Accordingly, issues of concern to other classes are, by joint agreement, left out of the agendas of the IMF, the WTO, and other international forums, and therefore out of the concerns of the global constitution. These include the rights of labor, the protection of the environment, public health, community stability … and of course, democracy and accountability.

These interests are championed by the minor party in the politics of global markets. Let us call it the Party of Porto Alegre, the original Brazilian site of the World Social Forum. This is the party of the opposition. It includes many labor unions, environmental organizations, religious and human rights activists, indigenous groups, and their many sympathizers around the world. They first came together at the WTO meeting in Seattle in the last weeks of the twentieth century when they crashed the party of the Party of Davos.

The often bizarre television images that the world sees of the street activists of the Party of Porto Alegre harassing the Party of Davos from one meeting of the IMF or the WTO to another are distortions designed to ridicule any opposition to Davos's hegemony. Yet, the images do capture an important part of their relationship. The goal of the Party of Davos is to escape popular constraints on capital and the goal of the Party of Porto Alegre is to constrain it—making it subject to democracy and accountability. This is why the constitution of the new world order is not being written at the United Nations. The UN is too unwieldy, too transparent, and too susceptible to Porto Alegre-ish sentiments.

Whatever separate goals its members might pursue, the common agenda of the Party of Davos is to break the bargaining power of labor. By labor, I do not mean just labor unions, but the vast majority of the people on this planet who must work in order to live—from industrial and service workers in advanced countries to rural laborers and marginal peasants in the most economically backward corners of the globe. The bargaining between labor and capital—which takes place within the firm and in a society's political life—is what makes up the "social contract" that is required in order to legitimize the unequal distribution of income, wealth, and power that markets generate.

Still, Davos makes a moral claim. It is that an emphasis on the distribution of wealth actually makes the poor worse off. In contrast, says Davos, deregulated capitalism makes for faster economic growth, and that growth improves life for everyone—especially the poor.

The Davos Record

We now have been at Davos's neoliberal program for twenty years, time enough to evaluate this claim. Of course, in a world of roughly two hundred separate nations and six billion people, measuring anything on a global scale is very tricky, particularly when the policies pursued by the different economies have not been uniform. But some things seem clear.

Most important, after two decades of neoliberalism, global economic growth has slowed from the previous twenty years. From 1960 to 1980, world gross domestic product grew at an average rate of 4.6 percent annually. In the following two decades, under increasing free trade and deregulation, growth in the world economy slowed to less than 2.9 percent annually. Moreover, those fast-growing countries that provide the most weight in the aggregate numbers—China and India—were the most resistant to the advice of the bankers, the international bureaucrats, and the army of consultants who work for the Party of Davos.

The trends on poverty and inequality are more difficult to sort out. But it appears that if one eliminates China and India—who represent 38 percent of the world's population—from the calculation, world poverty has not improved very much. Inequality among nations has certainly gotten worse. And inequality within nations seems to have increased in Latin America, Africa, Eastern and Central Europe, Central Asia. All but five industrialized countries (Denmark, Luxembourg, the Netherlands, Spain, and Switzerland) saw inequality increase while France saw no change in inequality. A recent analysis by Christian E. Weller, Robert E. Scott, and Adam S. Hersh of the Economic Policy Institute reports that the median income of the richest 10 percent of the world's people were 70 times that of the poorest 10 percent in 1980, and 122 times in 1999.

Competent scholars argue over these numbers, but one thing is obvious to all but the hopelessly ideological: the last twenty years have not produced the surge in living standards that neoliberalism's champions promised would flow from the liberation of capital from social constraints and the weakening of the bargaining power of the world's working people. Even then-World Bank president James Wolfensohn in 1999 was moved to admit, "At the level of people, the system isn't working," suggesting that there are other "levels" at which the system is working perfectly well.

The NAFTA Model

One place to see the process more clearly is here on our own continent, where in January the North American Free Trade Agreement (NAFTA) will be ten years old.

Like the WTO, NAFTA does more than just govern trade among its three members—Canada, Mexico, and the United States. If NAFTA had only been concerned with free trade, the agreement could have been written on a few pages. Instead, NAFTA is a thousand-page template for the constitution of an emerging continental economy.

In fact, NAFTA was a model for the WTO. It is the explicit template for the proposed Free Trade Agreement of the Americas, the Central American Free Trade Agreement, and the Asia-Pacific Economic Cooperation Forum. And it is the inspiration of the economic portion of the Bush administration's September 2002 National Security Strategy, openly referred to by its intellectual supporters as an agenda for "empire."

The vision of economic integration embodied in NAFTA differs from the vision of the other major model of regional market integration—the European Union. The development of the EU has been based on the understanding that common political institutions are the inevitable consequence of common economies. Every major step of the process was, and still is, transparent—subject to fiery public debates over the rules, particularly over the balance between individual rights, local sovereignty, and market efficiency.

In contrast, the constitution of the single North American market was merchandized to the citizens and legislators of each of the three countries as a simple, narrow, stand-alone agreement on foreign trade.

NAFTA does, of course, promote increased trade between Canada, Mexico, and the United States. Its text lays out a timetable for the elimination of customs barriers on everything from vegetables to truck transportation. But it is also as much an investment agreement as a trade agreement. The document binds each nation to extraordinary protection of the other member states' investors. It requires governments to guarantee the repatriation of profits in hard currency. Its Chapter 11 gives private investors the right to bring suit against governments over laws that might endanger future profits (defined as "tantamount to expropriation"). It inhibits efforts by national governments to liberalize the ownership of intellectual property. Disputes are settled in secret by tribunals of experts, many of whom are employed privately as corporate lawyers and consultants.

The result is a framework for the governance of the continental economy that curtails domestic powers of popularly elected government. NAFTA restricts the public sector's freedom of action in taxation, procurement, and capital market policies. Under NAFTA, corporations have forced state and provincial governments in each country to rescind environmental regulations. United Parcel Service is currently charging that Canada's government-owned postal service violates UPS's NAFTA-given right to provide private mail service. Little by little, policy proposals in all three nations now must pass the test of whether they are "NAFTA compatible."

In effect, NAFTA is a constitution that recognizes only one citizen—the multinational corporate investor. Governments will be punished for infringing on the rights of investors, whose protection is guaranteed. But governments may diminish, even abolish, the civil rights of workers or the claims of the environment with impunity. In contrast to the detailed protections for investors in NAFTA itself, the fig-leaf "side agreements" covering labor and the environment are weak and unenforceable.

Had this formula been proposed as the governing constitution of Canada, Mexico, or the United States, the electorates of each nation would have no doubt overwhelmingly rejected it. But, by defining the debate over its adoption as a dispute between abstract notions of "free trade" and "protectionism," the promoters of NAFTA diverted attention from the larger political significance of the agreement.

To be sure, there was protectionist opposition to NAFTA in all three nations. But the traditional politics of previous trade battles, in which industrial sectors—including employers, workers, and communities—who might lose from freer trade were pitted against industrial sectors that might win, was muted. The investor protections of NAFTA split off the interests of large U.S. employers from their workers by allowing firms to shift production to lower cost Mexico. Thus, U.S. auto firms' chief executive officers supported the treaty while U.S. auto workers opposed it.

The conflict over NAFTA thus reflected a new, class-based politics of trade. The opposition was led not by industrial "losers," but by the social movements—labor, environmentalists, consumers and nationalists in all three countries who were alarmed over the potential loss of national sovereignty and the domestic social contract.

The central claim for NAFTA was Davosian: the agreement would create a sustained economic boom in Mexico that would more than compensate for any social costs. One typical prediction, by a U.S. undersecretary of commerce, was that Mexico would grow, "between a super-charged 6 percent a year, worthy of Asia's tigers, and a startling 12 percent per year comparable to China's recent economic growth." The growth would lift the country's poor (more than 40 percent of Mexicans live on less than $2 a day) into the middle class.

The Mexican boom, in turn, would bring economic benefits to the United States and, to a lesser extent, Canada. First, the immigration of undocumented Mexican workers would diminish, if not disappear. In 1990, then-president of Mexico Carlos Salinas asked an American audience, "Where do you want Mexicans working, in Mexico or in the United States?" Second, NAFTA would create a new middle-class market in Mexico for the more expensive goods produced in the United States and Canada.

NAFTA at Ten

It is now painfully obvious that the promise of greater economic growth was not fulfilled. Over the last ten years, Mexico's growth has been at best half of what it needs to create enough jobs for its expanding labor force. Since 2000, Mexico has scarcely grown at all. The record would have been worse but for the unsustainable U.S. boom in the late 1990s which boosted Mexican exports. Since the mid-1980s, when the neoliberal reforms began, growth has fallen to less than a third of the 3.4 percent rate at which Mexico grew in the years of the 1960s and 1970s—the so-called "bad old days" of government industrial policies and import substitution.

While the economic benefits fell short, the human and social costs of the continent-wide reallocation of investment rose dramatically. These costs included the destruction of livelihood of millions of workers, particularly in Mexican agricultural labor and U.S. manufacturing. On both sides of the border, the promises made to these working populations were abandoned almost as soon as the ink was dry on the agreement. For example, Mexican farmers were promised that they would receive generous financial and technical assistance to help them meet competition from U.S. agribusiness. But after the treaty was signed, funding for farm programs dropped dramatically. Meanwhile, the U.S. government massively increased subsidies for corn, wheat, livestock, dairy products, and other farm

products exported to Mexico. This, "comparative advantage" enabled U.S. agribusiness to drive thousands of small Mexican farmers out of their own markets. When the displaced campesinos and their families arrived in nearby cities, few jobs were waiting. NAFTA concentrated growth along Mexico's northern border, where the Mexican government keeps unions out so that the *maquiladora* factories can process and assemble goods for export to the United States with workers who are desperate, pliable, and even cheaper than elsewhere in Mexico. Between 1994 and 2000, *maquiladora* employment doubled while employment in the rest of the country stagnated.

In the absence of labor and environmental protections, the expanding sweatshops of the north created a social and ecological nightmare. Rural migrants overwhelmed the already inadequate housing, health, and public-safety infrastructures, spreading shantytowns, pollution, and crime. *Maquiladora* managers often hire large numbers of women, whom they believe are more docile and more dexterous than men at assembly work. Earnings are typically about $55 a week for forty-five hours—not enough for survival in an area where acute shortages of basic services have raised the cost of living. Families break up as men cross the border in search of jobs, leaving women vulnerable to the social chaos.

An Amnesty International report on the border town of Ciudad Juárez, where hundreds of young women have been killed, quotes the director of the city's only rape crisis center (annual budget: $4,500): "This city has become a place to murder and dump women. [Authorities] are not interested in solving these cases because these women are young and poor and dispensable."

In the United States, workers were betrayed by major multinational firms that had assured the U.S. Congress that their interest in NAFTA was solely in the middle-class Mexican market. Once the agreement was signed, these same firms began to shift production south of the border, eliminating hundreds of thousands of jobs in the United States. Clearly, the object of their desire was the low-wage Mexican worker, not the mythical high-wage Mexican consumer.

The net effect was to undercut wage levels on both sides of the border. Indeed, despite the shift of manufacturing to Mexico, average real wages in Mexican manufacturing in January 2003 were some 9 percent below their January 1994 level. No doubt some Mexicans have benefited from cheaper prices of expensive U.S. and Canadian goods. But in a country where the poverty rate is above 50 percent, the basic cost of living for most people seems to have gotten worse. For example, in December 1994, the minimum wage (currently $4.20 per day) bought 44.9 pounds of tortillas. Today it buys 18.6 pounds. In December 1994, it bought 24.5 litres of gas for cooking and heating. Today it buys seven.

So the dangerous migration across the border continues. "If you're going to improve your life, you have to go to the United States," said a neighbor of one of the nineteen undocumented Mexican migrants found asphyxiated in a Houston-bound truck in May 2003.

The failure of NAFTA to produce sufficient growth to absorb its own labor force should not have been a surprise. The conventional economic argument for free trade is not that it promotes growth, but that the reallocation of capital among the lines of comparative advantage promotes efficiency gains in the form of lower prices. Freer trade can produce such gains, but most efforts to measure them consistently produce small numbers.

Recently, the World Bank estimated that the Doha round agenda would add roughly $160 billion in static gains—the gains consistent with free trade theory—to the GDP of the world's developing nations. The number was used in the chorus of recrimination against the third world nations for letting the meeting in Cancún fail. Yet, a closer look at the estimates shows that they completely ignore any costs of dislocation, unemployment, and the loss of markets by local producers. Even so, these "gross" gains represent an increase of only 1.5 percent of GDP by the year 2015. Harvard economist Dani Rodrik has observed that "no widely accepted model attributes to postwar trade liberalization more than a tiny fraction of the increased prosperity of advanced industrial countries."

Frightened by the disputed election of 1988 that almost installed a leftist president, elites in both countries wanted to make it much harder for a future populist Mexican government to pursue redistribution politics. It was a shared objective: inasmuch as the ownership of assets in a single market is commingled, there is little practical distinction between the rights of Canadian, U.S., or Mexican multinational investors. Moreover, NAFTA created new opportunities for Mexican business elites to broker privatized assets to foreign investors at enormous profit. For example, an investment group headed by the well-connected Roberto Hernandez bought Mexico's second largest commercial bank from the government for $3.1 billion and resold it to Citicorp for $12.5 billion. Foreign investors now own more than 85 percent of the Mexican banking system, yet credit available to Mexican business has actually shrunk.

The problem of Mexican growth will not disappear with the revival of the U.S. economy. Mexico's temporary faster growth in the late 1990s was a function of an extraordinary boom in the United States that we now know was unsustainable. With generous injections of fiscal stimulus, U.S. growth may accelerate for a while, but the chances of a return to those years of excessive speculation are remote. With the U.S. trade deficit now expanding to worrisome levels, policymakers may soon be looking for more ways to limit imports. The ominous shifting of production from Mexican *maquiladoras* to even lower cost China is further evidence that the assumption that Mexico's needed growth would automatically flow from free trade was naïve.

In many developing countries, the largest part of Mexico's economic problem lies not in restricted export markets, but in the stifling maldistribution of wealth and power that restricts internal growth. The rich pay hardly any taxes. Despite the image of Mexico as a country with a strong state, the public revenue is 19 percent of GDP, compared with the more than 30 percent that the presumably more conservative American public sector takes.

Seeking an Alternative

The alternatives thus far presented by the Party of Porto Alegre seem to be caught in a web of contradictions. For example, at the same time that demonstrators demand that the IMF and other world institutions respect local sovereignty and end efforts to impose the neoliberal model, they demand that a wide variety of their own rules—independence for indigenous tribes, gender and racial equality, priority for small farmers, environmental regulations—be imposed on sovereign nations

Moreover, the Party of Porto Alegre is caught in a Catch-22 situation:

- Social justice requires global political institutions to regulate the global market
- Global political institutions are dominated by the Party of Davos
- The Party of Davos is hostile to social justice

The Party of Porto Alegre is thus forced back into a defense of national sovereignty as the only available instrument for achieving social justice. Yet sovereignty is steadily eroding under the relentless pressure of global markets. Moreover, nationalist politics undercut the cross-border cooperation needed to balance the cross-border political reach of business and finance. Nationalism perpetuates the myth that national identity is the only factor in determining whether one wins or loses in the global economy. It obscures the common interests of working families in all countries when faced with the alliances of investors that now dominate the global marketplace.

Still, human rights and social justice will become part of the "constitution" of the global marketplace only when enough nation-states demand it. Therefore, the global opposition must pursue a common global program for working people that reinforces their national struggles for economic and social equity. Such a program would support national democratic movements and leaders who understand that national social contracts cannot be maintained in a global market that lacks one of its own, and that a global social contract cannot be established in the absence of effective social democracy at the national level.

The creation of a true global alternative requires a perspective through which the interests of workers in all countries are linked. In a global marketplace, workers' living standards increasingly rise and fall together. When workers in Brazil win a wage increase, it raises the bargaining power of workers in Germany. When workers in Indonesia improve their working conditions, workers in Nigeria benefit. Likewise, when the social safety net is strengthened in one country it helps those struggling for human economic and social rights in other countries as well.

In a world of countries desperate for investment, the development of a global political movement powerful enough to bring the investor class to the bargaining table is clearly a long way off. But, with a nod to Margaret Thatcher, there is no alternative.

I believe that it is time for us to concentrate on a feasible project—the building of a model of cross-border solidarity among the ordinary people of our own continent.

A Modest Continental Proposal

Despite the failure of NAFTA to deliver on the promises of its architects, it is here to stay. Every day more intracontinental connections in finance, marketing, production, and other business networks are being hardwired for a consolidated North American market. Almost 70 percent of U.S. imports from Mexico are within the same firm or related firms producing the same final product. Ford pick-up trucks are now assembled in Cuautitlan, Mexico, with engines from Windsor, Ontario, and transmissions from Livonia, Michigan. Labor markets are relentlessly merging, for professionals as well as migrant workers.

Post- 9/11 border security concerns in the U.S. slowed down the process. But commerce will prevail, and is now above pre- 9/11 levels. Ultimately, the War on Terrorism is more likely to constrict the political freedoms of North Americans than the freedom of money and goods to cross their borders.

Moreover, the writing of the North American constitution continues. Out of the public eye, trigovernmental task forces and committees are discussing proposals ranging from guest-worker programs to continental transportation systems and the privatization of Canadian water and Mexican oil. Think-tanks, new academic institutes, and business associations are debating ideas about the harmonization of taxes and regulation, monetary policies, and a single currency. As the former Canadian ambassador to the United States recently commented, "Few days go by without new ideas for keeping NAFTA." The shared assumption is that the necessary political governance of the North American economy can be achieved by stealth, by grafting new agreements onto the basic NAFTA framework without stirring up public concerns over sovereignty and accountability.

But, sooner rather than later, the question of NAFTA's future must become part of the domestic politics of each nation. We need a process in which electorates of all three countries share an honest dialogue over the common future that was denied them in the first NAFTA debate.

In all three countries, the sense that globalization is beyond the influence of the majority of people has disem-

powered the public discussion of how to shape a common future. A focus on the question, "What do we want North America to look like ten or twenty years from now?" might be a way to revive that discussion and eventually generate the basis for a new and more comprehensive bargain among all people of the three countries.

Shortly after his election, Mexican president Vicente Fox suggested that NAFTA countries adopt a version of the European Union's program for investment in poorer areas. Mexico—even more so than the poorest nations of Western Europe—needs substantial investment in education, health, and infrastructure to create sufficient jobs for its people.

Fox's proposal was rejected in both Washington and Ottawa. It may be time to revive that suggestion to create a new Grand Bargain. In return for long-term financial assistance for Mexico's public investment, the working people of Canada and the United States would get an agreement on enforceable labor and environmental standards, so that as Mexico grows, wage levels and working conditions will rise—creating a middle-class market in Mexico and preventing the undercutting of labor standards north of the border. It could also build a middle-class constituency for modern tax, legal and public administration systems. The credible prospect of widely shared prosperity in Mexico that is creating enough jobs for its people would, in turn, make it easier to achieve a satisfactory accord on migration.

Debate over a new bargain might also recognize that democracy is incompatible with Chapter 11 and other NAFTA provisions that undermine the authority of the local public sector. And it might initiate an honest effort to apply the principles of sustainability to the continent's economic growth.

A continent-wide project for economic and social justice has another great advantage. It could provide a way to work out a model for the governance of a global economy that reconciles the tension between the relentless drive of technology to expand the boundaries of the market and the human needs of a decent society. Focusing on building such a decent society in our own continental neighborhood could also help redirect our political energies away from the temptation of global empire.

Just perhaps, if we could achieve economic integration with social justice between two first world societies and one third world society on this continent, we might have something to contribute to the development of a just and prosperous global society.

JEFF FAUX was the founder, and is now Distinguished Fellow, of the Economic Policy Institute. This article is adapted from a paper delivered at the Villanova University Conference on Catholic Social Teaching and Globalization in November 2003.

Making the WTO More Supportive of Development

Bernard Hoekman

How to help developing countries integrate into the global trading system

IN WORLD trade negotiations there is a constant tension between attempting to establish a set of universally applicable rules and allowing certain opt-outs or exceptions, particularly for developing countries. The World Trade Organization (WTO) attempts to manage this tension through what is known as *special and differential treatment* (SDT). SDT spans promises by high-income countries to provide preferential access to their markets, the right to limit reciprocity in trade negotiating rounds to levels "consistent with development needs," and greater freedom to use otherwise restricted trade policies. The underlying premise is that industries in developing countries need assistance for some time in both their home market (protection) and in export markets (preferences).

But SDT is controversial. Many economists argue that the existing SDT "package" has not been very beneficial: preferences have been of limited value for most developing countries as a result of exceptions, nontrade conditionality, and supply capacity limits, while nonreciprocity and weaker disciplines on trade barriers have impeded more rapid integration into the world economy (as continued protection biases incentives against exporting and improving productivity). Others argue that preferences are needed because industrialized countries have consistently thwarted the development potential of the trading system by maintaining high barriers to developing country exports and that rich countries have historically intervened in trade in ways that the WTO now constrains. Thus, SDT is necessary to give developing countries the same opportunities.

Restrictive trade policies may help support the development of domestic industry. For such industries to become efficient, however, they need to be able to source inputs from the most competitive suppliers and confront competition in the markets for their products. Whatever one's views on the effectiveness of trade policy to support domestic industry, both theory and experience suggest that over time trade barriers should be lowered to ensure this. By establishing a mechanism through which countries negotiate the reduction of trade barriers, the WTO can be regarded as pro-development.

What then is the problem from a development perspective? First, the WTO is driven by mercantilism: the desire of members to improve their terms of trade through better access to the markets of other members. *The focus is not on the welfare or growth prospects of members, or on the identification of "good" policy, but on ways that national policies impose costs on other countries.* It may be the case, for example, that there is a rationale for subsidies (to offset a market failure), even if this is to the detriment of other countries.

Second, *the ambit of the WTO increasingly extends beyond trade policy.* Domestic regulatory policies (or their absence) may have a strong economic efficiency rationale even if they entail some negative spillovers on others. Intellectual property protection is an example—limited enforcement may well be the best option for poor countries (see Box 1). Regulatory disciplines may also give rise to high and asymmetric implementation costs, with the burden falling disproportionately on poorer countries. Longer transition periods—the basic instrument adopted in the Uruguay Round—are an inadequate response.

Third, *little effort is made to identify what the preconditions are for benefiting from specific WTO agreements or whether they have been satisfied.* Nor is there a mechanism to monitor the effectiveness of policies justified under SDT provisions or to identify alternative policies (including development assistance) that might be more efficient in attaining the objectives of a poor country. To return to the subsidy example, assuming there is a case for intervention, subsidies or taxes are generally more efficient than trade barriers in addressing market failures, but governments may not have the capacity to use them, resulting in the use of more distorting (costlier) trade policies.

Finally, *traditional SDT has resulted in significant discrimination among developing countries,* incentives by recipients of preferences to oppose liberalization, and less certainty and predictability of trade policy.

The current approach to SDT in the WTO places the primary focus on detailed negotiation of opt-outs, rules, and exemptions from specific agreements. An example is the Doha Round proposal that developing countries be permitted to designate special products and use special safeguard procedures for agricultural products. This approach requires poor countries to determine on an issue-by-issue basis the specific provisions that would be beneficial. What these are may not be clear, and the ability to get agreement from developed countries on such proposals is constrained by mercantilist calculus: the perceived cost to them of a proposal, not whether it makes sense from a developmental point of view. This article proposes a new approach that would imply major changes for *both* developed and developing countries. It would make the WTO more supportive of development and enable developing countries to better integrate into the global trading system by having all WTO members accept a set of core commitments while allowing latitude in other areas.

A development-friendly WTO

How can WTO trade agreements become more supportive of development? Arguably, such agreements should

• remove foreign barriers to trade for products that poor countries produce;

• lower domestic barriers that raise the prices and reduce the variety of goods and services that firms and households consume; and

• support the adoption of complementary regulations and institutions that enhance development.

Political economy forces constrain realization of the first objective. Small, poor countries have little to offer in the mercantilist WTO exchange to induce large countries to remove policies that harm them. The preferential access dimension of SDT was motivated in part by this observation. Many of the poorest countries today have not been able to use SDT to expand and diversify their exports. Moreover, preferences are not an enforceable commitment under the WTO. Instead, they are "best endeavor" promises that, in practice, have been subject to many restrictions and conditions. The second objective requires domestic reforms—here the question is how to mobilize political support for such reform, given fiscal constraints, industrial policy objectives, and the fact that nonreciprocal preference programs may imply that exporters already have free access to major markets. The third objective may be impeded by the fact that the rules adopted have often been developed in high-income countries. WTO rules on intellectual property protection are a good example.

The challenge is to introduce flexibility when it is desirable, while at the same time strengthening the trading system. An important role of the system should be the adoption of good policies—in part by increasing transparency and reducing uncertainty regarding the policies confronting traders. This function of the trading system is of great value to developing as well as developed economies.

Box 1

Cambodia: tripping up over TRIPS?

Recent case studies illustrate the potential payoffs of greater flexibility in the WTO's regulatory enforcement. One example is Cambodia, which has made significant efforts to adopt legislation consistent with the Trade-Related Aspects of Intellectual Property Rights (TRIPS) agreement and has trained government officials and the private sector in enforcement. The government drafted laws on trademarks, patents, copyright, protection of trade secrets, unfair competition, and plant variety protection. It set up training courses for lawyers, judges, law enforcement, and customs officers. And it expedited the drafting of laws and implementing regulations, publishing Khmer books on the subject.

While most of this was paid for by donors, it is questionable whether the benefits offset these costs, given that Cambodia is unlikely to be a producer of high-tech or pharmaceutical products for many years to come. Indeed, the economic price tag of strong enforcement of intellectual property rights may be a multiple of the direct administrative costs and the opportunity costs in human resource terms of devoting so much attention to this area. It is an open question whether these laws constituted priorities from a development perspective and whether the costs incurred would have passed a cost-benefit analysis. This was not undertaken, because complete adoption of TRIPS was seen as a requirement for accession to the WTO.

Proposed new approach

Making the WTO more supportive of development could involve three basic elements:

• First, unconditional acceptance by developing countries of a core set of disciplines relating to market access—including the most-favored-nation (MFN) principle, binding of tariffs and commitments to reduce tariffs in the future—as well as acceptance, in principle, of the WTO *as a whole.*

• Second, permitting countries not to implement "non-core" WTO rules on development grounds, in the context of multilateral consultations with representatives of the trade and development communities (donors, financial institutions) on the effectiveness and impact of the policies concerned. Assessments of these policies should consider negative spillovers and should be published in the relevant countries to increase the accountability of governments.

• Third, a shift away from discriminatory trade preferences as a form of "trade aid," coupled with strengthened grant-based financing targeted predominantly at the poorest countries to improve trade supply capacity and the competitiveness of local firms, and to redistribute some of the gains from trade liberalization.

The intention should *not* be to make the WTO a development organization. This is not desirable, even if it were feasible. Instead, the objective is to put in place an enabling mechanism to foster greater integration of developing countries into the WTO.

WTO is a binding contract: commitments are enforceable. This gives the WTO its value—traders have greater certainty regarding policy, and governments know what they are "buying" when they make commitments. Allowing for "policy space"—or leeway for countries to pursue policies that would otherwise be subject to multilateral discipline—will increase uncertainty and could reduce the willingness of major trading countries to make commitments in the first place. Agreement that a core set of WTO disciplines would constitute binding obligations on *all* members would help address this concern. Thus, violations of core rules would be enforced through existing dispute settlement mechanisms.

> "A trade-off for acceptance of the core principles by developing countries is that higher income countries augment and gradually replace preferences with expanded development assistance."

Negotiations would need to define what this core comprises. Arguably, it includes transparency, MFN treatment, the non-use of quotas, the binding of all tariffs, and the willingness to make commitments to lower such tariffs over time in the context of trade rounds. Why these? Because they constitute the fundamental principles on which the trading system is based, and are beneficial to all countries regardless of their level of development. If this were accepted, it would imply stronger multilateral commitments in the core areas than exist now. In particular it would imply the end of nonreciprocal trade preferences by developed countries. There are both systemic (the MFN principle) and developmental rationales for this. The evidence suggests that those countries that can benefit from trade preferences have already done so, while those that have not confront domestic constraints that impede them from fully exploiting these opportunities. The primary need is to address those constraints and to remove trade-distorting policies that affect developing countries disproportionately on a nondiscriminatory basis. Thus, a tradeoff for acceptance of the core principles by developing countries is that higher income countries augment and gradually replace preferences with expanded development assistance to bolster trade capacity. Another trade-off is that non-core disciplines become eligible for the policy flexibility mechanism.

How policy flexibility would work

Differentiation among developing countries in the application of SDT has been a sensitive topic in the WTO.

Many more advanced developing countries oppose suggestions that SDT be limited to a subset of poorer and more vulnerable countries. A major advantage of a development framework that is explicitly designed as an enabling mechanism is that assumptions about who is eligible can be avoided. One way to allow this would be for any (self-designated) developing country to be able to invoke the process, but to accept as well explicit consideration of a "spillover test" as part of the consultations—the extent to which a specific policy has negative effects on other countries. This would introduce differentiation on a de facto basis, and is discussed further below.

Judging non-core policies. Consultations would assess the impact and effectiveness of nonconforming policies (non-core). This first requires identification of such policies. Traditionally, this is left to dispute settlement and, in the case of small and poor countries, the dispute settlement procedure is unlikely to be invoked. (This is, in fact, a weakness of the status quo, in that poor countries are ignored.) Currently, there are only two WTO mechanisms that identify inconsistent policies of smaller countries: the Trade Policy Review and committees that oversee the operation of specific agreements. The former constitutes a valuable transparency exercise that arguably is underutilized because the secretariat is not permitted to form judgments regarding WTO consistency of observed policies, and impacts within and across countries are not considered. The process is infrequent (every 6+ years). Agreement-specific WTO committees focus mostly on (changes in) implementing legislation and its application—the focus is not on the economic rationale or effectiveness of policies. Moreover, attention centers predominantly on the larger markets.

An explicit link between a new development framework and an augmented Trade Policy Review could generate more information on the effects of developing country policies. Assessing whether instruments are achieving development objectives and whether less trade-distorting ones can be identified inherently require judgments regarding appropriate sequencing and the need for complementary reforms and investment. These must be made by the concerned government but can be informed by inputs from other members and from development and financing institutions. The involvement of the latter would be necessary and desirable for a number of reasons. First, they have the mandate, experience, local presence, and capacity to provide policy advice. Second, these organizations generally take the lead in developing and financing projects and programs in developing countries. The WTO should not move into project design and financing.

A major advantage of the proposed mechanism could be improved communication between the development and trade communities—identifying where development organizations should help and where WTO disciplines may not be optimal for a country. In any such process, development organizations must have a seat at the table. The launch of the first version of the Integrated Framework for

Trade-Related Technical Assistance to Least Developed Countries (IF) at the WTO's 1996 ministerial meeting in Singapore shows that initiatives by trade ministers are doomed to fail if they are not coordinated with (owned by) the development institutions that will be asked to provide assistance and the countries that will use it.

That said, if the membership of a monitoring mechanism were to span all WTO signatories as well as relevant international development institutions, it would probably not be effective. One option would be to build on the revamped Integrated Framework—which has now become a unique example of international collaboration in the trade area, following a major rethinking and redesign in 2000 (see Box 2). If extended beyond the least developed countries, the IF agencies and donors would overlap to a very great extent with the set of players one would expect to engage in any trade-related policy dialogue.

Recognizing spillovers. Whether a policy imposes significant negative financial costs on other countries should be part of the terms of reference of the consultation process, as is the identification of possible policies that are less trade distorting. For example, as mentioned, basic economics suggests that subsidies are more efficient instruments to address market failures than trade policies. If binding budget constraints in a developing country preclude the use of (temporary) subsidies, development assistance may be used to overcome them if there is agreement that it would help address a market failure. Linkage to aid may also help establish a credible exit mechanism, a key condition to prevent capture and control rent seeking.

Settling disputes. Although the process of determining the impact and effectiveness of a particular policy that is inconsistent with a non-core WTO discipline should enhance both transparency and accountability of governments, such policies may inflict substantial harm on other WTO members. If they are also developing countries, policy space may imply robbing Peter to pay Paul. This again points to the importance of identifying less trade-distorting policies to pursue the government's objective. If these do not exist or are not adopted, countries ultimately have recourse to the standard WTO remedy: dispute settlement.

Larger developing countries are more likely to impose relatively larger spillovers on trading partners, whether developed or developing. This suggests that the spillover assessment proposed above may be an effective way to differentiate between countries in terms of the extent to which they can invoke "policy space" for development purposes. While spillovers imposed by a small country on large WTO members will be small by definition, they may be relatively large for another small country. Thus, a development mechanism should complement dispute settlement and not replace it. In effect, developing countries would be granted immunity as long as policies do not create significant negative spillovers, with a higher threshold for the impact on higher-income countries.

Box 2

IF as a model
The Integrated Framework for Trade-Related Technical Assistance to Least Developed Countries (IF) is an unlikely name for a model of future cooperation. But it is a good basis on which to develop a policy dialogue on trade. It brings together key multilateral agencies working on trade development issues with donors and recipient countries. More than 40 of the least developed countries (LDCs) have applied for assistance under the scheme.

The basic purpose is to embed a trade agenda into a country's overall development strategy and ensure that trade-related adjustment and capacity building are in line with the trade policy aims of the country concerned and prioritized with other development assistance needs. Although the IF has raised awareness of trade issues within the LDCs, many countries need additional resources to implement the recommendations of their trade integration strategies.

The IF has a steering committee with rotating membership—spanning six multilateral agencies, contributing donors, and recipient countries, as well as an interagency working group that handles diagnostics and follow-up.

"There is a fundamental choice to be made regarding the development dimension of trade agreements. It is a choice that goes beyond the WTO, extending to North-South regional trade agreements as well."

Building capacity. The proposed mechanism should also help address supply capacity constraints in poor countries by going beyond their identification to include an expansion of aid funds for this purpose. In particular, consideration could be given to a binding commitment by richer countries to transfer a share of the gains realized from multilateral trade reforms (under the Doha Round) to developing countries. Such gains are potentially large, depending on the extent of liberalization commitments made. For example, part of the tariff revenue collected on goods that are due to be liberalized over time or part of the budgetary allocation for agricultural subsidies that is to be eliminated under a Doha agreement could be made available to fund trade capacity improvements in developing countries. Especially in small low-income countries that already have relatively free access to major markets, using aid to address constraints that reduce their competitiveness can have high payoffs. That said, a major lesson of World Bank experience with projects and programs in this area (and most others) is that country ownership and leadership at the highest levels are critical factors in ensuring concrete and sustained follow-up in removing constraints to trade expansion. As noted, the proposed mechanism could help mobilize such engagement within the context of overall poverty reduction strategies.

Worth the attempt

Would the establishment of a mechanism to allow greater flexibility on a country-specific basis, with all its complications, be of value? The potential upsides from the approach sketched out above are significant. Small developing countries are rarely subjected to litigation by developed economies due to their size. While that suggests a policy flexibility mechanism is not really needed, this fact in itself illustrates the need for change: greater engagement with poor countries on their trade policies would be beneficial. Acceptance of core principles by all developing countries, including MFN, and thus the (gradual) demise of trade preferences, and the explicit agreement by high-income countries to put greater weight on the policy objectives of developing countries by taking the supply-side capacity agenda seriously through an augmented IF-type mechanism would be other significant changes.

A major advantage of the WTO is that it is a single-issue organization: the focus is always on trade. This is not the case at other international organizations. Creating a focal point for a *constructive*, as opposed to adversarial, interaction among governments could do much to raise the domestic profile of the trade agenda in developing countries. It would also add to information on the effects of existing policy instruments—a necessary condition for adopting better policies—and ensure that trade-related policy actions and investments are considered by decision makers. Although there will certainly be greater human resource costs, much of the required work could be undertaken in the context of the activities and diagnostics of the Integrated Framework.

In a nutshell, there is a fundamental choice to be made regarding the development dimension of trade agreements. It is a choice that goes beyond the WTO, extending to North-South regional trade agreements as well. The same tension arises there, with asymmetries in both power and size that are often much greater than in the WTO. The type of mechanism proposed above could also be considered in the context of regional trade arrangements such as the Economic Partnership Agreements the EU is negotiating with the African, Caribbean, and Pacific countries.

Bernard Hoekman is a Senior Advisor in the World Bank's Development Research Group and a Visiting Professor, Groupe d'Economie Mondiale, Institut d'Etudes Politiques, Paris.

HOW TO RUN THE INTERNATIONAL MONETARY FUND

Jeffrey D. Sachs

To restore its credibility, the IMF must represent all its members,
not just the ones who chose its new director.

You are taking over the International Monetary Fund (IMF) at a critical moment. A decade ago, globalization looked like a sure winner. Expanding world markets and global cooperation, many thought, would extend prosperity and foster peace. But today, the world is at war. The gap between the richest and poorest people is wider than ever. The vaunted capacity of economic integration to mitigate extreme poverty and environmental degradation looks illusory. The world needs effective international institutions more than ever, yet the legitimacy of the IMF is at a low ebb in many parts of the world.

From the start, the IMF has lived with a particular tension. It is an international organization with 184 member countries, and the IMF Articles of Agreement call on it to represent all of its constituent members. Yet the fund is governed by rich nations, foremost among them the United States. How you handle this tension will determine your own success or failure as the new managing director, as well as the continued relevance of the IMF at a time of enormous international strain. On key issues such as foreign aid, debt relief, and exchange-rate policy, you must learn to represent the entire world, not just the U.S. and European governments that put you into your job.

MONEY TALKS (AND VOTES)

The way you arrived at the IMF speaks volumes about how the institution functions. Although your record as minister of economy in the last Spanish government is impressive, you were not an obvious candidate to lead a global financial institution with major operational responsibilities in the poorest countries. You were not a leading figure in the great debates of the past decade regarding the East Asian crisis, the initiative on highly indebted poor countries, capital market liberalization, African poverty, or other issues of central concern to the IMF. Indeed, you owe your job to a nontransparent process in which the richest countries dominate and most of humanity has little

say. Still, given your professional skills and the respect you command among your peers, there is widespread hope and anticipation that you will rise to the occasion.

The fund's governance starts with some basic arithmetic. The IMF operates on a voting system based on each country's quota at the fund, rather than a system of one person, one vote (or one country, one vote). The United States, with 5 percent of the population of IMF member countries, controls 17 percent of the vote. Europe has a remarkable 40 percent of the IMF vote, with just 13 percent of the population. China and India comprise 38 percent of the world's population, and just 5 percent of the vote at the fund. Little surprise that the United States and Europe jealously guard their voting powers.

Nothing happens at the fund without the say-so of the United States and Europe. If decisions are by consensus, it is only because developing countries long ago learned not to lock horns with rich nations on matters of financial diplomacy. The IMF, after all, can do great financial damage to unruly countries, causing them to lose not only the resources of the fund, but also those of the World Bank, regional development banks, the Paris Club, the London Club, and private creditors, all of which are influenced by IMF judgments.

The institution prioritizes the interests of rich countries, and especially those of the United States, at every turn. When the United States wisely sought to forestall a Mexican default in early 1995, the IMF was induced to make an emergency loan of unprecedented size. When, on the other hand, the U.S. Treasury was wary of lending to Ecuador in late 1999 following that country's default to private creditors, the IMF withheld a pending loan. That decision helped topple the financially strapped government in Quito in early 2000. When ideologues in the Bush administration wanted to punish an allegedly leftwing government in Haiti, the IMF obligingly froze lending in 2001. Eventually, the Haitian economy crumbled and the elected government was ousted. Either through action or deliberate inaction, the IMF has repeatedly influenced politically charged issues of privatization, trade, and financial market policy in emerging markets at the behest of rich nations.

Countries siding with U.S. geopolitics have a much easier time getting IMF loans and debt relief. Countries labeled ideological opponents, by contrast, have had funding frozen at tremendous cost to their poor citizens. Most shocking, the IMF has asked the poorest of the poor for unconscionable belt-tightening and debt servicing since the early 1980s, because the United States, Japan, and most of the leading creditor countries in Europe showed little interest in extending debt relief or increasing development assistance. Debt relief for the poorest countries has been slow, grudging, and inadequate. The IMF has helped by dressing up fiscal austerity as a macroeconomic necessity.

THE LIVES IN YOUR HANDS

Your new job grants you daily and pervasive influence over billions of people, especially the world's poorest. African governments cannot take a financial step without the blessing of the IMF. If you make mistakes regarding Africa's finances, people don't just suffer, they die. That is not hypothetical. For the last 20 years, the IMF has been the chief enforcer of inhuman austerity conditions imposed on Africa, because the United States could not rouse itself to give Africa more help. The rich countries collect debts from impoverished nations, while pandemic diseases cut life expectancies to half of those in the rich world. Yet wealthy nations have pretended that there is no alternative to this state of affairs.

In recent months, the IMF Executive Board has approved lending programs to Burkina Faso, Democratic Republic of the Congo, Madagascar, Nicaragua, Tanzania, and Sierra Leone, among others. The board knows very little about these countries. Most of the information from the IMF staff that goes to the board is about budget deficits, domestic credit expansion, exchange rates, and inflation—not about AIDS, malaria, malnutrition, deforestation, and drought. A basic disconnect exists between the work of the fund, which obsesses over financial indicators, and reality in much of the world, especially where people live in extreme poverty.

Poor countries need massive investment in the building blocks of economic growth—including physical infrastructure, health systems, and education systems. The IMF should help establish the financial and macroeconomic framework for these investments. I am arguing not for inflationary finance and macroeconomically irresponsible policies, but for sound strategies in which the rich countries contribute much more development assistance to poor and vulnerable nations. The IMF must appreciate that its policymaking is part of a larger reality and that its programs should be judged against a standard higher than whether they produce price stability or a balanced budget. They must be judged on how they support the escape from extreme poverty, the control of pandemic disease, and the positive evolution of the global economic and political system. On all of these counts, the United States and the other leading economic powers have failed, and they have used the IMF as a key instrument in that failure.

MORE TRUTH, LESS DEBT

The financial interests of rich and poor countries are likely to be at odds in at least four areas during your tenure. The most urgent involves the poorest countries in the world, especially in sub-Saharan Africa. The rich countries signed international treaties—including the Millennium Declaration of 2000 and the Monterrey Consensus and Johannesburg Plan of Action in 2002—committing more financial help to developing countries trying to cut extreme poverty and disease sharply by 2015, as embodied in the Millennium Development Goals (MDG). Without deeper debt cancellation and much greater development aid, especially from the United States, these goals will not be met.

Under your leadership, will the IMF continue to enforce unconscionable austerity, or will it fulfill its pledge to support the MDGs by telling the truth about the need for much greater aid from the United States and other rich countries? The key step will be for the IMF, in conjunction with the World Bank, the U.N. agencies, and the governments of the poor countries, to insist that the richest nations finally move toward 0.7 percent of gross national product (GNP) in development aid and support well-governed poor countries trying to achieve the MDGs. For the dozens of impoverished states within the IMF, your support for these goals will signal the institution's willingness to confront the greatest economic challenge of our time: the dramatic reduction of extreme poverty.

The debt of emerging-market economies will be your second test. In 2002, IMF management considered a valuable proposal for a new system known as the Sovereign Debt Restructuring Mechanism. This plan would have finally brought to cases of sovereign debt some of the worthy principles of bankruptcy settlements—including easing the collective action problems that arise when multiple creditors confront an insolvent debtor. When U.S. financial interests balked, however, the U.S. Treasury pulled the plug on the IMF 's draft proposals. It behooves you to reopen this discussion and bring it to a more satisfactory conclusion.

Your third area of concern will be theIMF's position on the exchange-rate systems in developing countries. The United States has often politicized exchange-rate questions on mercantilist grounds by calling on other countries to appreciate their currencies vis-à-vis the dollar in order to reduce foreign exports to U.S. markets. For example, Washington pushed Japan away from a much-needed depreciation of the yen during the 1990s, and it is now pressuring China to appreciate the yuan. You should discount the fevered political advice you will get from Capitol Hill and focus on how exchange-rate changes will affect the country in question.

TAMING THE UNITED STATES

Potentially the most dangerous problem you will face is the financial instability that will soon result from irresponsible U.S. macroeconomic policies. The Bush administration has nearly done the impossible, converting a budget surplus of more than 2 percent of GNP in 2000 to a budget deficit of 5 percent of GNP this year—a swing of $700 billion in just four years. This transformation occurred through the combination of irresponsible tax cuts, massive increases in military spending, and even a surprising splurge on some domestic social programs.

U.S. fiscal and monetary imbalances could eventually threaten global financial stability. To keep their currencies from appreciating against the dollar, Asian central banks have been accumulating massive foreign exchange reserves. Around $2 trillion now sits in the bulging central banks of China, Japan, Taiwan, South Korea, Hong Kong, Singapore, and India. Meanwhile, the U.S. Federal Reserve continues to run an irresponsibly loose monetary policy, seemingly politicking on behalf of the Bush administration in the run-up to the November elections. The United States is playing with fire: A sharp rise in long-term interest rates, higher inflation, or a plummeting dollar may result.

You must push the United States to quickly improve its fiscal situation. Realistically, Washington must raise taxes, since there is no public support for broad-based spending cuts. Another key step will be for the rest of the world to move progressively and smoothly to a multicurrency reserve system in which the U.S. dollar is no longer the single reserve currency or unit of account. The IMF can play an important advocacy role in this transition. The euro will come into its own as a major global reserve currency. East Asian countries as well will need to diversify their units of account into a more appropriate East Asian basket. South American countries should think more seriously about much closer monetary cooperation and perhaps eventual monetary union within Mercosur, the economic cooperation organization that includes Argentina, Brazil, Paraguay, and Uruguay.

I also urge you to take advantage of the IMF's superb professional staff by fostering internal debate, as well as more symposia, conferences, and outreach. You should support the work of the Independent Evaluation Office, a valuable recent addition to the fund's institutional design that objectively assesses IMF operations. You should also encourage your staff to forge closer relationships with the World Bank and with U.N. agencies such as the United Nations Development Programme, the World Health Organization, and the Food and Agriculture Organization. These organizations know vastly more about economic development and poverty alleviation than does the IMF staff. Finally, it is critical that you increase transparency of the organization, in part by opening board meetings to more scrutiny and public participation. The board is seen as a mysterious cabal—an image that is both unnecessary and debilitating.

The fund has urgent global fiscal and financial responsibilities at a time when globalization itself remains under extreme threat. More than 80 percent of humanity lives in the developing world, and half of the world's population lives on less than two dollars per day. The world wants to know whether globalization works for all or only for the most powerful nations. Representing global interests with professionalism and goodwill will be an enormously tall order. On this task you have my best wishes and strongest hopes for your success.

Jeffrey D. Sachs is director of the Earth Institute at Columbia University.

A regime changes

The World Bank's new president is famous for his commitment to "regime change".
The Bank is committed to a peaceful version of the same thing

O N ITS way to the Mekong river, the Nam Theun tributary flows uninterrupted across the Nakai plateau in Laos, the poorest country in South-East Asia. Not for much longer. In March, the World Bank backed a proposal to dam it. Hydroelectric turbines will generate up to 1,070 MW of electricity, 95% of which will be exported to neighbouring Thailand.

This is the World Bank's natural habitat, where its compulsions and capabilities are both shown to full advantage. The project is not just an exercise in hydrology. The Bank's grants will help to resettle villagers, including Vietic-speaking hunter-gatherers, from the inundated plateau behind the dam and to compensate inhabitants of the dried-out riversides below it. As the Bank's International Advisory Group reported earlier this year, the displaced are experimenting with new ways to make a living, from an organic mulch plant to eel breeding. The project will set aside a nature reserve, where wildlife, from pangolin to reticulated python, will be defended by village gamekeepers, their salaries paid out of the dam's revenues.

But this is not, it is safe to say, the natural habitat of Paul Wolfowitz, who took office as the Bank's new president on June 1st. The plight of the reticulated python and the Vietic-speaking peoples are unlikely to have crossed his desk in the Pentagon, where he previously served as America's deputy secretary of defence. Mr Wolfowitz has instead spent most of his career cogitating about America's power in the world, representing it abroad and lobbying to enlarge it, first in congressional back offices, most recently at the intellectual forefront of George Bush's foreign policy. He knows little about finance; only a little more about development, although, as ambassador to Indonesia for three years, he has lived in a populous, poor country. Behind him, he leaves the ongoing nightmare of reconstructing Iraq, a project that is certainly behind schedule and over budget.

The Bank which Mr Wolfowitz now heads has as many sides as the Pentagon he has left. Speaking on May 31st he said he would be willing to listen and experiment, but it will take him some time to get to grips with a complex organisation. The Bank's most prominent aspect is the International Development Association (IDA), which gives grants ($1.7 billion last year) and soft loans (another $7.3 billion) to 81 of the world's poorest countries. As important, but less widely understood, is the International Bank for Reconstruction and Development (IBRD), which lent about $11 billion last year. The IBRD has some claim to being a bank rather than a fund. Blessed with a AAA-credit rating, it can borrow cheaply on the capital markets, and lend, slightly less cheaply, to the aristocracy of the third world, such as China and Brazil.

The Bank also has third and fourth sides—two smaller agencies that take on some of the risk of private lending to poor countries—and a fifth that settles disputes between foreign lenders and sovereign borrowers. Dams in Laos notwithstanding, only 5% of the Bank's money went to the energy and mining sectors last year. Three times as much went to social services, such as health, while education received 8%. The Bank also performs a type of economic chiropractics, giving money to governments in need of an "adjustment" in their policies, fiscal or monetary.

Mr Wolfowitz may, in fact, discover much that is familiar to him at the Bank. It is first and foremost a formidable technocracy. But in its own bloodless idiom, the Bank now talks increasingly about politics, even if it does so in euphemisms such as "good governance", "capacity building", "voice" and "empowerment". It is committed to understanding the political institutions of the countries in which it operates. Haltingly, hesitantly, it is also committed to changing them.

In June 2000, for example, the Bank lent $190m to help finance a 1,000km pipeline from the oilfields of land-locked Chad to the port of Kribi in Cameroon. But laying the pipe was the easy bit. Much harder is managing the revenues, which threaten to overvalue Chad's currency and underwrite endemic corruption.

The Bank's answer was two-fold. It insisted that the pipeline revenues be paid into an offshore escrow account. About 10% of the money would be held aside for future generations. The rest would flow to the government's poverty-fighting efforts under the close supervision of a new body, commonly known as the Collège. Staffed by parliamentarians, judges and representatives from human-rights groups, the Collège was, in effect, a new institution of state. It was soon debating whether to withhold money from the government. Clearly then, even when it is in the business of erecting dams and laying pipelines, the Bank is also often building states and reforming regimes.

Naïfs no more

That is a big change. Until 1996, politics was the variable that dared not speak its name at the Bank. Country directors, who head its branch offices in borrowing countries, came to their jobs as "self-described political neophytes", according to a recent Bank publication that recounts their education in the ways of the world.

Their initial innocence was largely self-imposed. Basil Kavalsky, who served as the Bank's country director across eastern Europe, confesses that it was "an article of faith . . . that we did not take political considerations into account." Actually, it was more than an article of faith. The Bank's articles of agreement, its founding charter, enjoin its officers to remain studiously apolitical.

Of course, the neophytes soon learned all about the political character of their host countries. But, notes Mr Kavalsky, they treated corruption as "a given, a part of the environment to be factored into the calculation. We did not treat it as a variable—something which we should make a concerted effort to address."

That changed with James Wolfensohn, Mr Wolfowitz's predecessor. It was perhaps his most far-reaching innovation in a tumultuous ten-year reign. In May 1996, he visited Indonesia, where Mr Wolfowitz had been ambassador from 1986 to 1989. The brazen corruption of the country's ruling Suharto clan irked them both. Mr Wolfowitz broached the issue, albeit politely, as he prepared to leave

his ambassadorial post in the country in 1989. Seven years later Mr Wolfensohn was more forthright. "Let's not mince words," he said at the Bank's 1996 annual meeting in Washington, DC, "we need to deal with the cancer of corruption."

The following year, the *World Development Report,* written by a team led by Ajay Chhibber, was the first publication in which the Bank properly addressed the topic. It was the beginning of a thorough re-examination of the role of the state and political institutions in development.

Mr Chhibber is now given to quoting Napoleon: "institutions alone fix the destinies of nations". That dictum finds some support in the latest economic research on development. A number of economists believe the policies they advocated in the 1980s and 1990s—stabilise prices, liberalise trade, privatise industries—matter less than the institutions that stand behind those policies.

Leading the chorus are Daron Acemoglu, Simon Johnson and James Robinson of the National Bureau of Economic Research. As they point out, for example, the prescription of stable finances and sound money did little to help in Argentina. The state found itself chronically prone to profligacy, for deep institutional reasons. It had to appease the country's unruly outlying provinces, which contribute little to the economy but dominate parliament. Likewise, they argue, Ghana's wildly overvalued exchange rate in its post-independence decades was not a monetary blunder. It was a political strategy designed to redistribute resources from the country's cocoa exporters, who received artificially low prices for their exports, to the import-buying city dwellers, on whose support the regime depended.

Measure for measure

Testing such theories is fraught with difficulty. But the measurement of institutions has made some progress. Dani Kaufmann, at the World Bank, notes an explosion of indicators of good government, most

Renaissance and regression

Changes in governance in Africa, 1996-2004

Governance category	Number of countries	Countries
Voice And accountabiility		
Significantly improved	5	Ghana, Gambia, Nigeria, Sierra Leone, Tanzania
Improved	16	
Worsened	22	
Significantly worsened	4	Central African Republic, Côtevoire, Eritrea, Zimbabwe
Government effectiveness		
Significantly improved	4	Botswana, South Africa, Tanzania, Congo
Improved	16	
Worsened	21	
Significantly worsened	6	Central African Republic, Côtevoire, Comoros, Sierra Leone, Chad, Zimbabwe
Control of corruption		
Significantly improved	3	Gabon, Tanzania, Congo
Improved	13	
Worsened	12	
Significantly worsened	5	Côtevoire, Guinea, Gambia, Namibia, Zimbabwe

Source: "Governance Matters IV", by D. Kaufman, A. Kraay and M. Mastruzzi, World Bank

based on business surveys or expert perceptions, that offer measures of accountability, bureaucratic competence, the rule of law, and so on. By sorting and sifting these numbers, he and his colleagues believe that they can derive workable measures of misrule. Precise rankings between countries are not possible, but broad comparisons are, and changes over time can be discerned. Over the past eight years, for example, many governments in Africa have defied the Afro-pessimists (see table), although more have regressed.

Mr Kaufmann believes he and his colleagues can demonstrate a strong causal link between his indices of sound government and prosperity. If the rule of law in Somalia, for example, were to match even that prevailing in Laos, Somalia's income would rise two- to three-fold in the long run, Mr Kaufmann estimates.

These are powerful arguments. But even if it is true that institutions fix a nation's destiny, can the Bank fix a nation's institutions? Is there a reliable "transmission mechanism" between the levers the Bank can pull and the results it cares about?

By training and temperament, Bank staff have tended to view government as a practical art. But their efforts to date give comfort to those of a more fatalistic cast of mind, who believe good government cannot be engineered, but must evolve.

In 2000, the Bank unveiled its strategy for reforming public institutions and strengthening governments. Between 2000 and 2004, lending to promote economic reforms fell by 14% a year, but lending to improve governance rose by 11%. In the 2004 fiscal year the Bank committed 25% of its lending to law and public administration (see chart overleaf). It had 220 staff dedicated to the cause, and more than 840 professionals affiliated with it.

For the most part, its direct efforts were confined to poorer countries, dependent on IDA for grants and soft loans. The richer developing countries, such as Brazil or India, where the state apparatus was formidable, were reluctant to cede ground to outsiders. In China, where Edwin Lim once served as chief of mission for the Bank, "the economic dialogue was always," he admits, "within the Chinese ideological and political limits."

A review of the Bank's efforts to prune the lush bureaucracies of African states concluded that civil-service reform remains elusive and intractable. Elsewhere, anti-corruption commissions proliferated, but achieved little—indeed they were often set up in the wake of some scandal as an alternative to doing anything.

Part of the difficulty, as Dani Rodrik of Harvard University points out, is that typical measures capture institutional outcomes, not institutional forms. The "rule of law", for example, measures how secure an investor feels about his property. It tells us little about precisely what makes him feel that way. According to Michael Woolcock, of the Bank, and Lant Pritchett, of Harvard University, the development industry can agree on "objectives" (children should be taught, roads should be passable, the rule of law should prevail) and "adjectives" (government should be accountable, transparent and responsive). But that is about all. As a result, Mr Kavalsky notes, the Bank's prescriptions in this field often come "very close to a tautology". What is required for growth? Good governance. And what counts as good governance? That which promotes growth.

That "P" word again

But the main difficulty was the obvious one: politics. When the Bank moved in on examples of bad governance, it too often forgot to ask, bad for whom? Consider, says Mr Chhibber, Turkey's banking system prior to that country's financial crisis in 2001. In 1998, the government was advised to set up an independent financial regulator, styled on those of Britain and Canada. Instead it created a regulator that was packed with political appointees. To the Bank's technocrats, it was obvious that the country had too many banks, many of them state-owned, and that they were not serving the economy at all well. But in Turkey at that time, state banks had a different purpose. They were the playthings of politicians, given to them as the spoils of electoral victory.

In such a situation, Mr Chhibber points out, all the Bank can do is bide its time. After the 2001 financial crisis, political resistance to an independent regulator broke down. Once established, the regulator closed more than 20 private banks, and cleaned up the system, at a cost of 33% of GNP. Mr Chhibber argues that earlier failures contributed to the eventual success. The work undertaken in 1998 allowed Turkey, under a new economy minister, Kemal Dervis, himself an alumnus of the Bank, to take advantage of the opportunity for reform when it arose.

In a speech in 2000, Mr Wolfowitz reflected on the thawing of authoritarian regimes in South Korea, Taiwan and the Philippines—the last of them on his watch as assistant secretary of state for East Asia. In these regimes, he noted, America worked on institutional, rather than revolutionary, change. It once counted Ferdinand Marcos, the dictatorial president of the Philippines, as an ally. If it had written him off, it would have lost all influence over him, Mr Wolfowitz said. But America could not coddle Marcos indefinitely either.

Such dilemmas will almost certainly revisit Mr Wolfowitz in his new job. The Bank must continually choose whether to coddle bad governments, or to cut them off. If misrule matters so much for development, should it reserve its money for committed reformers, turning its back on the reform-shy? That would make its money go further; it might also encourage laggards to reform. David Dollar and Victoria Levin, two Bank economists, reckon that since 1995 the Bank's soft-loan arm, IDA, has become much choosier about its clients. Broadly speaking, money flows to countries based on two main criteria: how well run is it? And how poor?

IDA may be pickier than it once was, but the Bank as a whole is not quite as discriminating as this study suggests. Richer countries, even if badly run, can still unlock money from the IBRD, the Bank's commercial-loan arm. And disastrously run countries are never entirely shunned by IDA. Each gets a small allocation regardless of its performance, and some qualify for money from the Bank's £25m trust fund for failed states, which it calls "low-income countries under stress".

Some think that, if it were to confine itself to the well-governed parts of the globe, the World Bank would scarcely warrant its title. But the Bank is learning that every unfit government is unfit in its own way. In

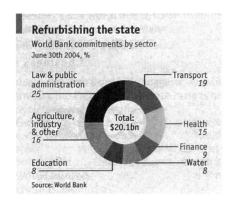

Refurbishing the state
World Bank commitments by sector
June 30th 2004, %

Law & public administration 25
Transport 19
Agriculture, industry & other 16
Total: $20.1bn
Health 15
Finance 9
Education 8
Water 8

Source: World Bank

some countries, citizens cannot hold policymakers to account (China); in others, policymakers cannot bend the bureaucracy to their will (Armenia). In some cases, the state is captured by venal interests—either wealth captures power (Russia under Yeltsin), or power captures wealth (Russia under Putin). In others, the state is so weak there is nothing worth capturing.

The Bank must pitch itself accordingly. If the state is honest, but weak, the Bank can try to train judges and equip civil servants. But there is no point investing in the machinery of a captured state. A project to strengthen the fiscal apparatus of Mobutu Sese Seko, the kleptocratic former ruler of Zaire, counts as the most misguided Bank project ever, in the opinion of Susan Rose-Ackerman, a corruption expert at Yale University.

If there is no will for reform on the part of government leaders, the Bank can try to go over their heads, stimulating demand for reforms in the public at large. Sometimes this works. When Thailand slipped in the Bank's ratings of good government, Mr Kaufmann recalls, the prime minister had to go on the radio to explain himself.

Truman's show

Some will argue, of course, that foreign aid has been political since its inception. The World Bank owes its existence to America's strategic commitment to rebuild post-war Europe. And many think the modern aid business and the cold war were twin-born at the moment of President Harry Truman's inaugural address in 1949. That speech is famous for Truman's vow to strengthen the freedom-loving nations of the world against the false philosophy of communism. But in it he also promised to share America's know-how and some of its resources with those parts of the world threatened by the "ancient enemies—hunger, misery and despair."

Mr Wolfowitz, of all people, is not one to disavow Truman's commitment to strengthen freedom. But if the ends Truman sought were deeply political, the means were mostly technocratic. The Bank which Mr Wolfowitz now leads is in a different game. The ends it pursues are primarily technocratic—it wants to fight poverty, not a false philosophy. But the means it employs have to be canny, opportunistic and, yes, political.

Ranking the Rich 2004

The second annual CGD/FP Commitment to Development Index ranks 21 rich nations on how their aid, trade, investment, migration, environment, security, and technology policies help poor countries. Find out who's up, who's down, why Denmark and the Netherlands earn the top spots, and why Japan—once again—finishes last.

The world's poor countries are ultimately responsible for their own development—and for years, rich countries have measured, categorized, scored, advised, and admonished them to cut their budget deficits, invest more in education, or liberalize their financial markets. Last year, the Center for Global Development (CGD) and FOREIGN POLICY turned the tables: We created the Commitment to Development Index (CDI), a ranking of rich nations according to how their policies help or hinder social and economic development in poor countries. One year and much additional data later, we unveil a second edition of the CDI that brings into sharper focus which governments lead the global community in the challenge of development.

Why should rich countries care about development in poor ones? For reasons both pragmatic and principled. In a globalizing world, rich countries cannot insulate themselves from insecurity. Poverty and weak institutions are breeding grounds for public-health crises, violence, and economic volatility. Fairness is another reason to care. No human being should be denied the chance to live free of poverty and oppression, or to enjoy a basic standard of education and health. Yet rich nations' current trade policies, for example, place disproportionate burdens on poor countries, discriminating against their agricultural goods in particular. Finally, the countries ranked in the CDI are all democracies that preach concern for human dignity and economic opportunity within their own borders. The index measures whether their policies promote these same values in the rest of the world.

In order to rank rich nations as accurately as possible, this year the aid, trade, and environment components of the index were revised, a technology component added, and the sections on investment, migration, and security (formerly called peacekeeping) overhauled. Australia gains most from these improvements in method, surging from 19th place in 2003 to 4th place this year, due in part to changes in the investment and security components. The new measure of security also helps boost the

United States 13 slots; Australia, the United States, and Canada all gain from improved data on migration. Amid all the jockeying, however, the same stalwarts anchor first and last place: Japan remains at the bottom of the CDI while the Netherlands stays at the top, though it now shares that position with last year's number two, Denmark.

In 2001, the United States collected more in import duties from Bangladesh than it did from France, despite importing 12 times as much from France.

Some governments got the CDI's message last year. For example, the Dutch government has adopted the CDI as one of its external performance standards for development and is now drafting a report on how it can improve its score. But despite such encouraging signs, underlying realities appear to have changed little. True, most donor countries gave more aid in 2002—the last year for which data are available—than in 2001. And under the aegis of the World Trade Organization (WTO), rich nations came to agreement on permitting poorer countries to import "generic" copies of patented pharmaceuticals, thus opening the door to cheaper AIDS drugs for Africa. However, rich countries—led by the United States, Japan, and France— remained intransigent on removing their agricultural tariffs and subsidies, contributing to the collapse of WTO negotiations in Cancun, Mexico in September 2003. And international efforts to reduce harmful greenhouse gas emissions also suffered when Russia joined the United States in blocking the passage of the Kyoto Protocol.

Ultimately, for all the CDI's focus on winners and losers, no wealthy country lives up to its potential to help poor countries. Generosity and leadership remain in short supply.

How the Index Is Calculated

Each country's overall score on the CGD/*FP* Commitment to Development Index is average of its scores in seven categories: trade, technology, environment, migration, investment, and aid.

Three fourths of the **trade** score on barriers to developing countries—tariffs, quotas, and subsidies for farmers in rich countries. Higher barriers yield lower scores. The remainder measures how much rich nations import from developing countries. Imports from the world's poorest nations receive greater weight, as do manufactured imports from all developing countries.

The **technology** component measures government support for research and development (R&D) as a percentage of GDP, including direct spending and tax subsidies. Defense-related R&D is discounted by half.

The **security** score rewards participation in peacekeeping operations and forcible humanitarian interventions during 1993–2002, counting only military operations approved by international bodies such as the U.N. Security Council or NATO Military operations are assessed

and converted to dollar terms based on the size of countries' defense budgets and share of standing forces committed.

Two thirds of the **environment** component reflects harm done to the global commons, factoring in consumption of ozone-depleting substances, subsidies for fishing, emissions of greenhouse gasses, and low gasoline taxes. The remaining third rewards contributions to international initiatives, such as ratification of major environmental treaties and donations to funds that help developing countries meet international environmental goals.

The net inflow of people from developing countries to wealthy ones between 1995 and 2000 accounts for 65 percent of the **migration** score. The percentage of students from developing countries among the total foreign-student population in rich countries counts for 15 percent, and aid to refugees and asylum seekers counts for 20 percent.

The **investment** component rewards policies that encourage helpful investment into developing countries. Eighty percent of the score recognizes policies promoting appropriate foreign

direct investment, such as political risk insurance and rules preventing double taxation. The remaining points reward long-term portfolio investment.

Finally, the **aid** component assesses total official assistance—grants and low-interest loans—as a percentage of the donor country's gross domestic product (GDP). It discounts by 20 percent aid that is "tied" to the purchase of goods or services from the donor nation, and subtracts debt payments received on past aid. The index penalizes donors based on the share of aid commitments made in amounts less than $100,000, which tend to overburden poor governments. Rich countries with tax incentives that encourage private charitable contributions gain points. They also win points simply for having lower taxes (relative to Sweden, the country with the highest taxes as a share of GDP), because higher post-tax income leads to more private giving.

A comprehensive explanation of the index's methodology is available on the Center for Global Development's Web site at `www.cgdev.org`.

THE ELEMENTS OF DEVELOPMENT

The CDI assesses seven major domains of government action: foreign aid, trade, investment, migration, environment, security, and—new this year—technology policy. How much foreign aid do countries give and to whom? Do rich nations erect high trade barriers to products made in the developing world? How do they treat the global environmental commons? Each country is scored in each area and averaged to arrive at a country's final ranking. See sidebar for a detailed description of the CDI's methodology. The index ranks Australia, Canada, Japan, New Zealand, the United States, and most of "Western Europe in their effort and leadership in promoting development in poorer countries—not their absolute impact. For instance, one cannot expect Denmark to give as much foreign aid to poor countries as Japan (whose economy is 20 times bigger), but one can ask Japan to give as large a share of its gross domestic product (GDP) as Denmark does.

Aid Foreign aid is the national policy most commonly associated with development efforts. In 2002, total aid flows from rich countries to poor ones reached $58 billion. Rich countries provide poor ones with grants, loans, food, and technical advice to support everything from massive infrastructure projects to immunization programs in tiny rural villages. Most comparisons of aid examine simple measures such as total assistance as a percentage of the donor's GDP. The CDI goes further, by considering the quality and not just the quantity of aid provided.

For starters, the CDI discounts "tied aid," whereby donors require recipient countries to spend their aid on goods and services such as tractors or educational consultants from the donor nation. (Tying aid can raise the costs of any given development project by 15 to 30 percent by preventing recipients from shop-

ping around for the best deal.) The index also subtracts all debt payments received from developing countries on aid loans, rewarding donors that forgive poor countries' debts. Choice of recipient countries is considered too: The CDI rewards aid to countries that are relatively poor yet relatively uncorrupt and accountable to their citizens.

The CDI also penalizes donors for overloading governments in poor nations with onerous aid reporting requirements and countless "mission" visits from foreign aid officials. For example, Mozambique, with its combination of high poverty and relatively good governance, has attracted much donor interest in recent years, resulting in 1,413 new aid project commitments between 2000 and 2002. That amount is more than India (1,339 new projects) and China (1,328), countries with vastly more administrative staff to manage relationships with donors. Rich nations help Mozambique more when they jointly fund a few large-scale programs in, say, education or health. Last year, Tanzania even declared a four-month "mission holiday," during which the country received only the most urgent visits by donor officials. Evidently Tanzanian officials needed some peace so they could get work done.

This year, the CDI rewards governments for allowing their citizens to write off charitable contributions on their income taxes—and for taxing their citizens less, leaving more money in private hands for charity. Some of those contributions go to humanitarian organizations such as Oxfam and CARE that do important work in developing countries. Currently, all index countries except Austria, Finland, and Sweden offer tax deductions or credits for such contributions. However, even in the United States—often considered a stingy government donor and generous source of charity—private giving is small compared to public giving U.S. government aid in 2002 was $13.3

The Rankings

The CGD/*FP* Commitment to Development Index ranks 21 of the world's richest countries based on how much their policies help or hinder development in poorer nations. The index examines seven policy categories: foreign aid, investment, openness to immigration, responsible environmental practices, contributions to internationally approved security operations, support for technology development, and openness to international trade. Dark cells indicate a particularly favorable score, and light cells indicate poor performance. Substantial improvements in method explain most changes in rank since 2003.

Rank	Country	Aid	Investment	Migration	Environment	Security	Technology	Trade	Average	2003 Rank*
1	Netherlands	11.2	6.7	5.9	5.3	6.4	5.5	5.9	6.7	1
1	Denmark	12.3	4.8	6.1	5.7	7.1	5.0	5.8	6.7	2
3	Sweden	12.4	3.8	5.1	5.8	4.5	5.7	5.8	6.1	8
4	Australia	2.9	6.5	8.8	3.3	9.0	6.4	4.4	5.9	19
4	United Kingtom	4.8	6.4	4.4	5.8	9.1	4.7	5.8	5.9	11
6	Canada	3.6	6.3	11.2	2.9	4.3	6.6	5.7	5.8	18
7	United States	1.9	5.6	10.5	2.3	4.9	5.5	6.7	5.3	20
7	Germany	3.9	6.7	6.1	6.1	2.9	5.6	5.8	5.3	6
7	Norway	10.6	5.3	4.9	4.0	9.3	5.5	-2.7	5.3	10
7	France	6.0	4.7	2.7	5.9	5.6	6.1	5.8	53	14
11	Finland	5.0	5.1	2.6	5.0	6.7	6.3	5.7	5.2	17
12	Austria	3.7	4.4	2.9	6.1	3.1	6.9	5.8	4.7	9
13	Belgium	6.0	4.3	2.6	5.9	4.0	3.4	5.8	4.6	12
14	Portugal	2.3	5.6	2.8	5.4	5.2	4.5	5.8	4.5	3
14	Italy	2.8	5.3	3.6	5.5	3.6	4.7	5.9	4.5	15
16	New Zealand	0.8	2.9	5.0	4.7	6.7	4.1	5.9	4.3	4
17	Greece	1.8	4.1	6.2	4.7	4.0	2.5	5.8	4.1	13
18	Ireland	3.0	2.7	5.8	2.8	5.5	2.0	5.8	3.9	15
18	Switzerland	5.8	4.7	3.6	7.9	0.7	4.5	0.3	3.9	5
20	Spain	2.0	4.5	2.3	5.5	2.0	4.0	5.8	3.7	6
21	Japan	2.4	4.6	1.9	4.5	0.4	5.4	3.4	3.2	21

*2003 ranks calculated using 2003 method.

billion, or 13 cents a day per U.S. citizen. U.S. private giving to developing countries was another $5.7 billion, less than six cents a day, two cents of which is attributed to U.S. tax policy as opposed to individuals' own decisions. In the end, factoring in tax policy only lifts the U.S. aid rank from 20th to 19th.

Sweden led the aid component this year, followed closely by neighbors Denmark, Norway, and the Netherlands. All four governments scored above 10 (On a scale of 0 to 10; see the "Off the Scale?" sidebar.) by virtue of the sheer amount given. Many of the CDI nations increased their foreign aid in 2002, especially the United States, which favored geopolitically important actors such as Turkey, Indonesia, Russia, and Afghanistan. But although the United States gives more aid than any other country in absolute terms, it still gives less aid in proportion to its size than any other rich country, and so finished near the bottom in this category. However, due to the penalty for overloading countries with projects, Greece and New Zealand scored below the United States. Evidently these countries spread their modest aid thinly, covering many countries with small projects and overburdening local administrators. [See chart: Throwing Money Around.]

Trade The WTO negotiations in Cancun collapsed last September after an alliance of developing countries challenged rich nations over their agricultural subsidies and tariffs. Agriculture typically comprises between 17 and 35 percent of GDP in developing countries, compared with less than 3 percent in rich na-

tions. When high-income countries tax food and subsidize their own farmers' production, they destroy markets for developing country farmers who lack such protection.

Wealthy nations' tariffs on industrial goods also tend to hurt the poor, with high rates for the labor-intensive products that are the mainstay of developing countries. In 2001, the United States collected more in import duties from Bangladesh ($331 million, mainly on clothing) than it did from France ($330 million), despite importing 12 times as much from France in dollar terms. The index penalizes all such barriers. However, a few may soon fall: Rich countries must abolish their quotas on textiles and clothing made in developing countries on December 31, 2004, under the 1994 treaty that created the WTO. China may reap big benefits from greater access to Western clothing markets, but Bangladesh—which receives a large share of the quotas—may be hurt. Still, on balance, if rich countries eliminated all their trade barriers, the ranks of the global poor would shrink by more than 270 million over 15 years, estimates CGD Senior Fellow William Cline.

Latin American and Caribbean economies received $32 billion in remittances in 2002, six times what they received in foreign assistance.

Off the Scale?

For the second year in a row, the United States tops the index's trade score. Norway repeats its poor performance, due to extremely high agricultural tariffs. Although Norway supports poor countries with a generous foreign aid budget, it undermines that support with its high trade barriers.

Investment Foreign investment can distort development and feed corruption and violence. For example, Angola's government, which reaps massive oil revenues from foreign firms, has reportedly filched or misspent $4.2 billion in five years, equivalent to nearly a tenth of its annual GDP. However, foreign investment can also be a significant driver of development in poor countries. In China, India, and Mexico, foreign investors have brought not only money but technical and managerial know-how.

This year's CDI investment component surveys what governments in rich countries are doing to facilitate investment flows to developing nations as well as ensure that investment promotes development. The index looks at two kinds of investment. The first is foreign direct investment, where a company builds factories or buys large stakes in companies overseas. Do governments offer political risk insurance to encourage companies to invest in poor countries whose political climate would otherwise be deemed too insecure? Do they avoid investment projects likely to harm the environment or exploit workers? Do governments help investors avoid double-taxation of profits earned in developing countries? The second type is portfolio investment, where foreigners buy securities traded on open exchanges outside their home country. Do countries help create securities markets and institutions? Do they allow domestic pension funds to invest in developing countries?

Ireland places last on the investment ranking this year in part because it does not provide political risk insurance or help investors avoid double-taxation. In contrast, the top-ranked Netherlands does both, although its insurance program does not screen for environmental and labor problems.

Migration Rich countries frequently tout free trade's positive impact on economic development. The basic arguments for freer trade apply to migration as well. People who move from poor countries to rich ones usually earn more in their new homes and send money back to support their families. For example, Latin American and Caribbean economies received $32 billion in remittances in 2002, six times what they received in foreign assistance. Remittances accounted for nearly 30 percent of Nicaragua's GDP in 2002 and 25 percent of Haiti's. The impact on poor countries when professionals leave—the so-called "brain drain"—is more complex. For instance, the exodus of doctors and nurses from Ghana and South Africa has devastated these countries. However, sometimes professionals gain skills abroad and then move back home: Returning Indian expatriates are playing a big role in that country's software and services boom. Even when professionals remain abroad, they often retain links with industry and research at home.

Unfortunately, insufficient data prevent the index from distinguishing between skilled and unskilled migrant flows. So, the CDI relies on the core conviction that freer movement of people, including those with considerable professional skills, benefits development overall.

The migration component has improved since last year; it now measures not only how many migrants come to rich countries but how many leave, better reflecting whether immigrants stay long enough to put down roots, send home substantial sums, and pick up real skills. The 2004 CDI estimates the net flow of immigrants from developing countries in a five-year period, 1995 to 2000. Because of this change, Canada bumped Switzerland and New Zealand off their perch as the most migrant-friendly countries. Switzerland in particular admits many people from developing countries—but many of them leave soon after. In contrast, the migrant population is growing steadily in Canada, as well as in Australia and the United States. The CDI also considers openness to students, refugees, and asylum seekers from poor countries.

Environment Citizens in rich countries often think of environmental protection in terms of preserving the world for their children and grandchildren—people who do not participate in today's environmental degradation but who will suffer its consequences. Yet today's global poor are already harmed by irresponsible environmental policies. Rich countries are the

Throwing Money Around

The Commitment to Development Index penalizes wealthy nations for overloading poor nations with aid projects that strain governments' administrative capacities and dilute the impact of the projects.

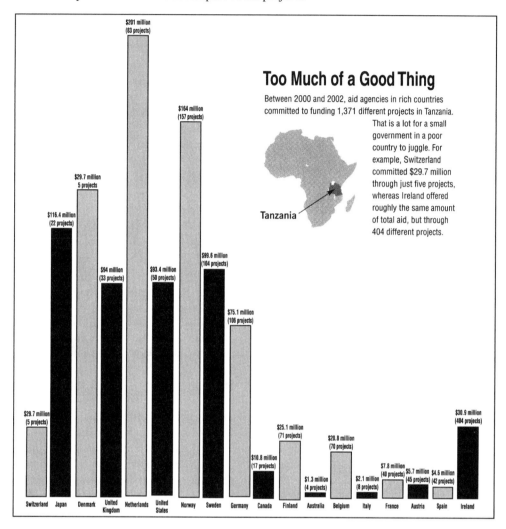

Too Much of a Good Thing

Between 2000 and 2002, aid agencies in rich countries committed to funding 1,371 different projects in Tanzania. That is a lot for a small government in a poor country to juggle. For example, Switzerland committed $29.7 million through just five projects, whereas Ireland offered roughly the same amount of total aid, but through 404 different projects.

Tanzania

primary users of scarce global resources, but poor countries are the most likely to be hurt by ecological deterioration and the least capable of adapting. These countries typically have weak infrastructures and social services, making them particularly vulnerable to the floods, droughts, and spread of infectious diseases that global climate change could bring.

In the end, no wealthy country lives up to its potential to help poor countries. Generosity and leadership will remain in short supply.

The index assesses whether countries are reducing their depletion of shared resources and contributing to multilateral efforts to protect the environment, such as the Montreal Protocol fund established in 1990 to help developing countries phase out ozone-

depleting chemicals. This year, Switzerland retained its hold on first place in this category because of its low greenhouse gas emissions. The United States remained in last place because of its high emissions and low taxes on gasoline. Denmark, France, and the United States picked up about half a point each for ratifying the latest international agreement on protecting the ozone layer, known as the Beijing Amendment, in 2003.

Security As recent events in Liberia and Haiti demonstrate, rich countries' military power can protect developing countries' citizens from the violent upheaval that political instability and civil conflict too often create. Internal instability can take a terrible toll on people's well-being: Children forced into armed rebel groups in Sierra Leone and Uganda during the 1990s lost their childhood, chance for education, and in many cases, their lives. Instability also undermines economic and political development, robbing entire countries of their future.

Want to Know More?

For the details of the 2004 CGD/*FP* Commitment to Development Index, see **"The Commitment to Development Index: 2004 Edition,"** by David Roodman, available at **www.cgdev.org.** Background papers are available for each of the index's componentsa: Roodman on foreign aid and environmental performance, Alicia Bannon and Roodman on technology, William Cline on trade, Theodore Moran on investment, Elizabeth Grieco and Kimberly A. Hamilton on the Migration Policy Institute on migration, and Michael O'Hanlon and Adriana Lins de Alburquerque on security.

William Esterly offer cricital views on ast aid efforts in *The Elusive Quest for Growth: Economist' Adventures and Misadventures in the Tropics* (Cambridge: Massachusetts Institute of Technology Press, 2001). For an analysis of the United States' new aid programs, consult Steven Radelet's *Challenging Foreign Aid: A Policymaker's Guide to the Millennium Challenge Account* (Washington: Center for Global Development, 2003). Nancy Birdsall and Brian Deese

discuss the overburdening or poor countries in *"Hard Currency"* (*Washington Monthly,* March 2004).

Visit the Web site of the **Migration Policy Institute** for excellent data and analysis on migration issues. See also Devesh Kapur and John McHale's article **"Migration's New Payoff"** (FOREIGN POLICY, November/December 2003) for a look at the social and economic impact of remittances in poor countries. Moran's *Beyond Sweatshops: Foreign Direct Investment and Globalization in Developing Countries* (Washington: Brookings Institution Press, 2002) analyzes how government policies can shape foreign direct investment (FDI), and how FDI can, in turn, shape development. O'Hanlon's *Expanding Global Military Capacity for Humananitarian Intervention* (Washington: Brookings Institution Press, 2003) examines the challenges government face when preparing humanitarian interventions of the kind the CDI rewards.

Carol C. Adelman offers an alternative approach to comparing donors in **"The Privatization of Foreign Aid: Reassess-**

ing National Largesse" (*Foreign Affairs,* November/December 2003). The British humanitarian group Oxfam assails rich nations for their trade practices in **"Rigged Rules and Double Standards: Trade, Globalization, and the Fight Against Poverty"** (Oxford: Oxfam, 2002). Arvind Panagariya's **"Think Again: International Trade"** (FOREIGN POLICY, November/December 2003) dissects the conventional wisdom on how trade affects poor nations. Also see the United Nations Development Programme's **"Human Development Report 2001: Making New Technologies Work for Human Development"** (New York: Oxford University Press, 2001) for a discussion of the impact of technological innovations on economic development.

For links to relevant Web sites, access to the *FP* Archive, and a comprehensive index of related FOREIGN POLICY articles, go to **www.foreignpolicy.com.**

The CDI tallies the financial and personnel contributions that governments have made to peacekeeping operations and (new this year) forcible humanitarian interventions. Because the merits and motives of such interventions are often highly controversial, the CDI filters out operations lacking approval from international bodies such as the U.N. Security Council or the African Union. After extensive data gathering, the CDI now considers a country's history of contributions over a decade—instead of two years—to assess its current willingness and ability to participate.

Australia's 4,500-troop intervention to stop the Indonesian military's oppression of the East Timorese in 1999 (a large deployment for a country the size of Australia) earns that country third place. Surprisingly, the United States comes in only 11th, despite contributing more than 50,000 personnel to interventions in Haiti, Somalia, Kosovo, and Bosnia. By the standards of its peers, this is not a large contribution after adjusting for economic, size. Because data are incomplete for 2003, U.N.-approved postwar security aid for Iraq is not included this year—but it could be next year. The invasion of Iraq, however, will not be counted because no major international body approved it. Japan and Switzerland rank at the bottom of the component, due to Switzerland's traditional neutrality and Japan's constitutional limits on military interventions.

Technology Arguably, the most profound long-term effect of rich countries on poorer countries' development comes from new technologies. East Asian countries have enjoyed near-miraculous growth, halving poverty rates between 1975 and 1995, thanks in part to their production of electronic goods originally invented in rich countries. Moreover, vaccines and antibiotics

led to major gains in life expectancy in Latin America and East Asia in the 20th century; these regions achieved in only four decades improvements that took Europe almost 150 years. Cell phones have revolutionized communications in poor countries such as Nigeria. The Internet also helps developing countries access and disseminate information, form civil-society movements, and trade with rich economies.

To capture the government's role in encouraging globally beneficial innovation, the CDI's new technology measure counts total government subsidies for research and development (R&D)—whether delivered through spending or tax breaks—as a share of GDP. Unfortunately, few data are available on the amount of R&D funding in areas most relevant to the world's poorest populations, such as malaria vaccines and tropical agriculture. The index discounts military spending on R&D by 50 percent because, while some military innovations have useful civilian spin-offs (including the Internet), much military R&D does more to improve the destructive capacity of rich countries than the productive capacity of poor ones.

Austria and Canada come out on top of the CDI technology component, with their governments spending 0.9 percent of GDP on R&D (discounting for military R&D). Greece and Ireland finish last with 0.3 percent. The U.S. government actually devotes the most to R&D as a share of GDP, but half of that is military, so the defense penalty pulls the country to seventh place.

ROOM FOR IMPROVEMENT

A quick scan down the final column of the 2004 CDI rankings reveals big changes in standings since last year. But changes in rankings reflect almost entirely improvements in methods and measures. The bottom line, as was the case last year, is that every country's performance is mediocre or worse in at least one area. Even the highest-ranking countries could do much better.

When rich countries improve the health of other countries, their own outlook will improve as well. A prescription for the CDI countries in the coming year would include abolishing agricultural subsidies and tariffs, legalizing more migrant flows, and giving more aid to countries based on their needs and prospects, not on narrow geopolitical interests. There is tremendous room for all rich countries to demonstrate true leadership in support of global development.

Calculating the Benefits of Debt Relief

How cutting the external debt burden can boost growth in low-income countries

Rina Bhattacharya and Benedict Clements

A LARGE number of low-income countries are now receiving debt relief under the Heavily Indebted Poor Countries (HIPC) and Enhanced HIPC initiatives, launched by the IMF and the World Bank in 1996 and 1999, respectively. These initiatives aim to cut the debt burdens of some of the world's poorest countries to help them combat poverty. But how do lower debt burdens and reduced debt service payments translate into higher growth and better living standards?

Economists have often argued that high external debt makes it more difficult for countries to achieve the Millennium Development Goals (MDGs). High debt service absorbs resources that could be used for essential spending on poverty reduction, and diverts resources away from public investment. However, despite substantial research into the impact of external debt on growth in general, surprisingly few studies have focused on low-income countries, and the HIPCs in particular. Because most low-income countries do not have access to international capital markets, the impact of external debt on growth can be different in low-income and emerging market countries. The channels through which debt affects growth may also differ. Further, low-income countries are usually net recipients of concessional loans and aid, even when debt service is high, suggesting that the adverse impact of debt service on growth may not be large.

We assessed the impact of external debt on growth in low-income countries and the channels through which these effects are realized. Special attention was given to the indirect effects of external debt on growth through its impact on public investment because of the statistically significant influence of public investment on economic growth.

Debt and growth in theory

What does economic theory have to say about the relationship between the stock of external debt and growth? External debt can potentially help foster higher economic growth, provided that it is used to help finance investment. In light of the diminishing returns to capital, however, the net benefits of additional investment could decline as debt increases. In addition, high levels of debt may hamper growth through the effects of "debt overhang." When there is a debt overhang, a country's debt exceeds its expected ability to repay, and expected debt service is likely to be an increasing function of the country's output level. Thus, some of the returns from investing in the domestic economy are effectively "taxed away" by foreign creditors. As a result, investment by both domestic and foreign investors—and thus economic growth-is discouraged. A high level of external debt can reduce a government's incentive to carry out important structural and fiscal reforms if it anticipates that foreign creditors will reap most of the benefits. Debt overhang can also depress growth by increasing uncertainty about the actions and policies that the government will resort to in order to meet its debt service obligations.

The theoretical literature thus suggests that foreign borrowing has a positive impact on investment and growth up to a certain threshold level; beyond this level, however, its impact is adverse, giving rise to a "Laffer curve"-type relationship between external debt, on the one hand, and investment and per capita income growth on the other (see "External Debt and Growth," *F&D*, June 2002, p. 32).

External debt service (in contrast to the total debt stock) can also potentially affect growth by crowding out private investment or altering the composition of public spending. Other things being equal, higher debt service can raise the government's interest bill and the budget deficit, reducing public savings; this, in turn, may either raise interest rates or crowd out credit available for private investment. Higher debt service payments can also squeeze the amount of resources available for infrastructure and human capital formation, with further negative effects on growth.

Filling a gap

But existing empirical research does not provide a clear picture of how debt affects growth, particularly in low-income countries. Our research attempts to fill this gap. We start by estimating an empirical equation for per capita income growth, based on data for 1970-99 for a group of 55 low-income countries classified as eligible to receive funds under the IMF's Poverty Reduction and Growth Facility (PRGF). The variables in the equation include lagged per capita GDP; the secondary school enrollment rate; private investment as a share of GDP; public investment as a share of GDP; and a measure of the openness of the economy to foreign trade.

> ## "Debt appears to affect growth via its effect on how efficiently resources are used, rather than by discouraging private investment."

We add to the traditional growth equation different measures of external public and publicly guaranteed debt to assess the effect of this debt on growth. These measures are the face value of the stock of external debt as a share of GDP; the net present value of the stock of external debt as a share of GDP; the face value of the stock of external debt as a share of exports of goods and services; and the net present value of this debt as a share of exports of goods and services. The net present value takes into account the degree of concessionality of the debt. To capture the interaction between growth and debt, an appropriate econometric technique known as *generalized method of moments* is used.

Our results suggest that high levels of debt can indeed depress economic growth in low-income countries, but only after it reaches a certain threshold. This threshold is estimated at about 50 percent of GDP for the face value of external debt and about 20-25 percent of GDP for its estimated net present value. For the external debt indicators expressed as a ratio to exports, the results are somewhat weaker, but they indicate a threshold level for the net present value of external debt of about 100-105 percent of exports. Moreover, debt appears to affect growth via its effect on how efficiently resources are used, rather than by discouraging private investment, since the results indicated that the latter does not have a statistically significant impact on growth in this group of countries. The empirical estimates thus provide some support for the debt overhang hypothesis.

Our results also suggest that debt service has no direct effect on real per capita GDP growth. One reason why debt service may be insignificant is that its effect is realized through its impact on public investment, which is included as an explanatory variable in the model and is thus held constant. We explored this possibility in greater detail by estimating a public investment equation and

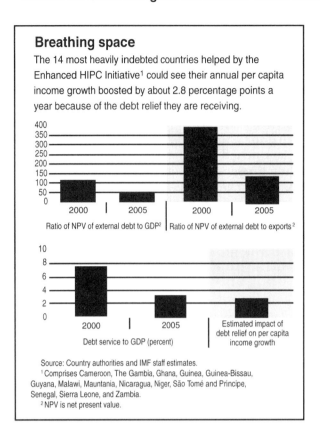

Breathing space

The 14 most heavily indebted countries helped by the Enhanced HIPC Initiative[1] could see their annual per capita income growth boosted by about 2.8 percentage points a year because of the debt relief they are receiving.

Ratio of NPV of external debt to GDP[2] | Ratio of NPV of external debt to exports[2]

Debt service to GDP (percent)

Estimated impact of debt relief on per capita income growth

Source: Country authorities and IMF staff estimates.
[1] Comprises Cameroon, The Gambia, Ghana, Guinea, Guinea-Bissau, Guyana, Malawi, Mauritania, Nicaragua, Niger, São Tomé and Príncipe, Senegal, Sierra Leone, and Zambia.
[2] NPV is net present value.

looking at the impact of both the stock of external debt and the external debt service ratio.

The empirical results provide support for the hypothesis that higher debt service crowds out public investment: under most formulations of the model, debt service has a statistically significant negative effect on public investment. The relationship appears to be nonlinear, with the crowding-out effect intensifying as the ratio of debt service to GDP rises.

How significant is the crowding-out effect? Under linear formulations of the model, the results indicate that for every 1 percentage point of GDP increase in debt service, public investment declines by about 0.2 percent of GDP. The modest size of this coefficient is somewhat surprising and indicates that high debt burdens have not had a very large effect on public investment in low-income countries. These results suggest that debt relief on its own cannot be expected to lead to large increases in public investment. In most cases, it instead leads either to greater public consumption, or—if used for deficit reduction or lower taxes—to higher private consumption or investment.

Policy implications

A high stock of debt tends to depress economic growth in low-income countries. This has important implications for the impact of debt relief on growth in the HIPCs. But how big are these effects?

Consider the case of the 14 most heavily indebted HIPCs (in terms of debt service payments-to-GDP ratios) in 2000 (see chart). The net present value of external debt for these countries is projected by the IMF to fall from over 113 percent of GDP in 2000 to just under 45 percent of GDP in 2005. Our results suggest that this sharp reduction would directly add about 2.8 percentage points to annual per capita income growth. At the same time, the average ratio of debt service to GDP in these countries is projected to fall from 7.5 percent to 3.3 percent over the same time period. Calculations using the results from the best-fitting regression suggest that this would increase public investment by 0.5-0.8 percent of GDP and indirectly raise real per capita GDP growth by 0.1-0.2 percent annually. Moreover, if a larger share of debt relief were channeled to public investment, the impact on annual per capita income growth would be correspondingly higher. Under all scenarios, greater public investment only bolsters growth if matched by other revenue and expenditure measures that keep the budget deficit from rising.

These results have important implications for the design of adjustment programs in countries receiving debt relief. Reducing the stock of debt alone can have a significant positive direct impact on per capita income growth in the most heavily indebted poor countries. Cutting debt service obligations can also provide breathing space for raising public investment. To further strengthen the link between debt relief and growth, countries could consider allocating a larger share of this relief to productive public investment than in the past.

Rina Bhattacharya is an Economist in the African Department and Benedict Clements is a Deputy Division Chief in the IMF's Fiscal Affairs Department.

Recasting the case for aid

A new report for the UN says that aid can work, and demands lots more of it

NOBODY ever accused Jeffrey Sachs of lacking purpose, thinking small or keeping his ideas to himself. The United Nations shrewdly chose the Columbia University professor to lead a team of "more than 250 experts" (they must have lost count) charged with producing a grand new proposal on development. This week Mr Sachs and his army delivered.

"Investing in Development: A Practical Plan to Achieve the Millennium Development Goals" is, in many ways, an impressive—one might say heroic—piece of work. The document in full runs to ten supporting volumes and more than 3,000 pages (this writer lost count). The overview paper is packed with high-octane analysis and recommendations, no waffle, not a sentence wasted. Its aim is no less than to dispel the prevailing pessimism on aid—a deeply entrenched attitude, based on years of disappointment—and to mobilise hundreds of billions of dollars in new help for the developing world.

In this, it might succeed. Whether it deserves to is another question.

The Millennium Development Goals (MDGs) were adopted at a United Nations summit in 2000. They have already begun to shape the policies, as well as the rhetoric, of institutions such as the World Bank and the International Monetary Fund. For the first time, in thinking about development, governments are having to frame their policies around specific intended outcomes rather than policy inputs. It is a big-

ger change than you might suppose. Because of the MDGs, allocations of aid, or targets for budget deficits or other policy settings, are no longer front and centre. These and other inputs are increasingly being tested against the final goals. The question now—and it is the right question—is what policy inputs will be required to hit the targets[?]

For the least successful regions of the world, the targets are extremely ambitious: halve the proportion of people living in extreme poverty (ie, on less than a dollar a day) between 1990 and 2015; achieve universal primary education by 2015; reduce the under-five mortality rate by two-thirds; and more. Altogether the MDGs encompass 18 targets arranged under eight broad headings. The table shows how one region, sub-Saharan Africa, is doing so far (measured against a smaller set of quantitative, as opposed to qualitative, targets).

Plainly, it is doing very badly. Unlike some other regions, notably East Asia and the former Soviet empire, sub-Saharan Africa has met or is on track to meet not a single MDG. Even where it is making some progress (the blue panels), the change is too slow to bring the MDG within reach. Mr Sachs boldly claims that even sub-Saharan Africa can achieve the MDGs by 2015—but only with a huge new commitment of aid, designed and delivered in new ways. The document published this week spells out the amounts that Mr Sachs and his team believe will be required, not just for Africa but for all of the devel-

oping world, and the methods to be used.

Given what is at stake, Mr Sachs's passion and ambition are entirely warranted—but does the approach he advocates make sense? To persuade rich-country governments to give more, he first reminds them that they have been promising for the past 35 years to spend 0.7% of GDP on aid. Very few countries have kept that promise, or even come close. But there is a reason for this: often in the past aid has failed, or at least has been understood to fail. The report devotes a good deal of attention to arguing that this perception is wrong.

Many of the studies which found that aid fails were flawed, says Mr Sachs. For instance, if you simply compare, country by country, growth rates and aid volumes, you will find that aid and slow growth are correlated. But this is because a lot of aid is given to countries recovering from natural disasters, famines or other humanitarian emergencies. You would expect countries battered down by such calamities to grow more slowly than the average, so the correlation between aid and slow growth is false. Looking only at development aid, the report argues, you find that aid works: it spurs growth.

Also, the report says that a lot of aid in the past has been badly designed: too much reliance on short-term-funding, for instance, causing projects to fail because recurrent expenses cannot be covered. Well-designed aid, delivered in a sustained

Missing the target

Major trends in Millennium Development Goals in sub-Saharan Africa

Goal 1: Eradicate extreme poverty and hunger	■ Progress, but too slow ▲ No or negative change
Reduce extreme poverty by half	▲ high, no change
Reduce hunger by half	▲ very high, little change

Goal 2: Achieve universal primary education	
Universal primary schooling	■ progress, but lagging

Goal 3: Promote gender equality and empower women	
Girl's equal enrolment in primary school	■ progress, but lagging
Girl's equal enrolment in secondary school	■ progress, but lagging
Literacy parity between young men and women	■ lagging
Women's equal representation in national parliaments	■ progress, but lagging

Goal 4: Reduce child mortality	
Reduce mortality of under-five-year-olds by two-thirds	▲ very high, no change
Measles immunisation	▲ low, no change

Goal 5: Improve maternal health	
Reduce maternal mortality by three-quarters	▲ very high

Goal 6: Combat HIV/AIDS, malaria and other diseases	
Halt and reverse spread of HIV/AIDS	■ stable
Halt and reverse spread of malaria	▲ high
Halt and reverse spread of tuberculois	▲ high, increasing

Goal 7: Ensure environmental sustainability	
Reverse loss of forests	▲ declining
Halve proportion without improved drinking water in urban areas	▲ no change
Halve proportion without improved drinking water in rural areas	■ progress, but lagging
Halve proportion without sanitation in urban areas	▲ low access, no change
Halve proportion without sanitation in rural areas	▲ no change
Improve the lives of slum dwellers	▲ rising numbers

Goal 8: A global partnership for development	
Youth unemployment	▲ high, no change

Source: UN Millennium Project

way to countries with reasonably good governments, does what it is supposed to.

That good-government precondition is crucial, however, and causes the team some difficulty. The development literature emphasises the importance of sound institutions and clean, effective government. Countries with those things can put aid to good use. The trouble is, countries with those things tend to make progress unassisted. The poorest countries, including the basket-cases of sub-Saharan Africa, are the most deserving by the test of need, but tend to be the worst governed.

Potentially rich

The report challenges this thinking in two ways. First, it insists that some of the world's poorest countries are in fact pretty well-governed: they are poor for other reasons, to do with geography, history, incidence of disease, and so forth. Identify the good governments, the thinking goes, and give them aid "at scale".

The table suggests that standards of governance in sub-Saharan Africa are indeed far from uniform. It shows countries and their governance ratings according to a variety of measures—as judged by peers in the African Union; by the IMF and World

Bank as part of the Heavily Indebted Poor Countries initiative; by the American government for the purposes of its Millennium Challenge scheme (well-governed countries can qualify for assistance or, in the case of "threshold countries", be deemed on track to qualify); according to whether countries have prepared a recognised poverty-reduction strategy paper; and finally as judged by the World Bank in a study of countries' capacity to absorb aid. Ghana, Mozambique, Senegal, Tanzania and Uganda score well. Countries such as these, says the report, should be fast-tracked for more aid, and plenty of it.

Too many blanks on the scoresheet

Potential candidates for MDG* fast-tracking in sub-Saharan Africa

Country	African Peer Review Mechanism	HIPC completion poing	Millennium Challenge Corporation		Poverty Reduction Strategy Paper	World Bank Absorptive Capacity Study
			Qualifier	*Threshold*		
Angola	✔					
Benin	✔	✔	✔		✔	✔
Burkina Faso	✔	✔		✔	✔	✔
Cameroon	✔				✔	
Cape Verde			✔			
Chad					✔	
Congo-Brazzaville	✔					
Djibouti					✔	
Ethiopia	✔	✔			✔	✔
Gabon	✔					
Gambia					✔	
Ghana	✔	✔	✔		✔	
Guinea					✔	
Kenya	✔			✔		
Lesotho	✔		✔			
Madagascar		✔	✔		✔	✔
Malawi	✔			✔	✔	
Mali	✔	✔	✔		✔	✔
Mauritania		✔			✔	✔
Mozambique	✔	✔	✔		✔	✔
Niger		✔			✔	
Nigeria	✔					
Rwanda	✔				✔	
São Tomé and Principe				✔		
Senegal	✔	✔	✔		✔	
Sierra Leone	✔					
South Africa	✔					
Tanzania	✔	✔		✔	✔	✔
Uganda	✔	✔		✔	✔	✔
Zambia				✔	✔	

Source: UN Millennium Project
*Millennium Development Goals

Unfortunately, and perhaps as a matter of political necessity, the report wants lots of aid to go to the other countries too. Even countries such as Chad and Nigeria, regarded by Transparency International as among the most corrupt in the world, are described as "potentially well-governed"—much as Mike Tyson is potentially well-behaved—and deserving of consideration for fast-tracking. The report draws the line only at cases as egregious as Zimbabwe, Myanmar and North Korea: not as bravely discriminating as one might wish.

The report has a second way of dealing with lack of good governance: it argues that aid can be spent on remedying this. But that may be wishful thinking. The problem with aid to bad governments is that it can help to keep them in place. Donors have tried before to invest in improved governance. The record is not good.

The report has recommendations by the dozen on how aid should be patterned and delivered. It argues that the development needs of countries vary a great deal from case to case: strategies need to be carefully tailored, with policies designed and "owned" by the country itself. Well and good. But many would question the very idea of a top-down development strategy. William Easterly, for instance, formerly of the World Bank, now at New York University, author of "The Elusive Quest for Growth" and a leading authority on development, says he is sceptical. He regards the proliferation of goals and recommendations in the report as hugely over-ambitious, as tending towards a kind of international central planning, and as placing far too great a strain on the puny resources of "global bureaucrats and dysfunctional administrations".

Mr Easterly makes a further point, echoed by others this week. The report advances great, not to say extravagant, claims about what aid can achieve, even in regions such as sub-Saharan Africa, over the next ten years. Even if Mr Sachs succeeds in his chief purpose of inspiring big new commitments of aid from the West, Africa is very likely to fail by the daunting standards he has set in this report. Over the longer term, a more sober and guarded assessment of what can be achieved—one less inclined to end once more in disappointment—might better serve the cause of development. Driven and effective, yes: sober and guarded, Mr Sachs is not.

Microfinance and the Poor

Breaking down walls between microfinance and formal finance

Elizabeth Littlefield and Richard Rosenberg

CONTRARY TO a common impression, poor people need and use a variety of financial services, including deposits, loans, and other services. They use financial services for the same reasons as anyone else: to seize business opportunities, improve their homes, deal with other large expenses, and cope with emergencies. For centuries, the poor have used a wide range of providers to meet their financial needs. While most poor people lack access to banks and other formal financial institutions, informal systems like moneylenders, savings and credit clubs, and mutual insurance societies are pervasive in nearly every developing country. The poor can also tap into their other assets, such as animals, building materials, and cash under the mattress, when the need arises. Or, for example, a poor farmer may pledge a future season's crops to buy fertilizer on credit from commercial vendors.

However, the financial services usually available to the poor are limited in terms of cost, risk, and convenience. Cash under the mattress can be stolen and can lose value as a result of inflation. A cow cannot be divided and sold in parcels to meet small cash needs. Certain types of credit, especially from moneylenders, are extremely expensive. Rotating savings and credit clubs are risky and usually don't allow much flexibility in amount or in the timing of deposits and loans. Deposit accounts require minimum amounts and may have inflexible withdrawal rules. Loans from formal institutions usually have collateral requirements that exclude most of the poor.

Microfinance institutions (MFIs) have emerged over the past three decades to address this market failure and provide financial services to low-income clients. Most of the early pioneer organizations in the modern microfinance movement operated as nonprofit, socially motivated nongovernmental organizations. They developed new credit techniques: instead of requiring collateral, they reduced risk through group guarantees, appraisal of household cash flow, and small initial loans to test clients. Experience since then has shown that the poor repay uncollateralized loans reliably and are willing to pay the full cost of providing them: access is more important to them than cost.

The poor need and use a broad range of financial services, including deposit accounts, insurance, and the ability to transfer money to relatives living elsewhere. Experience has shown that the poor can be served profitably, on a long-term basis, and in

some cases on a large scale. Indeed, well-run MFIs can outperform mainstream commercial banks in portfolio quality. The top-performing MFIs in some countries are more profitable than the top-performing local commercial bank.

In turbulent times, microfinance has been shown to be a more stable business than commercial banking. During Indonesia's 1997 crisis, for example, commercial bank portfolios deteriorated, but loan repayment among Bank Rakyat Indonesia's 26 million microclients barely declined. And, during the recent Bolivian banking crisis, MFIs' portfolios suffered but remained substantially healthier than commercial bank portfolios.

Today, microfinance is reaching only a small fraction of the estimated demand for financial services by poor households. While a few hundred institutions have proved the poor can be served sustainably and on a large scale, most of the institutions are weak, heavily donor-dependent, and unlikely to ever reach scale or independence. Only financially sound, professional organizations have a chance to compete effectively, access commercial loans, become licensed to collect deposits, and grow to reach significant scale and impact. To achieve its full potential, microfinance must become a fully integrated part of a developing country's mainstream financial system rather than being confined to a niche of the development community.

Encouraging signs of integration are beginning to emerge. In some countries, the walls between microfinance and the formal financial sector are coming down. The commercial success of some MFIs has begun to attract new, mainstream actors. Partnerships are forming, and public and private sector infrastructure and knowledge are being shared and leveraged. New technologies are also driving costs and risks down to provide services to poorer clients more cost effectively. The quality and comparability of financial reporting, ratings, and audits are improving, and domestic and international commercial investors are investing in the sector.

What's at stake

Reliable measurement of the impact of financial services on household welfare is expensive and methodologically difficult. However, an increasing number of serious studies are suggesting that microfinance can produce improvements in a range of welfare measures, including income stability and growth, school attendance, nutrition, and health. Microfinance has been widely credited with empowering women by increasing their contribution to household income and assets and, thus, their control over decisions affecting their lives. Of course, microfinance has generated considerable enthusiasm—not just in the development community but also politically—with the predictable result that some of its merits have been oversold.

Microfinance alone is not a magic solution that will propel all of the poor—particularly the very poorest people—out of poverty. But there is no doubt that poor clients themselves value microfinance very highly, as evidenced by their strong demand for such services, their willingness to pay the full cost of those services, and high loan repayment rates that are motivated mainly by a desire to have access to future loans. Moreover, because microfinance can be delivered sustainably, its benefits can be made available for the long term—well beyond the duration of donor or government subsidies.

MFIs form one part of a much broader spectrum of socially oriented financial institutions (SOFIs) that includes state-owned development, postal, agricultural, and savings banks, as well as smaller entities like savings and loan cooperatives. These institutions are considered socially oriented because, for the most part, they were created not to be profit maximizers but, rather, to reach clients who were not being well served by the commercial banking system. SOFIs represent a vast infrastructure and clientele: a recent, far from exhaustive survey identified well over 600 million accounts in these institutions. Although no concrete data are available on the proportion of SOFI clients who are poor, average account sizes suggest that this proportion is substantial.

Despite their extensive outreach and infrastructure, SOFIs also have significant limitations. Some of them—especially the state-owned ones—provide inferior services, are highly inefficient, and generate large, continuing losses. In many countries, financial authorities do not consider SOFIs part of the mainstream financial system and do not supervise them as seriously as commercial banks. Except in a few countries, SOFIs account for a small percentage of financial system assets and may not pose systemic risk. But in many countries, a large proportion—sometimes the majority—of households using financial services access them through SOFIs. The SOFI share of total financial system accounts is, for example, 50 percent in Bolivia and 65 percent in Côte d'Ivoire.

When large SOFIs can be turned around and run on a commercial basis, the results can be dramatic. In Mongolia, for example, the state agricultural bank restructured, moved into microfinance, and has been privatized. The bank, which serves half of all rural households in Mongolia through 375 points of sale, is now profitable. Bank Rakyat Indonesia is another restructured SOFI that now provides high-quality services to massive numbers of poor people and generates very healthy profits.

Increasingly commercial orientation

Most leading MFIs operate today on a commercial basis using the techniques and disciplines of commercial finance. They are investing in more sophisticated management and information systems, applying international accounting standards, contracting annual audits from mainstream auditing firms, and seeking ratings from commercial rating agencies. Last year, rating agencies, including industry leaders Standard & Poor's and Moody's Investors Service, carried out over 100 credit ratings of MFIs.

"To achieve its full potential, microfinance must become a fully integrated part of a developing country's mainstream financial system.

There is growing awareness that building financial systems for the poor means building sound domestic financial intermediaries that can mobilize and recycle domestic savings. Foreign donor and social investor capital diminishes as individual institutions and entire markets mature. For this reason, increasing numbers of MFIs are getting licensed as banks or specialized finance companies, allowing them to finance themselves by accessing capital markets and mobilizing deposits from large institutional investors as well as poor clients. Several MFIs, mainly in Latin America, have tapped local debt markets, largely by issuing private placements taken up by local financial institutions.

Dozens of countries are considering legislation to create new types of financial licenses, usually with lower minimum capital requirements, designed for specialized microfinance intermediaries. Although generally positive, this trend does pose risks. Supervisory authorities who are already stretched thin trying to monitor commercial banks can find it difficult to take on responsibility for a new group of small institutions. Moreover, a move toward specialized MFIs sometimes overlooks opportunities to involve mainstream commercial banks in microfinance.

In countries as different as Haiti, Georgia, and Mexico, partnerships between commercial banks and MFIs are an alternative to MFIs seeking their own financial licenses. These partnerships enable MFIs to cut costs and extend their reach, while banks can benefit from the opportunity to tap new markets, diversify assets, and increase revenues. Partnerships vary in their degree of engagement and risk sharing, ranging from sharing or renting front offices to banks making actual portfolio and direct equity investments in MFIs (see figure).

In Africa, Asia, and Latin America, some local financial institutions are pursuing lower-end retail banking directly, as financial globalization heightens the competition posed by international banks for larger corporate customers. Banque du Caire in Egypt, for example, entered the market two years ago and now delivers microfinance alongside traditional products through its 230 branches. It is still too early to tell whether large numbers of commercial banks will move into microfinance. Well-run MFIs have proved to be profitable, but serving this market requires changes in systems, staffing, and culture that are not easy for traditional banks.

Lower-income customers have smaller account and transaction sizes, which underscores the importance of reducing transaction costs. Credit scoring and computerization have

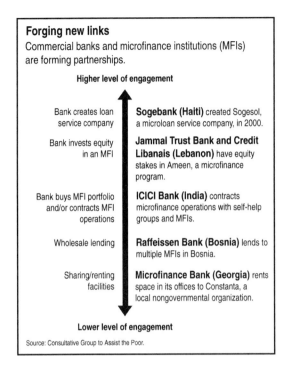

Forging new links
Commercial banks and microfinance institutions (MFIs) are forming partnerships.

Higher level of engagement

Bank creates loan service company — **Sogebank (Haiti)** created Sogesol, a microloan service company, in 2000.

Bank invests equity in an MFI — **Jammal Trust Bank and Credit Libanais (Lebanon)** have equity stakes in Ameen, a microfinance program.

Bank buys MFI portfolio and/or contracts MFI operations — **ICICI Bank (India)** contracts microfinance operations with self-help groups and MFIs.

Wholesale lending — **Raffeissen Bank (Bosnia)** lends to multiple MFIs in Bosnia.

Sharing/renting facilities — **Microfinance Bank (Georgia)** rents space in its offices to Constanta, a local nongovernmental organization.

Lower level of engagement

Source: Consultative Group to Assist the Poor.

with Garanti Bankasi, a leading private bank, to gain access to the national credit bureau to screen loan applicants for credit card and other debt. New regulations on microfinance issued by Rwanda's central bank require that MFIs communicate information about their borrowers to a credit bureau.

Creative new delivery channels and new information technology also hold promise for reducing risk and cutting delivery costs. Microfinance providers are now exploring ways to piggyback financial services delivery onto existing infrastructure, such as retail shops, Internet kiosks, post offices, and even lottery outlets. Existing distribution systems may make it possible to provide financial services more cost effectively and, thus, to poorer and more sparsely populated areas. On the technology side, companies in southern Africa are developing low-cost, cell phone-based banking services for poor clients. MFIs in Bolivia, Mexico, India, and South Africa are also making use of smart cards, fingerprint readers, and personal digital assistants to improve efficiency and expand into rural areas. Not surprisingly, the actual performance of such new technologies and strategies does not always match the initial level of enthusiasm generated. But some of these new approaches have proved themselves already, and others will no doubt continue to emerge.

Twenty years ago, the main challenge in microfinance was methodological: finding techniques to deliver and collect uncollateralized loans to "microentrepreneurs" and poor households. After notable successes on that front, the challenge today is a more systemic one: finding ways to better integrate a full range of microfinance services with mainstream financial systems and markets. While it is not yet clear how far that integration will go, the early signs are encouraging. Advances taking place around the world would have been dismissed as unthinkable just a decade or two ago.

underpinned many of the important new down-market opportunities, with the result that the boundary between microfinance and consumer finance is now blurring in many places. Retailers, consumer finance institutions, and building societies in some developing countries are adapting microfinance methodologies so they can use their infrastructure to tap the new market of uncollateralized, character-based lending to the self-employed or to households, more generally.

MFIs are beginning to tap into mainstream credit bureaus—an important strategy to increase productivity, improve portfolio quality, and reduce spreads between borrowing and saving rates. This not only reduces risk for the MFIs but also allows their clients to build a public credit history that makes them more attractive to mainstream banks and retailers. For example, more than 80 MFIs in Peru are registered to use Infocorp, a private credit bureau. In Turkey, Maya Enterprise for Microfinance negotiated

Elizabeth Littlefield is the Chief Executive Officer of the Consultative Group to Assist the Poor (CGAP), a multidonor organization created to help build a large-scale microfinance industry providing flexible, high-quality financial services to the poor on a sustainable basis. Richard Rosenberg is a Senior Advisor on policy issues at CGAP.

The real digital divide

Encouraging the spread of mobile phones is the most sensible and effective response to the digital divide

IT WAS an idea born in those far-off days of the internet bubble: the worry that as people in the rich world embraced new computing and communications technologies, people in the poor world would be left stranded on the wrong side of a "digital divide". Five years after the technology bubble burst, many ideas from the time—that "eyeballs" matter more than profits or that internet traffic was doubling every 100 days—have been sensibly shelved. But the idea of the digital divide persists. On March 14th, after years of debate, the United Nations will launch a "Digital Solidarity Fund" to finance projects that address "the uneven distribution and use of new information and communication technologies" and "enable excluded people and countries to enter the new era of the information society". Yet the debate over the digital divide is founded on a myth—that plugging poor countries into the internet will help them to become rich rapidly.

The lure of magic

This is highly unlikely, because the digital divide is not a problem in itself, but a symptom of deeper, more important divides: of income, development and literacy. Fewer people in poor countries than in rich ones own computers and have access to the internet simply because they are too poor, are illiterate, or have other more pressing concerns, such as food, health care and security. So even if it were possible to wave a magic wand and cause a computer to appear in every household on earth, it would not achieve very much: a computer is not useful if you have no food or electricity and cannot read.

Yet such wand-waving—through the construction of specific local infrastructure projects such as rural telecentres—is just the sort of thing for which the UN's new fund is intended. How the fund will be financed and managed will be discussed at a meeting in September. One popular proposal is that technology firms operating in poor countries be encouraged to donate 1% of their profits to the fund, in return for which they will be able to display a "Digital Solidarity" logo. (Anyone worried about corrupt officials creaming off money will be heartened to hear that a system of inspections has been proposed.)

This sort of thing is the wrong way to go about addressing the inequality in access to digital technologies: it is treating the symptoms, rather than the underlying causes. The benefits of building rural computing centres, for example, are unclear (see the article in our *Technology Quarterly* in this issue). Rather than trying to close the divide for the sake of it, the more sensible goal is to determine how best to use technology to promote bottom-up development. And the answer to that question turns out to be remarkably clear: by promoting the spread not of PCs and the internet, but of mobile phones.

Plenty of evidence suggests that the mobile phone is the technology with the greatest impact on development. A new paper finds that mobile phones raise long-term growth rates, that their impact is twice as big in developing nations as in developed ones, and that an extra ten phones per 100 people in a typical developing country increases GDP growth by 0.6 percentage points (see Economics focus, page 94).

And when it comes to mobile phones, there is no need for intervention or funding from the UN: even the world's poorest people are already rushing to embrace mobile phones, because their economic benefits are so apparent. Mobile phones do not rely on a permanent electricity supply and can be used by people who cannot read or write.

Phones are widely shared and rented out by the call, for example by the "telephone ladies" found in Bangladeshi villages. Farmers and fishermen use mobile phones to call several markets and work out where they can get the best price for their produce. Small businesses use them to shop around for supplies. Mobile phones are used to make cashless payments in Zambia and several other African countries. Even though the number of phones per 100 people in poor countries is much lower than in the developed world, they can have a dramatic impact: reducing transaction costs, broadening trade networks and reducing the need to travel, which is of particular value for people looking for work. Little wonder that people in poor countries spend a larger proportion of their income on telecommunications than those in rich ones.

The digital divide that really matters, then, is between those with access to a mobile network and those without.

The good news is that the gap is closing fast. The UN has set a goal of 50% access by 2015, but a new report from the World Bank notes that 77% of the world's population already lives within range of a mobile network.

And yet more can be done to promote the diffusion of mobile phones. Instead of messing around with telecentres and infrastructure projects of dubious merit, the best thing governments in the developing world can do is to liberalise their telecoms markets, doing away with lumbering state monopolies and encouraging competition. History shows that the earlier competition is introduced, the faster mobile phones start to spread. Consider the Democratic Republic of Congo and Ethiopia, for example. Both have average annual incomes of a mere $100 per person, but the number of phones per 100 people is two in the former (where there are six mobile networks), and 0.13 in the latter (where there is only one).

Let a thousand networks bloom

According to the World Bank, the private sector invested $230 billion in telecommunications infrastructure in the developing world between 1993 and 2003—and countries with well-regulated competitive markets have seen the greatest investment. Several firms, such as Orascom Telecom and Vodacom, specialise in providing mobile access in developing countries. Handset-makers, meanwhile, are racing to develop cheap handsets for new markets in the developing world. Rather than trying to close the digital divide through top-down IT infrastructure projects, governments in the developing world should open their telecoms markets. Then firms and customers, on their own and even in the poorest countries, will close the divide themselves.

UNIT 3
Conflict and Instability

Unit Selections

Key Points to Consider

- What are the causes of civil war?

- How do failing states pose an international security threat?

- Why has the danger of nuclear proliferation increased?

- What is the alternative to the establishment of a Palestinian state in the Middle East?

- Why does the Democratic Republic of Congo continue to teeter on the brink of war?

- What is the source of the new conflict in the Darfur region of Sudan?

- What conditions characterize the plight of the world's refugees?

Student Website
www.mhcls.com/online

Internet References
Further information regarding these websites may be found in this book's preface or online.

The Carter Center
   http://www.cartercenter.org
Center for Strategic and International Studies (CSIS)
   http://www.csis.org/
Conflict Research Consortium
   http://www.Colorado.EDU/conflict/
Institute for Security Studies
   http://www.iss.co.za
PeaceNet
   http://www.igc.org/peacenet/
Refugees International
   http://www.refintl.org

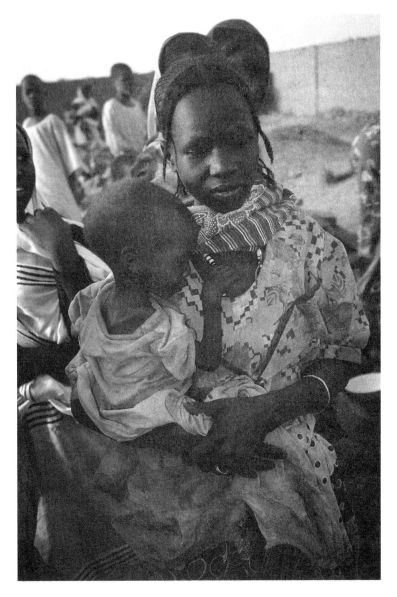

Conflict and instability in the developing world remain major threats to international peace and security. Conflict stems from a combination of sources including ethnic and religious diversity, nationalism, the struggle for state control, and competition for resources. In some cases, boundaries that date from the colonial era encompass diverse groups. A state's diversity can increase tension among groups competing for scarce resources and opportunities. When some groups benefit or are perceived as enjoying privileges at the expense of others, ethnicity can offer a convenient vehicle to organize and mobilize around. Moreover, ethnic politics lends itself to manipulation both by regimes that are seeking to protect privileges, maintain order, or retain power and those that are challenging existing governments. In a politically and ethnically charged atmosphere, conflict and instability often result as groups vie to gain control of a state apparatus that can extract resources and allocate benefits. While ethnicity has certainly played a role in many recent conflicts, conflicts over power and resources may be mistakenly viewed as ethnic in nature. Recent conflicts in Africa and Asia have increasingly centered on valuable resources. Conflicts like the war in the Democratic Republic of Congo generate economic disruption, population migration, environmental degradation, and may also draw other countries into the fighting. Failing states also contribute to the potential for conflict. They can offer a haven for terrorists and criminals to operate and the spill-over from conflict in these states can widen instability.

Early literature on modernization and development speculated that as developing societies progressed from traditional to modern, primary attachments such as ethnicity and religious affiliation would fade and be replaced by new forms of identification. Clearly, however, ethnicity remains a potent force, as does religion. Ethnic politics and the emergence of religious radicalism demonstrate that such attachments have survived the drive toward modernization.

Initially inspired and encouraged by the theocratic regime in Iran, radical Muslims have not only pushed for the establishment of governments based on Islamic law but have engaged in a wider struggle with what they regard as the threat of western cultural dominance. Radical Islamic groups that advocate a more rigid and violent interpretation of Islam have increasingly challenged more mainstream Islamic thought. These radicals are driven by hatred of the West and the United States in particular. They were behind the 1993 New York City World Trade Center bombing, the United States embassy bombings in Kenya and Tanzania in 1998, the devastating attacks that destroyed the World Trade Center and damaged the Pentagon in September 11, 2001, and the bombing of a night club in Bali in 2002 and a hotel in Jakarta in 2003. Although there is a tendency to equate Islam with terrorism, it is a mistake to link the two. A deeper understanding of the various strands of Islamic thought is required to separate legitimate efforts to challenge repressive regimes and forge an alternative to western forms of political organization from radicals who pervert Islam and use terrorism.

Whatever the cause, there is no shortage of tension and conflict around the world. Prospects for peace in the Middle East faded in the midst of Palestinian suicide bombings, Israeli retaliation, and acrimony over the establishment and nature of a Palestinian state. The climate has improved recently but hardliners on both sides threaten to disrupt efforts to dismantle Israeli settlements and extend Palestinian control. The war in Iraq has entered yet another new phase with the installation of an elected government. Attacks continue against both the U.S. and Iraqi government forces, and the civilian death toll continues to rise. Remnants of the Taliban pose a renewed security threat in Afghanistan. Nepal's conflict between Maoist rebels and government forces demonstrates that ideology remains a source of conflict. North Korea's nuclear weapons program has not only heightened tensions with the United States but has also made regional neighbors increasingly wary and illustrates the growing problem of nuclear proliferation. Parts of Africa also continue to be conflict prone. The Democratic Republic of Congo's tenuous peace has been marred by sporadic fighting in the eastern part of the country. Zimbabwe continues its slide into economic disaster and instability. Although a tentative peace agreement between the Sudanese government and rebels in the South has been negotiated, the Darfur region has become a focus of worldwide concern amid controversy over whether the conflict amounts to genocide. Arab militias supported by the government have carried out deadly attacks against black residents causing tens of thousands of deaths and creating what the UN has called the world's worst humanitarian crisis. In South America, the Colombian government continues to face a serious challenge by leftist rebels, a conflict made more complicated by ties between the rebels and drug traffickers as well as the activities of right-wing paramilitaries.

Although conflict may have declined worldwide, the threats to peace and stability in the developing world remain complicated and dangerous and clearly have the potential to threaten international security. These circumstances require a greater effort to understand and resolve conflicts, whatever their source.

THE MARKET FOR CIVIL WAR

Ethnic tensions and ancient political feuds are not starting civil wars around the world. A groundbreaking new study of civil conflict over the last 40 years reveals that economic forces—such as entrenched poverty and the trade in natural resources—are the true culprits. The solution? Curb rebel financing, jump-start economic growth in vulnerable regions, and provide a robust military presence in nations emerging from conflict.

By Paul Collier

Every time a civil war breaks out, some historian traces its origin to the 14th century and some anthropologist expounds on its ethnic roots. Don't buy into such explanations too quickly. Certain countries are more prone to civil war than others, but distant history and ethnic tensions are rarely the best explanations for a conflict. Look instead at a nation's recent past and, most important, its economic conditions.

Once a country has reached a per capita income rivaling that of the world's richest nations, its risk of civil war is negligible. Today, about 900 million people live in such societies. Four billion more live in countries that are either already middle income or on track to becoming so, thanks to rapidly growing and diversifying economies. This group, which includes the economic success stories of the post–World War II era, faces fairly low risk of civil war. The potential for conflict is concentrated among the countries inhabited by the world's remaining 1.1 billion people. These countries typically have poor and declining economies and rely on natural resources—such as diamonds or oil—for a large proportion of national income. As the British, French, Portuguese, and Soviet empires successively dissolved during the last century, the number of such countries increased in waves.

Such at-risk countries are engaged in a sort of Russian roulette. Every year that their dismal economic conditions persist increases the odds that their societies will fall into armed conflict. Whether by luck or prudence, many such nations have so far escaped civil war. Others have not. And once civil war has started, the decline in income and the accumulation of arms, fighting skills, and military capabilities greatly increase the risks of further conflict.

To date, academics and policymakers alike have misdiagnosed the nature of the problem; little surprise, then, that their efforts to prevent civil wars have been ineffective. When the world's leaders can identify the real factors most likely to drive such conflicts, they will have a better chance of preventing future wars.

THE MYTH OF ETHNIC STRIFE

Between 1960 and 1999, there were 52 major civil wars for which comprehensive data is available on social, political, historical, economic, and geographic circumstances. Such wars spanned the developing regions, with the typical conflict lasting around seven years and leaving a legacy of persistent poverty and disease in its wake. To understand the causes of these conflicts, economist Anke Hoeffler and I studied each five-year-period from 1960 to 1999 and identified preexisting conditions that helped predict the outbreak of war.

> If a country is mountainous and has a large, lightly populated hinterland, it faces an enhanced risk of rebellion... Nepal is therefore more at risk of civil war, geographically speaking, than Singapore.

For example, income inequality and ethnic-religious diversity are frequently cited as causes for conflict. Yet surprisingly, inequality—either of household incomes or of land owner-ship—does not appear to increase systematically the risk of civil war. Brazil got away with its high inequality; Colombia didn't. And, in fact, ethnic and religious diversity actually reduces the risk of civil conflict. One important exception: Where the largest ethnic group constitutes a majority but lives alongside a substantial minority, such as in Sri Lanka and Rwanda, the risk of civil war roughly doubles. Once wars start, they also tend to last much longer if the nation in question displays two or three dominant ethnic groups.

Conflicts in ethnically diverse countries may be ethnically patterned without being ethnically caused. International media coverage of civil wars often focuses on history and ethnicity because rebel leaders adopt this sort of discourse. Grievances are to a rebel organization what image is to a business. The rebel group needs to stimulate a sense of collective grievance to build cohesion in its army and to attract funding from its diaspora living in rich countries.

Much to the dismay of democratization activists, democracy fails to reduce the risk of civil war, at least in low-income countries. Indeed, politically repressive societies have no greater risk of civil war than full-fledged democracies. Countries falling between the extremes of autocracy and full democracy—where citizens enjoy some limited political rights—are at a greater risk of war. Low-income societies with new democratic institutions are often at enhanced risk: Just consider the current catastrophe in Ivory Coast, where uncertainty over who could stand for the presidential election in 2000 triggered violent clashes and ongoing political instability.

Wherever a civil war occurs, observers will invariably find some deep history of conflict. But overwhelmingly, conflicts in the distant past are not generating civil wars in the present. The history that matters is recent history, not that of the 14th century. If a country recently experienced a civil war, it is much more likely to have another one. The risk fades the longer peace endures.

Civil war is self-perpetuating, partly because it changes the balance of interests within countries. Groups engaged in conflict invest in armaments, skills, and infrastructure that are only good for violence. These groups' leaders, and indeed all those who gain from lawlessness, prosper during war, even though society as a whole suffers. The part of the elite that prefers peace will have shifted much of its wealth outside the country. Hence, as a result of the conflict, the balance of elite interests shifts toward further conflict.

Geography matters, too. If a country is mountainous and has a large, lightly populated hinterland, it faces an enhanced risk of rebellion. Presumably, rebels are harder to find and defeat in such terrain. Nepal is therefore more at risk of civil war, geographically speaking, than Singapore.

DIAMONDS, A REBEL'S BEST FRIEND

All these factors notwithstanding, economic conditions remain paramount in explaining civil wars. For the average country in our study, the risk of a civil war in each five-year period was around 6 percent, but the risks increased alarmingly if the economy was poor, declining, and dependent on natural resource exports. For a country with conditions like those in the Democratic Republic of the Congo (formerly Zaire) in the late 1990s—with deep poverty, a collapsing economy, and huge miner exploitation—the risk reaches nearly 80 percent.

> When valuable natural resources are discovered in a particular region of a country, the people living in such localities suddenly have an economic incentive to secede, violently if necessary.

Once started, wars last longer in low-income countries, which are prone to rebellion for many reasons: Recruits have less of a stake in the status quo, and central governments are typically weak. Each additional percentage point in the growth rate of per capita income shaves off about 1 percentage point of conflict risk; conversely, wars are more likely to follow periods of economic collapse—such as the conflicts that have surged in Indonesia since the East Asian economic crisis of the late 1990s. If a country's per capita income doubles, its risk of conflict drops by roughly half. Simply put, economic growth matters because opportunities for youth depend upon a robust economy.

Conflict is also more likely in countries that depend heavily on natural resources for their export earnings, in part because rebel groups can extort the gains from this trade to finance their operations. Diamonds funded the National Union for the Total Independence of Angola (UNITA) rebel group during Angola's long civil war, as well as the Revolutionary United Front (RUF) in Sierra Leone; timber funded the Khmer Rouge in Cambodia. Indeed, methods of extortion abound. For example, multinational corporations that extract natural resources must often pay huge sums to ransom kidnapped workers and to protect infrastructure from sabotage at the hands of rebel groups. Laughably, such payments are sometimes charged to the companies' "corporate social responsibility" budgets.

Natural resources also fuel war because they make secession more likely. When valuable natural resources are discovered in a particular region of a country, the people living in such localities suddenly have an economic incentive to secede, violently if necessary. Since most countries are ethnically diverse, the lucky, resource-rich locality is likely to be ethnically distinct as well. Often, the weak political force of ethnic romanticism latches on to the stronger force of economic self-interest so that secessionist movements voice ethnic grievances—Biafra (Nigeria), Cabinda (Angola), and Aceh (Indonesia) come to mind. The incentive to secede is probably compounded by the corrupt, incompetent way to which governments commonly use natural resource wealth: The greed of a resource-rich locality can seem ethically less ugly if a corrupt national elite is already hijacking the resources.

THE WAR DIVIDEND

One striking lesson from these patterns is that the motivations for rebellion generally matter less than the conditions that make a rebellion financially and militarily viable. Civil wars only occur if a rebel organization can build and sustain a private army. These organizations are unlike traditional opposition groups such as political parties or protest movements. They are hierarchical, authoritarian, expensive, and usually small. Where such organizations are financially militarily feasible, rebellions are likely to emerge, promoting whatever political agenda their leaders happen to support.

Global efforts to curb civil war should therefore focus on reducing the viability—rather than just the rationale—of rebellion. Of course, policies should address legitimate grievances, not because addressing them paves a royal road to peace but because they are legitimate.

Botswana and Sierra Leone were similarly poor countries, both sitting on vast diamond deposits… Botswana harnessed this opportunity, becoming the fastest-growing economy in the world. Sierra Leone used the same resources to impoverish itself…

Those nations currently at war and those that have recently emerged from civil war constitute the core of the problem. Many countries have fallen into a conflict trap: a damaging war that sharply increases the risk of further conflict, followed by a fragile peace, and then back to war. The expected duration of a civil war is currently about eight years—double what it was before the 1980s. Wars therefore do more damage now and thus more powerfully provoke further conflict.

No one knows why wars last longer now. Perhaps global markets in both natural resources and arms make rebellion easier to finance and equip. Rebel groups can now sell the future rights to mineral extraction (conditional on rebel victory) to raise funds for weapons purchases. A similar arrangement reputedly helped French oil Giant Elf Aquitaine (now TotalFinaElf) gain its current access to oil in the Republic of the Congo.

So, what specific measures can countries take to reduce the occurrence and likelihood of civil war?

For developing countries already growing rapidly, the most significant risk may be episodes of economic crisis, such as that experienced by Indonesia in the late 1990s. The opportunity to prevent war in such cases only strengthens the justification for international efforts to avert economic crises. In this light, initiatives to reform and rethink the workings of the global financial system are not merely an academic debate or an effort to ease investors' concerns, but rather a much more serious matter with immediate life-and-death consequences.

Poor countries that are not developing but have so far escaped civil war, such as Zambia and Malawi, are also racing against time. If they do not find ways to accelerate their economic growth and development, they will likely stumble into conflict. Recent casualties include Ivory Coast and Nepal. Nations in these conditions should get the message that change is urgent. Often the remedy should go beyond the standard package of market access, debt relief, and aid programs from the developed countries to include credible policy reform and honest governance within vulnerable countries.

Due to their heavy dependence on natural resource revenues, the governments in many at-risk nations face acute problems of corruption and exposure to international price shocks. But natural resources need not be a curse. Twenty-five years ago, Botswana and Sierra Leone were similarly poor countries, both sitting on vast diamond deposits. Over the ensuing quarter century, Botswana harnessed this opportunity, becoming the fastest-growing economy in the world. Sierra Leone used the same resources to impoverish itself, experiencing the most rapid sustained decline of any country; it now ranks at the bottom of the Human Development Index put out by the United Nations Development Programme. These contrasting examples show that good policy and governance are especially vital where natural resources are discovered.

So far, the record has been dismal: There are many more Sierra Leones than Botswanas. But some encouraging signs are emerging. The "Fowler Report" to the U.N. Security Council in 2000 detailed how UNITA evaded U.N. sanctions against arms smuggling and diamond-based financing; as a result, scrutiny of the international diamond trade increased. Such attention may well have contributed to the demise of UNITA in Angola and the RUF in Sierra Leone, two highly durable, diamond-dependent rebel organizations. Moreover, diamonds are now being tracked through the new Kimberley Process certification scheme, making it harder for rebel groups to obtain financing from these goods. Transparency is the first step toward effective national scrutiny.

And if such strategies work with diamonds, why not replicate the process for timber? The Group of Eight industrialized nations flagged the problems posed by natural resources in its 2002 meeting and may take up the issue again in June 2003 at the Evian summit. Specific political leaders have also begun taking action. British Prime Minister Tony Blair has launched an initiative for greater transparency in the reporting of resource revenues by multinational companies in extractive industries; by nationally owned oil companies such as Angola's Sonangol, which are often a state within a state; and by recipient governments. And in a complementary effort, French President Jacques Chirac has recently called for better mechanisms to cushion low-income countries when the prices of their commodity exports plummet. A new compact could emerge: Rich nations take action to cut rebel financial and cushion adverse shocks, while low-income nations adopt better governance of their revenues from natural resources.

ESCAPING THE CONFLICT TRAP

From 1960 to 1999, international interventions—whether economic or military—intended to shorten civil wars were disappointing. Some strategies may have succeeded in individual cases (as in the recent destruction of the Taliban regime in Afghanistan), but no type of intervention has worked regularly.

More effective interventions could target the systems that finance and equip rebel organizations, beyond solely focusing on the trade in diamonds and other commodities. Many rebel movements also receive illicit support from neighboring governments. Such support can be exposed and penalized, and the penalties should outweigh the benefits of rebel alliances. Moreover, governments can discourage huge ransom payments by corporations—such as the $20 million reputedly paid in 1984 by the German engineering company Mannesmann-Anlagenbau AG to the Colombian rebel group ELN, or National Liberation Army, for the release of three of the company's staff. Should such payments be tax deductible and so, in effect, subsidized? Governments could ban the insurance arrangements that facilitate and inevitably inflate such payments. National authorities have started to improve scrutiny of national and international banking systems; recent evidence that al Qaeda is shifting its assets into diamonds suggests that this strategy is becoming effective. National and multilateral policymakers should also look again at drug policy. Wouldn't it be easier and more effective to curb the demand for criminally supplied drugs? And certainly, the flow of arms can be curtailed, especially if efforts are made earlier to catch the big operators, like suspected gunrunner Victor Bout, who is though to have supplied arms to rebel groups in several African nations, including Angola.

The best way to break out of the conflict trap is to ensure that countries that have just ended one conflict do not quickly become enmeshed in another. In some nations, the risks of renewed conflicts are so high that an external military peacekeeping force is normally necessary. The operative word is "external" because high military spending by a post-conflict government actually increases the risk of another war. That external military presence must be credible. In Sierra Leone, the RUF took hostage a large U.N. force that it sensed would not fight, yet when confronted by a smaller British force, the rebel group collapsed.

Unfortunately, peacekeeping missions normally do not last long enough to allow economic recovery to take hold and help keep the peace. The peak time for economic recovery is usually during the middle of the first post-conflict decade. That is also when aid is most effective in promoting economic growth. Unfortunately, international aid is frequently mistimed. It pours in during the first year of peace, when the country's institutions are too weak for the money to be used effectively, then tapers out just when it would be most useful.

Governments in countries recovering from civil war also must give greater priority to economic reform: The post-conflict period is a good time to reform because vested interests are loosened up. For example, after the end of civil conflict in Uganda in 1986, the country's economic policies moved from

Want to Know More?

The arguments in this article draw on **Breaking the Conflict Trap: Civil War and Development Policy** (Washington: World Bank and Oxford University Press, 2003) by Paul Collier. See also the studies and data from **The Economics of Civil War, Crime and Violence** project, available on the World Banks' Web site. Additional research from the project is available in a special issue of the **Journal of Conflict Resolution** (Vol. 46, No. 1, February 2002), edited by Collier and Nicholas Sambanis. See also the International Peace Academy's (IPA) program on **Economic Agendas in Civil War,** available on IPA's Web site.

On the links between natural resources and conflict, see Michael T. Klare's **Natural Resource Wars: The New Landscape of Global Conflict** (New York: Henry Holt, 2001), Michael Renner's **The Anatomy of Resource Wars** (Washington: Worldwatch Institute, 2002), and the Web site of the nongovernmental organization **Global Witness.** Yahya Sadowski contends that ethnic conflict is not as widespread or as "ethnic" as it seems in **"Think Again: Ethnic Conflict"** (FOREIGN POLICY, Summer 1998). For information on the British initiative on transparency in extractive industries, see the Web site of the United Kingdom's **Department of International Development.** And for views on how rebel groups and local movements promote their causes around the globe, see Clifford Bob's **"Merchants of Morality"** (FOREIGN POLICY, March/April 2002).

For political and historical perspectives on the underlying causes of civil wars, see Monty G. Marshall and Ted Robert Gurr's report **"Peace and Conflict 2003: A Global Survey of Armed Conflicts, Self-Determination Movements, and Democracy"** (College Park: Center for International Development and Conflict Management, 2003). See also Chester A. Crocker, Fen Osler Hampson, and Pamela Aall's, eds., **Turbulent Peace: The Challenges of Managing International Conflict** (Washington: United States Institute of Peace Press, 2001). For economic views on the sources of conflict, see Herschel Grossman's **"A General Equilibrium Model of Insurrections"** (American Economic Review, Vol. 81, No. 4, September 1991), where the author does not distinguish rebels and revolutionaries from "bandits or pirates." And Jack Hirshleifer postulates the Machiavellian theorem that no opportunity for profitable violence will go unexploited in **The Dark Side of the Force: Economic Foundations of Conflict Theory** (Cambridge: Cambridge University Press, 2001).

For links to relevant Web sites, access to the FP Archive, and a comprehensive index of related FOREIGN POLICY articles, go to www.foreignpolicy.com.

among the worst in Africa to among the best in the following decade.

Finally, diasporas in rich countries pose a particular danger in post-conflict situations. They tend to be more extreme than the populations they leave behind, and they finance extremist and violent organizations. For example, the Tamil and Irish diasporas in North America have both been gullible financiers of murder in the past. Diaspora organizations can play an important role in economic recovery; their networks of skills and businesses are potentially valuable. Afghanistan is now trying constructively to deploy its diaspora, for example. But governments of rich nations should help keep the behavior of diaspora organizations in their borders within legitimate bounds.

LOCAL WARS, GLOBAL CASUALTIES

Civil war is not just disastrous for the countries directly affected; it hurts the surrounding regions and often poses risks for even remote, seemingly unaffected nations. Within the country at war, combat-related deaths represent just a small if gruesome part of the costs: War-related economic ruin also intensifies poverty and disease. Throughout the region, economic growth declines and investment flows dry up. Disease spreads across borders through the flow of refuges. And higher military spending induced by real or potential civil war can fuel pointless regional arms races.

Finally, civil war creates territories beyond the control of recognized governments. These no-go areas can be damaging to the international community. Around 95 percent of the global production of hard drugs is located in civil war countries. Witness how sources of supply shift in response to the changing pattern of conflict: As the Shining Path guerrillas were defeated in Peru in the early 1990s, drug production shifted to territory held by the FARC, or the Revolutionary Armed Forces of Co-lombia. These lawless areas also provide safe havens and training for international terrorists.

Over the last several decades, national, regional, and global organizations seeking to end or prevent civil wars have often focused on the wrong challenges, or on the right challenges but at the wrong time. Certainly, no single, magic policy will fix the problem; a range of initiatives is urgently required across a broad front. But if governments and multilateral organizations can help curb rebel financing and armament, accelerate the economic development of the countries most at risk, and provide an effective military presence in post-conflict settings, the global incidence of civil war will decline dramatically. These are viable objectives, and they are likely much cheaper than the long-term consequences of continued conflict and neglect.

Paul Collier is director of the World Bank's development research group.

The End of War?

Explaining 15 years of diminishing violence.

GREGG EASTERBROOK

DAILY EXPLOSIONS IN Iraq, massacres in Sudan, the Koreas staring at each other through artillery barrels, a Hobbesian war of all against all in eastern Congo—combat plagues human society as it has, perhaps, since our distant forebears realized that a tree limb could be used as a club. But here is something you would never guess from watching the news: War has entered a cycle of decline. Combat in Iraq and in a few other places is an exception to a significant global trend that has gone nearly unnoticed—namely that, for about 15 years, there have been steadily fewer armed conflicts worldwide. In fact, it is possible that a person's chance of dying because of war has, in the last decade or more, become the lowest in human history.

Five years ago, two academics—Monty Marshall, research director at the Center for Global Policy at George Mason University, and Ted Robert Gurr, a professor of government at the University of Maryland—spent months compiling all available data on the frequency and death toll of twentieth-century combat, expecting to find an ever-worsening ledger of blood and destruction. Instead, they found, after the terrible years of World Wars I and II, a global increase in war from the 1960s through the mid-'80s. But this was followed by a steady, nearly uninterrupted decline beginning in 1991. They also found a steady global rise since the mid-'80s in factors that reduce armed conflict—economic prosperity, free elections, stable central governments, better communication, more "peace-making institutions," and increased international engagement. Marshall and Gurr, along with Deepa Khosla, published their results as a 2001 report, *Peace and Conflict*, for the Center for International Development and Conflict Management at the University of Maryland. At the time, I remember reading that report and thinking, "Wow, this is one of the hottest things I have ever held in my hands." I expected that evidence of a decline in war would trigger a sensation. Instead it received almost no notice.

"After the first report came out, we wanted to brief some United Nations officials, but everyone at the United Nations just laughed at us. They could not believe war was declining, because this went against political expectations," Marshall says. Of course, 2001 was the year of September 11. But, despite the battles in Afghanistan, the Philippines, and elsewhere that were ignited by Islamist terrorism and the West's response, a second edition of *Peace and Conflict*, published in 2003, showed the total number of wars and armed conflicts continued to decline. A third edition of the study, published last week, shows that, despite the invasion of Iraq and other outbreaks of fighting, the overall decline of war continues. This even as the global population keeps rising, which might be expected to lead to more war, not less.

In his prescient 1989 book, *Retreat from Doomsday*, Ohio State University political scientist John Mueller, in addition to predicting that the Soviet Union was about to collapse—the Berlin Wall fell just after the book was published—declared that great-nation war had become "obsolete" and might never occur again. One reason the Soviet Union was about to collapse, Mueller wrote, was that its leaders had structured Soviet society around the eighteenth-century assumption of endless great-power fighting, but great-power war had become archaic, and no society with war as its organizing principle can endure any longer. So far, this theory has been right on the money. It is worth noting that the first emerging great power of the new century, China, though prone to making threatening statements about Taiwan, spends relatively little on its military.

Last year Mueller published a follow-up book, *The Remnants of War*, which argues that fighting below the level of great-power conflict—small-state wars, civil wars, ethnic combat, and clashes among private armies—is also waning. *Retreat from Doomsday* and *The Remnants of War* are brilliantly original and urgent books. Combat is not an inevitable result of international discord and human malevolence, Mueller believes. War, rather, is "merely an idea"—and a really bad idea, like dueling or slavery. This bad idea "has been grafted onto human existence" and can be excised. Yes, the end of war has been predicted before, prominently by H.G. Wells in 1915, and horrible bloodshed followed. But could the predictions be right this time?

FIRST, THE NUMBERS. The University of Maryland studies find the number of wars and armed conflicts worldwide peaked in 1991 at 51, which may represent the most wars happening simultaneously at any point in history. Since 1991, the number has fallen steadily. There were 26 armed conflicts in 2000 and 25 in 2002, even after the Al Qaeda attack on the United States and the U.S. counterattack against Afghanistan. By 2004, Marshall and Gurr's latest study shows, the number of armed conflicts in the world had declined to 20, even after the invasion of Iraq. All told, there were less than half as many wars in 2004 as there were in 1991.

Marshall and Gurr also have a second ranking, gauging the magnitude of fighting. This section of the report is more subjective. Everyone agrees that the worst moment for human conflict was World War II; but how to rank, say, the current separatist fighting in Indonesia versus, say, the Algerian war of independence is more speculative. Nevertheless, the *Peace and Conflict* studies name 1991 as the peak post-World War II year for totality of global fighting, giving that year a ranking of 179 on a scale that rates the extent and destructiveness of combat. By 2000, in spite of war in the Balkans and genocide in Rwanda, the number had fallen to 97; by 2002 to 81; and, at the end of 2004, it stood at 65. This suggests the extent and intensity of global combat is now less than half what it was 15 years ago.

How can war be in such decline when evening newscasts are filled with images of carnage? One reason fighting seems to be everywhere is that, with the ubiquity of 24-hour cable news and the Internet, we see many more images of conflict than before. A mere decade ago, the rebellion in Eritrea occurred with almost no world notice; the tirelessly globe-trotting Robert Kaplan wrote of meeting with Eritrean rebels who told him they hoped that at least spy satellites were trained on their region so that someone, somewhere, would know of their struggle. Today, fighting in Iraq, Sudan, and other places is elaborately reported on, with a wealth of visual details supplied by mini-cams and even camera-enabled cell phones. News organizations must prominently report fighting, of course. But the fact that we now see so many visuals of combat and conflict creates the impression that these problems are increasing: Actually, it is the reporting of the problems that is increasing, while the problems themselves are in decline. Television, especially, likes to emphasize war because pictures of fighting, soldiers, and military hardware are inherently more compelling to viewers than images of, say, water-purification projects. Reports of violence and destruction are rarely balanced with reports about the overwhelming majority of the Earth's population not being harmed.

Mueller calculates that about 200 million people were killed in the twentieth century by warfare, other violent conflicts, and government actions associated with war, such as the Holocaust. About twelve billion people lived during that century, meaning that a person of the twentieth century had a 1 to 2 percent chance of dying as the result of international war, ethnic fighting, or government-run genocide. A 1 to 2 percent chance, Mueller notes, is also an American's lifetime chance of dying in an automobile accident. The risk varies depending on where you live and who you are, of course; Mueller notes that, during the twentieth century, Armenians, Cambodians, Jews, kulaks, and some others had a far higher chance of death by war or government persecution than the global average. Yet, with war now in decline, for the moment men and women worldwide stand in more danger from cars and highways than from war and combat. World Health Organization statistics back this: In 2000, for example, 300,000 people died in combat or for war-related reasons (such as disease or malnutrition caused by war), while 1.2 million worldwide died in traffic accidents. That 300,000 people perished because of war in 2000 is a terrible toll, but it represents just .005 percent of those alive in that year.

This low global risk of death from war probably differs greatly from most of the world's past. In prehistory, tribal and small-group violence may have been endemic. Steven LeBlanc, a Harvard University archeologist, asserts in his 2003 book about the human past, *Constant Battles*, that warfare was a steady feature of primordial society. LeBlanc notes that, when the aboriginal societies of New Guinea were first observed by Europeans in the 1930s, one male in four died by violence; traditional New Guinean society was organized around endless tribal combat. Unremitting warfare characterized much of the history of Europe, the Middle East, and other regions; perhaps one-fifth of the German population died during the Thirty Years War, for instance. Now the world is in a period in which less than one ten-thousandth of its population dies from fighting in a year. The sheer number of people who are *not* being harmed by warfare is without precedent.

NEXT CONSIDER a wonderful fact: Global military spending is also in decline. Stated in current dollars, annual global military spending peaked in 1985, at $1.3 trillion, and has been falling since, to slightly over $1 trillion in 2004, according to the Center for Defense Information, a nonpartisan Washington research organization. Since the global population has risen by one-fifth during this period, military spending might have been expected to rise. Instead, relative to population growth, military spending has declined by a full third. In current dollars, the world spent $260 per capita on arms in 1985 and $167 in 2004.

The striking decline in global military spending has also received no attention from the press, which continues to promote the notion of a world staggering under the weight of instruments of destruction. Only a few nations, most prominently the United States, have increased their defense spending in the last decade. Today, the United States accounts for 44 percent of world military spending; if current trends continue, with many nations reducing defense spending while the United States continues to increase such spending as its military is restructured for new global anti-terrorism and peacekeeping roles, it is not out of the question that, in the future, the United States will spend more on arms and soldiers than the rest of the world combined.

Declining global military spending is exactly what one would expect to find if war itself were in decline. The peak year in global military spending came only shortly before the peak year for wars, 1991. There's an obvious chicken-or-egg question, whether military spending has fallen because wars are rarer or whether wars are rarer because military spending has fallen. Either way, both trend lines point in the right direction. This is an extremely favorable development, particularly for the world's poor—the less developing nations squander on arms, the more they can invest in improving daily lives of their citizens.

WHAT IS CAUSING war to decline? The most powerful factor must be the end of the cold war, which has both lowered international tensions and withdrawn U.S. and Soviet support from proxy armies in the developing world. Fighting in poor nations is sustained by outside supplies of arms. To be sure, there remain significant stocks of small arms in the developing world—particularly millions of assault rifles. But, with international arms shipments waning and heavy weapons, such as artillery, becoming harder to obtain in many developing nations, factions in developing-world conflicts are more likely to sue for peace. For example, the long, violent conflict in Angola was sustained by a weird mix of Soviet, American, Cuban, and South African arms shipments to a potpourri of factions. When all these nations stopped supplying arms to the Angolan combatants, the leaders of the factions grudgingly came to the conference table.

During the cold war, Marshall notes, it was common for Westerners to say there was peace because no fighting affected the West. Actually, global conflict rose steadily during the cold war, but could be observed only in the developing world. After the cold war ended, many in the West wrung their hands about a supposed outbreak of "disorder" and ethnic hostilities. Actually, both problems went into decline following the cold war, but only then began to be noticed in the West, with confrontation with the Soviet empire no longer an issue.

Another reason for less war is the rise of peacekeeping. The world spends more every year on peacekeeping, and peacekeeping is turning out to be an excellent investment. Many thousands of U.N., NATO, American, and other soldiers and peacekeeping units now walk the streets in troubled parts of the world, at a cost of at least $3 billion annually. Peacekeeping has not been without its problems; peacekeepers have been accused of paying very young girls for sex in Bosnia and Africa, and NATO bears collective shame for refusing support to the Dutch peacekeeping unit that might have prevented the Srebrenica massacre of 1995. But, overall, peacekeeping is working. Dollar for dollar, it is far more effective at preventing fighting than purchasing complex weapons systems. A recent study from the notoriously gloomy RAND Corporation found that most U.N. peacekeeping efforts have been successful.

Peacekeeping is just one way in which the United Nations has made a significant contribution to the decline of war. American commentators love to disparage the organization in that big cereal-box building on the East River, and, of course, the United Nations has manifold faults. Yet we should not lose track of the fact that the global security system envisioned by the U.N. charter appears to be taking effect. Great-power military tensions are at the lowest level in centuries; wealthy nations are in-

creasingly pressured by international diplomacy not to encourage war by client states; and much of the world respects U.N. guidance. Related to this, the rise in "international engagement," or the involvement of the world community in local disputes, increasingly mitigates against war.

The spread of democracy has made another significant contribution to the decline of war. In 1975, only one-third of the world's nations held true multiparty elections; today two-thirds do, and the proportion continues to rise. In the last two decades, some 80 countries have joined the democratic column, while hardly any moved in the opposite direction. Increasingly, developing-world leaders observe the simple fact that the free nations are the strongest and richest ones, and this creates a powerful argument for the expansion of freedom. Theorists at least as far back as Immanuel Kant have posited that democratic societies would be much less likely to make war than other kinds of states. So far, this has proved true: Democracy-against-democracy fighting has been extremely rare. Prosperity and democracy tend to be mutually reinforcing. Now prosperity is rising in most of the world, amplifying the trend toward freedom. As ever-more nations become democracies, ever-less war can be expected, which is exactly what is being observed.

FOR THE GREAT-POWER nations, the arrival of nuclear deterrence is an obvious factor in the decline of war. The atomic bomb debuted in 1945, and the last great-power fighting, between the United States and China, concluded not long after, in 1953. From 1871 to 1914, Europe enjoyed nearly half a century without war; the current 52-year great-power peace is the longest period without great-power war since the modern state system emerged. Of course, it is possible that nuclear deterrence will backfire and lead to a conflagration beyond imagination in its horrors. But, even at the height of the cold war, the United States and the Soviet Union never seriously contemplated a nuclear exchange. If it didn't happen then, it seems unlikely for the future.

In turn, lack of war among great nations sets an example for the developing world. When the leading nations routinely attacked neighbors or rivals, governments of emerging states dreamed of the day when they, too, could issue orders to armies of conquest. Now that the leading nations rarely use military force—and instead emphasize economic competition—developing countries imitate that model. This makes the global economy more turbulent, but reduces war.

In *The Remnants of War,* Mueller argues that most fighting in the world today happens because many developing nations lack "capable government" that can contain ethnic conflict or prevent terrorist groups, militias, and criminal gangs from operating. Through around 1500, he reminds us, Europe, too, lacked capable government: Criminal gangs and private armies roamed the countryside. As European governments became competent, and as police and courts grew more respected, legitimate government gradually vanquished thug elements from most of European life. Mueller thinks this same progression of events is beginning in much of the developing world. Government and civil institutions in India, for example, are becoming more professional and less corrupt—one reason why that highly populous nation is not falling apart, as so many predicted it would. Interstate war is in substantial decline; if civil wars, ethnic strife, and private army fighting also go into decline, war may be ungrafted from the human experience.

Is it possible to believe that war is declining, owing to the spread of enlightenment? This seems the riskiest claim. Human nature has let us down many times before. Some have argued that militarism as a philosophy was destroyed in World War II, when the states that were utterly dedicated to martial organization and violent conquest were not only beaten but reduced to rubble by free nations that initially wanted no part of the fight. World War II did represent the triumph of freedom over militarism. But memories are short: It is unrealistic to suppose that no nation will ever be seduced by militarism again.

Yet the last half-century has seen an increase in great nations acting in an enlightened manner toward one another. Prior to this period, the losing sides in wars were usually punished; consider the Versailles Treaty, whose punitive terms helped set in motion the Nazi takeover of Germany. After World War II, the victors did not punish Germany and Japan, which made reasonably smooth returns to prosperity and acceptance by the family of nations. Following the end of the cold war, the losers—the former Soviet Union and China—have seen their national conditions improve, if fitfully; their reentry into the family of nations has gone reasonably well and has been encouraged, if not actively aided, by their former adversaries. Not punishing the vanquished should diminish the odds of future war, since there are no generations who suffer from the victor's terms, become bitter, and want vengeance.

Antiwar sentiment is only about a century old in Western culture, and Mueller thinks its rise has not been given sufficient due. As recently as the Civil War in the United States and World War I in Europe, it was common to view war as inevitable and to be fatalistic about the power of government to order men to march to their deaths. A spooky number of thinkers even adulated war as a desirable condition. Kant, who loved democracy, nevertheless wrote that war is "sublime" and that "prolonged peace favors the predominance of a mere commercial spirit, and with it a debasing self-interest, cowardice and effeminacy." Alexis De Tocqueville said that war "enlarges the mind of a people." Igor Stravinsky called war "necessary for human progress." In 1895, Oliver Wendell Holmes Jr. told the graduating class of Harvard that one of the highest expressions of honor was "the faith … which leads a soldier to throw away his life in obedience to a blindly accepted duty."

Around the turn of the twentieth century, a counterview arose—that war is usually absurd. One of the bestselling books of late-nineteenth-century Europe, *Lay Down Your Arms!*, was an antiwar novel. Organized draft resistance in the United Kingdom during World War I was a new force in European politics. England slept during the '30s in part because public antiwar sentiment was intense. By the time the U.S. government abolished the draft at the end of the Vietnam War, there was strong feeling in the United States that families would no longer tolerate being compelled to give up their children for war. Today, that feeling has spread even to Russia, such a short time ago a totalitarian, militaristic state. As average family size has decreased across the Western world, families have invested more in each child; this should discourage militarism. Family size has started to decrease in the developing world, too, so the same dynamic may take effect in poor nations.

There is even a chance that the ascent of economics to its pinnacle position in modern life reduces war. Nations interconnected by trade may be less willing to fight each other: If China and the United States ever fought, both nations might see their economies collapse. It is true that, in the decades leading up to World War I, some thought rising trade would prevent war. But today's circumstances are very different from those of the fin de siècle. Before World War I, great powers still maintained the grand illusion that there could be war without general devastation; World Wars I and II were started by governments that thought they could come out ahead by fighting. Today, no major government appears to believe that war is the best path to nationalistic or monetary profit; trade seems much more promising.

The late economist Julian Simon proposed that, in a knowledge-based economy, people and their brainpower are more important than physical resources, and thus the lives of a country's citizens are worth more than any object that might be seized in war. Simon's was a highly optimistic view—he assumed governments are grounded in reason—and yet there is a chance this vision will be realized. Already, most Western nations have achieved a condition in which citizens' lives possess greater economic value than any place or thing an army might gain by combat. As knowledge-based economics spreads throughout the world, physical resources may mean steadily less, while life means steadily more. That's, well, enlightenment.

In his 1993 book, *A History of Warfare*, the military historian John Keegan recognized the early signs that combat and armed conflict had entered a cycle of decline. War "may well be ceasing to commend itself to human beings as a desirable or productive, let alone rational, means of reconciling their discontents," Keegan wrote. Now there are 15 years of positive developments supporting the idea. Fifteen years is not all that long. Many things could still go badly wrong; there could be ghastly surprises in store. But, for the moment, the trends have never been more auspicious: Swords really are being beaten into plowshares and spears into pruning hooks. The world ought to take notice.

THE FAILED STATES INDEX

About 2 billion people live in countries that are in danger of collapse. In the first annual Failed States Index, FOREIGN POLICY and the Fund for Peace rank the countries about to go over the brink.

America is now threatened less by conquering states than we are by failing ones." That was the conclusion of the 2002 U.S. National Security Strategy. For a country whose foreign policy in the 20th century was dominated by the struggles against powerful states such as Germany, Japan, and the Soviet Union, the U.S. assessment is striking. Nor is the United States alone in diagnosing the problem. U.N. Secretary-General Kofi Annan has warned that "ignoring failed states creates problems that sometimes come back to bite us." French President Jacques Chirac has spoken of "the threat that failed states carry for the world's equilibrium." World leaders once worried about who was amassing power; now they worry about the absence of it.

Failed states have made a remarkable odyssey from the periphery to the very center of global politics. During the Cold War, state failure was seen through the prism of superpower conflict and was rarely addressed as a danger in its own right. In the 1990s, "failed states" fell largely into the province of humanitarians and human rights activists, although they did begin to consume the attention of the world's sole superpower, which led interventions in Somalia, Haiti, Bosnia, and Kosovo. For so-called foreign-policy realists, however, these states and the problems they posed were a distraction from weightier issues of geopolitics.

About 2 billion of the world's people live in insecure states, with varying degrees of vulnerability to widespread civil conflict.

Now, it seems, everybody cares. The dangerous exports of failed states—whether international terrorists, drug barons, or weapons arsenals—are the subject of endless discussion and concern. For all the newfound attention, however, there is still uncertainty about the definition and scope of the problem. How do you know a failed state when you see one? Of course, a government that has lost control of its territory or of the monopoly on the legitimate use of force has earned the label. But there can be more subtle attributes of failure. Some regimes, for example, lack the authority to make collective decisions or the capacity to deliver public services. In other countries, the populace may rely entirely on the black market, fail to pay taxes, or engage in large-scale civil disobedience. Outside intervention can be both a symptom of and a trigger for state collapse. A failed state may be subject to involuntary restrictions of its sovereignty, such as political or economic sanctions, the presence of foreign military forces on its soil, or other military constraints, such as a no-fly zone.

How many states are at serious risk of state failure? The World Bank has identified about 30 "low-income countries under stress," whereas Britain's Department for International Development has named 46 "fragile" states of concern. A report commissioned by the CIA has put the number of failing states at about 20.

The Rankings

In the table, the columns highlight the 12 political, economic, military, and social indicators of instability. Higher scores (in black) represent more instability; lower scores (in white) suggest less.

To present a more precise picture of the scope and implications of the problem, the Fund for Peace, an independent research organization, and FOREIGN POLICY have conducted a global ranking of weak and failing states. Using 12 social, economic, political, and military indicators, we ranked 60 states in order of their vulnerability to violent internal conflict. (For each indicator, the Fund for Peace computed scores using software that analyzed data from tens of thousands of international and local media sources from the last half of 2004. For a complete discussion of the 12 indicators, please go to www.ForeignPolicy.com or www.fundforpeace.org.) The resulting index provides a profile of the new world disorder of the 21st century and demonstrates that the problem of weak and

			Indicators of Instability											
Rank	Total	Country	Demographic Pressures	Refugees and Displaced Persons	Group Grievance	Human Flight	Uneven Development	Economic Decline	Delegitimization of State	Public Services	Human Rights	Security Apparatus	Factionalized Elites	External Intervention
1	106.0	Ivory Coast	8.0	8.0	7.7	8.8	9.0	7.7	9.8	9.5	9.4	9.0	9.1	10.0
2	105.3	Dem. Rep. of the Congo	9.0	9.4	9.0	7.0	9.0	8.0	8.0	9.0	9.1	8.7	9.1	10.0
3	104.1	Sudan	8.6	9.4	7.8	9.1	9.0	8.5	9.2	8.7	8.0	9.8	8.7	7.3
4	103.2	Iraq	8.0	9.4	8.3	6.3	8.7	8.2	8.8	8.9	8.2	8.4	10.0	10.0
5	102.3	Somalia	9.0	8.0	7.4	6.3	9.0	8.3	9.8	10.0	7.8	10.0	8.7	8.0
6	102.1	Sierra Leone	9.0	8.0	7.5	8.9	8.7	10.0	7.5	9.1	8.7	6.3	8.6	9.8
7	100.9	Chad	8.0	9.1	7.1	8.3	9.0	8.0	8.9	9.0	9.1	7.0	9.4	8.0
8	99.7	Yemen	7.8	8.0	6.4	8.2	9.0	8.8	9.8	9.3	6.4	9.0	9.4	7.6
9	99.5	Liberia	9.0	7.8	7.3	8.1	9.0	10.0	7.5	8.2	8.2	6.5	7.9	10.0
10	99.2	Haiti	8.8	8.0	7.7	3.4	9.0	8.1	9.4	9.8	8.7	7.8	8.5	10.0
11	99.0	Afghanistan	9.0	8.0	8.0	7.4	8.8	7.5	8.1	8.1	7.9	8.2	8.0	10.0
12	96.5	Rwanda	9.0	7.8	8.0	8.6	9.0	9.2	9.5	5.0	8.3	5.0	8.9	8.2
13	95.7	North Korea	8.0	6.0	7.2	8.1	9.0	9.6	9.8	9.7	9.0	8.3	8.0	3.0
14	95.0	Colombia	9.0	8.0	6.9	9.2	9.0	7.1	9.8	4.2	8.2	5.4	9.2	9.0
15	94.9	Zimbabwe	9.0	8.0	6.4	7.7	9.0	7.3	7.9	8.5	7.5	9.0	7.9	6.7
16	94.7	Guinea	9.0	6.0	6.1	10.0	9.0	4.5	9.7	7.5	8.1	8.1	9.2	7.5
17	94.3	Bangladesh	8.4	7.0	7.6	6.0	9.0	7.4	9.5	8.2	8.5	8.0	8.7	6.0
18	94.3	Burundi	9.0	7.2	7.1	3.8	8.8	7.8	7.2	9.0	8.3	7.5	8.6	10.0
19	94.2	Dominican Republic	9.0	8.0	7.1	8.5	9.0	6.8	6.8	9.6	9.2	7.0	9.2	4.0
20	93.7	Central African Republic	9.0	5.0	8.8	3.0	7.0	9.0	9.7	8.0	8.2	9.0	10.0	7.0
21	93.5	Bosnia and Herzegovina	7.0	8.0	8.6	5.7	9.0	5.7	8.5	6.0	7.3	9.0	8.7	10.0
21	93.5	Venezuela	8.0	8.0	6.8	7.6	9.0	4.5	9.8	8.2	9.1	7.8	7.2	7.5
23	93.4	Burma	8.9	8.0	6.3	8.0	9.0	6.9	9.2	8.0	9.6	9.0	7.5	3.0
24	93.2	Uzbekistan	6.5	8.0	6.8	6.8	9.0	6.0	9.1	5.0	9.6	9.0	9.4	8.0
25	92.7	Kenya	9.0	8.0	6.7	8.3	8.8	6.3	8.9	7.4	8.5	8.4	8.4	4.0
26	92.0	Bhutan	8.0	8.0	5.5	8.0	9.0	8.0	9.8	5.0	8.0	6.0	10.0	6.7
27	91.7	Uganda	9.0	7.6	6.9	5.7	8.4	6.0	8.0	8.4	8.3	8.0	8.1	7.3
28	91.5	Laos	9.0	6.7	6.3	8.8	9.0	6.5	7.9	2.5	9.4	9.0	9.7	6.7
28	91.5	Syria	9.0	8.0	7.5	6.8	9.0	5.0	9.0	5.0	7.6	9.0	8.2	7.4
30	91.1	Ethiopia	8.7	8.0	6.0	7.3	9.0	8.5	7.9	5.5	6.3	9.0	8.9	6.0
31	91.0	Guatemala	9.0	6.0	7.4	7.5	9.0	7.7	9.5	5.0	8.7	8.1	9.1	4.0
31	91.0	Tanzania	9.0	7.2	7.6	6.7	8.9	4.5	8.2	7.8	8.6	7.9	7.5	7.1
33	90.9	Equatorial Guinea	8.0	6.0	6.3	9.0	9.0	5.1	9.9	8.0	7.8	7.0	9.8	5.0
34	89.4	Pakistan	5.0	5.0	6.9	8.0	9.0	3.3	9.8	7.5	8.1	9.0	9.3	8.5
35	89.0	Nepal	9.0	8.0	5.6	4.0	9.0	7.1	8.9	6.0	9.1	7.6	8.0	6.7
36	88.9	Paraguay	4.0	5.0	6.9	8.3	9.0	7.8	9.9	7.0	8.3	8.0	8.7	6.0
36	88.9	Lebanon	8.0	8.0	7.5	7.1	7.0	4.7	8.7	4.3	7.3	8.1	9.2	9.0
38	88.8	Egypt	9.0	8.0	7.8	5.0	9.0	3.8	9.5	7.3	7.7	8.5	8.2	5.0
38	88.8	Ukraine	9.0	7.0	6.9	8.8	9.0	7.3	8.9	5.5	8.5	2.0	9.1	6.8
40	88.1	Peru	6.0	7.0	6.6	9.0	8.5	5.0	9.6	4.4	7.1	9.0	8.9	7.0
41	87.6	Honduras	9.0	6.0	5.3	9.7	9.0	5.4	9.9	3.0	7.2	8.0	9.1	6.0
42	87.5	Mozambique	9.0	8.0	5.7	9.0	8.8	7.8	8.1	6.7	7.4	3.8	8.2	5.0
43	87.3	Angola	7.9	8.6	6.3	3.8	9.0	4.4	7.9	7.2	8.3	7.0	8.1	8.8
43	87.3	Belarus	9.0	8.0	7.0	2.4	9.0	5.4	8.5	7.0	7.3	6.8	9.4	7.5
45	87.1	Saudi Arabia	7.6	6.3	7.8	8.8	9.0	2.2	9.8	4.3	8.6	9.0	8.3	5.4
46	87.0	Ecuador	9.0	6.0	5.6	6.9	9.0	5.0	9.5	7.5	7.9	8.0	8.6	4.0
46	87.0	Indonesia	8.6	7.0	6.3	8.9	9.0	4.0	9.2	4.0	8.6	7.6	8.8	5.0
48	86.7	Tajikistan	9.0	5.0	6.2	6.7	9.0	5.3	8.6	5.0	9.4	8.0	9.5	5.0
49	86.1	Turkey	9.0	8.0	7.3	5.0	9.0	4.2	9.7	4.8	5.0	8.0	9.1	7.0
50	85.7	Azerbaijan	8.0	6.0	6.0	5.8	9.0	4.1	9.7	5.0	8.5	7.0	9.6	7.0
51	85.6	Bahrain	6.0	5.0	6.7	9.0	9.0	1.7	9.7	4.0	8.4	9.0	9.6	7.5
52	84.9	Vietnam	8.6	8.0	5.6	8.5	8.9	3.4	7.6	4.3	8.4	8.0	6.4	7.2
53	84.6	Cameroon	9.0	7.0	5.1	8.6	9.0	4.2	6.4	7.5	6.6	8.0	8.2	5.0
54	84.3	Nigeria	7.2	3.0	6.5	8.7	8.9	5.8	8.8	6.9	6.7	9.0	8.3	4.5
55	84.1	Eritrea	8.0	8.0	5.4	4.0	9.0	8.8	9.0	7.0	5.7	5.0	9.2	5.0
56	83.9	Philippines	7.0	7.0	6.5	8.2	9.0	4.7	9.3	3.8	8.2	7.0	9.2	4.0
57	83.8	Iran	5.0	8.0	7.3	6.0	9.0	3.3	9.1	4.8	8.8	7.3	9.1	6.1
58	83.7	Cuba	5.0	8.0	6.3	5.4	8.8	5.7	7.8	3.8	9.0	9.0	8.6	6.3
59	83.5	Russia	9.0	6.0	7.5	2.3	9.0	3.8	9.4	6.7	9.0	7.6	9.2	4.0
60	82.4	Gambia	7.0	7.0	5.4	4.0	9.0	6.7	8.1	7.0	7.9	6.0	8.3	6.0

failing states is far more serious than generally thought. About 2 billion people live in insecure states, with varying degrees of vulnerability to widespread civil conflict.

The instability that the index diagnoses has many faces. In the Democratic Republic of the Congo or Somalia, state failure has been apparent for years, manifested by armed conflict, famine, disease outbreaks, and refugee flows. In other cases, however, instability is more elusive. Often, corrosive elements have not yet triggered open hostilities, and pressures may be bubbling just below the surface. Large stretches of lawless territory exist in many countries in the index, but that territory has not always been in open revolt against state institutions.

Conflict may be concentrated in local territories seeking autonomy or secession (as in the Philippines and Russia). In other countries, instability takes the form of episodic fighting, drug mafias, or warlords dominating large swaths of territory (as in Afghanistan, Colombia, and Somalia). State collapse sometimes happens suddenly, but often the demise of the state is a slow and steady deterioration of social and political institutions (Zimbabwe and Guinea are good examples). Some countries emerging from conflict may be on the mend but in danger of backsliding (Sierra Leone and Angola). The World Bank found that, within five years, half of all countries emerging from civil unrest fall back into conflict in a cycle of collapse (Haiti and Liberia).

The 10 most at-risk countries in the index have already shown clear signs of state failure. Ivory Coast, a country cut in half by civil war, is the most vulnerable to disintegration; it would probably collapse completely if U.N. peacekeeping forces pulled out. It is followed by the Democratic Republic of the Congo, Sudan, Iraq, Somalia, Sierra Leone, Chad, Yemen, Liberia, and Haiti. The index includes others whose instability is less widely acknowledged, including Bangladesh (17th), Guatemala (31st), Egypt (38th), Saudi Arabia (45th), and Russia (59th).

Weak states are most prevalent in Africa, but they also appear in Asia, Eastern Europe, Latin America, and the Middle East. Experts have for years discussed an "arc of instability"—an expression that came into use in the 1970s to refer to a "Muslim Crescent" extending from Afghanistan to the "Stans" in the southern part of the former Soviet Union. Our study suggests that the concept is too narrow. The geography of weak states reveals a territorial expanse that extends from Moscow to Mexico City, far wider than an "arc" would suggest, and not limited to the Muslim world.

The index does not provide any easy answers for those looking to shore up countries on the brink. Elections are almost universally regarded as helpful in reducing conflict. However, if they are rigged, conducted during active fighting, or attract a low turnout, they can be ineffective or even harmful to stability. Electoral democracy appears to have had only a modest impact on the stability of states such as Iraq, Rwanda, Kenya, Venezuela, Nigeria, and Indonesia. Ukraine ranks as highly vulnerable in large part because of last year's disputed election.

What are the clearest early warning signs of a failing state? Among the 12 indicators we use, two consistently rank near the top. Uneven development is high in almost all the states in the index, suggesting that inequality within states—and not merely poverty—increases instability. Criminalization or delegitimization of the state, which occurs when state institutions are regarded as corrupt, illegal, or ineffective, also figured prominently. Facing this condition, people often shift their allegiances to other leaders—opposition parties, warlords, ethnic nationalists, clergy, or rebel forces. Demographic factors, especially population pressures stemming from refugees, internally displaced populations, and environmental degradation, are also found in most at-risk countries, as are consistent human rights violations. Identifying the signs of state failure is easier than crafting solutions, but pinpointing where state collapse is likely is a necessary first step.

From *Foreign Policy*, July/August 2005, pp. 56-65. © 2005 by the Carnegie Endowment for International Peace. www.foreignpolicy.com Permission conveyed through Copyright Clearance Center, Inc.

Gaza: Moving Forward by Pulling Back

David Makovsky

THE OPPORTUNITY

AFTER FOUR and a half years of terror and violence, the proverbial stars seem to be aligned for a new push for peace between the Israelis and the Palestinians. Unlike his predecessor, the newly elected Palestinian Authority (PA) president, Mahmoud Abbas, stresses the importance of peaceful problem solving and has condemned suicide bombing (in Arabic and in English) as counterproductive. On the Israeli side, Prime Minister Ariel Sharon, the onetime architect of the settlement movement, is leading the drive to evacuate all settlers from Gaza and the northern West Bank. At Sharm-el-Sheikh earlier this year, he and Abbas committed to a cease-fire, an important step even if rejectionists on both sides are certain to try to exploit it. In Washington, meanwhile, Condoleezza Rice is as close to the commander in chief as any secretary of state has been since James Baker teamed up with George W. Bush's father, guaranteeing that she speaks with the president's authority.

But even under such relatively favorable conditions, it is wrong to assume that the Israelis and the Palestinians can simply return to the summer of 2000, when Washington thought that an end to the conflict was within sight. Since then, trust between the parties has been shattered by violence, and rebuilding it will not be quick or easy. Reaching for too much too soon will turn the current opening into one more lost opportunity.

Optimists—arguing that the time is right to work out compromises on such thorny issues as the borders of a Palestinian state, the status of Jerusalem, and the rights of Palestinian refugees—want to move shortly to negotiations on a final-status agreement. Rushing to an endgame approach, however, will energize hard-liners in both camps and undermine the leadership of Abbas and Sharon. Abbas, despite his victory in the January elections, does not yet have the authority to veer from Yasir Arafat's legacy on the conflict's most sensitive issues. Sharon, for his part, has won domestic support for his plan to disengage from Gaza and the northern West Bank, but an overly ambitious focus would be equally damaging for him. There is no evidence that he is either willing or politically able to strike a grand deal. Any attempt to do so would lead to his ouster in Likud (to the benefit of his hard-line opponent Binyamin Netanyahu), threaten the survival of his government,

and reverse the favorable political dynamic set off by pulling Israeli forces out of Gaza.

Before negotiating a final agreement and before the United States issues its own blueprint for a final outcome both sides need to provide tangible evidence that they are willing to compromise, thereby restoring trust and reinvigorating the peace process. Israel's disengagement from Gaza and the northern West Bank provides the perfect opportunity for doing so, and Washington should focus its immediate energy and resources on coordinating this endeavor. A successful withdrawal will shatter old taboos, undermine extremists, embolden moderates, and facilitate further withdrawals. A failed effort, meanwhile, will condemn both the Israelis and the Palestinians to many more years of violence and despair.

THE RISK OF OVERREACH

THERE ARE two variants of the case for moving shortly to a final-status agreement. Those who are most optimistic believe that the United States should jettison the three-phase "road map"—a set of mutual guidelines endorsed by the international community and meant to serve as markers on the way to a two-state solution and fast-forward to discussions of the most sensitive issues: namely, Jerusalem, refugees, and territory. Others call for Washington to issue a blueprint for final status now, regardless of the ripeness of the Israelis and the Palestinians themselves. (Both groups note that Bush has already gone part of the way toward defining his vision of a final agreement—which is not inconsistent with that offered by President Bill Clinton.)

The premise behind these arguments is that since opportunities in the Middle East are fleeting, it is best to go for a grand deal as soon as there is an opening. (Some in Washington offer an additional rationale: that issuing a blueprint will improve the United States' standing in the Arab world.) But in reality, such a high-risk move would be counterproductive, undermining peacemakers on both sides of the conflict as well as in Washington. And any blueprint that neglects conditions on the ground and raises expectations only to see them dashed again will trigger a new wave of violence and despair.

Opponents of Abbas will inevitably accuse him of betrayal on two key issues: refugees and Jerusalem. While campaigning,

Abbas veered from Arafat's legacy on violence and on the issue of coexistence. He refused, however, to contradict Arafat on the questions of the "right of return"—whether all Palestinian refugees should have the option of returning to Israel and exclusive Palestinian sovereignty over holy sites in Jerusalem. The implication is that Abbas does not yet have enough strength to compromise on such sensitive issues, as would be necessary in any discussion of final status, leaving him vulnerable to attack from rejectionists.

On the Israeli side, jumping to final status would undermine positive short-term dynamics. The Gaza disengagement plan has created a right-wing opposition to Sharon. If the United States or the international community were to present a blueprint for final status and press Israel to accept it, there is little doubt that Sharon's government would collapse and his disengagement plan would be thwarted. Sharon, a vociferous critic of the 2000 Camp David talks and the Clinton parameters for a final peace agreement, might have to come out against Washington. If forced to confront controversial issues when distrust of Palestinian intentions still runs high, he could lose the broad support of the public and his national unity government.

From Washington's perspective, meanwhile, presenting a U.S. backed blueprint for final status before calm and confidence have been restored on the ground would be counterproductive on several fronts. The experience of 2000 demonstrates that for any agreement to succeed, Arab states must give vocal support to it so that the Palestinian leader will have the political cover he needs to compromise on questions central to Palestinian identity. Since Arab governments have been unwilling to do this, it is preferable to focus on practical steps, especially a successful disengagement from Gaza.

The Bush administration has already used language to guide the next steps in the peace process. President Bush is the first U.S. leader to have articulated support for a solution with two states, Israel and Palestine, and he has emphasized the importance of a Palestinian state being independent and contiguous. In a June 24, 2002, speech he declared:

> Ultimately, Israelis and Palestinians must address the core issues that divide them if there is to be a real peace, resolving all claims and ending the conflict between them. This means that the Israeli occupation that began in 1967 will be ended through a settlement negotiated between the parties, based on UN Resolutions 242 and 338, with Israeli withdrawal to secure and recognized borders.

Palestinians and Arabs positively interpreted this reference to 1967 as signaling a commitment to ensuring the viability of West Bank territories in the future Palestinian state.

Instead of making a high-risk move to final status, Washington should use this moment of opportunity to facilitate confidence-building measures between the Israelis and the Palestinians. One way of signaling progress and bringing the sides closer together would be to formally activate the first phase of the road map. The road map certainly has its disadvantages. Since the Israelis and the Palestinians did not negotiate the document themselves, they do not feel a sense of ownership over the process it prescribes. But it is the only diplomatic framework broadly acceptable to both parties and backed by the international community. Recognizing that shattered trust is the legacy of the last several years, its guidelines are rooted in gradualism but focused on the ultimate goal of guaranteeing security for the Israelis and a state for the Palestinians. Following UN Security Council Resolution 242—the document that has traditionally guided the peace process—the road map's three-phase approach demands parallel performances, generating compromises and obligations on both sides.

The first phase of the road map calls for initial confidence-building measures. As demonstrated by the recent Palestinian elections and Abbas' public commitment to reforming the PA and restructuring its security services, this phase is essentially being implemented already. Since his election, Abbas has talked about: the road map as the way forward and—in line with the document's initial requirements-taken some preliminary measures to eliminate incitement to violence. (These steps can be reinforced by a review of the PA educational curriculum and a crackdown on imams who use incendiary language.) Despite some reservations, Israel has expressed support for the mad map's guidelines as well. Israel's willingness to move forward will. of course be contingent on the PA'S willingness to fight terrorism, but Israel should start by honoring its commitment under the road map to remove unauthorized settlement outposts in the West Bank and curb settlement activity.

A formal activation of the first phase would demonstrate that Washington remains committed to the road map's "performance-based approach" and to preserving the balance inherent in that idea. More important, it would show that a withdrawal from Gaza will not be an isolated step. For the Palestinians, this would provide evidence that Abbas has succeeded in ensuring U.S. involvement in the peace process—which would help prevent "Gaza first" from becoming "Gaza last."

GET GAZA RIGHT

SO FAR, Washington's approach has been to provide security and economic assistance but avoid micromanaging the disengagement process. In his most recent State of the Union address, Bush announced that he would give the PA $350 million in aid. And, as announced by Secretary of State Rice, he will dispatch Lieutenant General William "Kip" Ward to help restructure the PA's security services and facilitate cooperation between Israel and the PA.

Apart from reactivating the road map, Washington should focus on a number of short-term tasks to facilitate a successful withdrawal from Gaza ensuring that the Israeli pullout is complete and that Israel's safety is not compromised in the wake of it. To this end, Washington should work to revive Israeli-Palestinian security coordination, ensure a viable cease-fire, and maintain calm on the ground by training Palestinian security personnel and working with Egypt to reconfigure the Multinational Force and Observers (MFO) currently stationed in Sinai to patrol the border between Gaza and Egypt. These measures would build much-needed confidence between the Israelis and the Palestinians.

A successful revival of security coordination after several years of violence would have multiple benefits for both sides. Security is the cornerstone of coexistence, and sustained coordination would help rebuild trust between the two parties. Moreover, a better security environment would improve conditions on the ground for Palestinians by reducing the need for checkpoints. Israeli Defense Minister Shaul Mofaz has also said that Israel is willing to pull out of major West Bank cities as soon as the PA is ready to accept responsibility for controlling these areas.

Most important, security coordination would facilitate a successful withdrawal from Gaza. A coordinated pullout is more likely than a unilateral move by Israel to lead to a smooth transfer of authority. To avoid destabilization after the withdrawal, it is important that the PA be committed to the terms of the pullout.

Both sides have taken encouraging steps toward cooperation, showing greater sensitivity to the other side's domestic concerns than in the past. Sharon has released several hundred prisoners and agreed to halt targeted killings of suspected terrorists and the demolition of homes belonging to the families of suicide bombers. Abbas, for his part, has sacked Palestinian security chiefs who were considered corrupt Arafat cronies and ordered his troops to disrupt arms smuggling and rocket attacks on Israel.

If there is to be a viable cease-fire, the United States must take the lead in ensuring that it takes hold and is maintained. In order to avoid ambiguity, Washington should get the terms of the cease-fire in writing. And if the United States pressures Israel to uphold its half of the bargain, then perhaps Egypt can help Abbas secure Palestinian support as well.

Abbas must take on rejectionists with a blend of confrontation and competition. Confrontation does not have to be massive to be effective. Abbas should start with an array of important measures: closing rocket labs, arresting some key operatives, and removing certain imams who are inciting violence. He should also make clear that suicide bombing is counterproductive. Having campaigned on this issue, he has a mandate to act on it.

In addition to such measures, Abbas needs to neutralize Hamas and other militant groups with a new political and economic strategy. Two weeks after Abbas won a majority in Gaza in presidential elections, Hamas swept municipal elections there thanks to its track record of providing essential social services, from health care to education, not being delivered by the government. The PA, accordingly, must be capable of supporting both private- and public-sector groups that can replace Hamas as providers of such services. It must also show clear economic results, proving the financial benefits of nonviolence, and reduce corruption in the mainstream Fatah Party.

A key security challenge will be stopping weapons smuggling into Gaza from Sinai. The MFO deployed along the Egyptian side of the Gaza border under Washington's leadership could work with Egypt's newly upgraded 750-member border police. Egyptian aid is crucial, as the Egyptians know the local geography and culture. Weapons smuggling has proved to be one of the most difficult issues surrounding the Gaza withdrawal. But the combined efforts of the Egyptian border police

and the MFO could prove effective in stemming the flow of weapons. In the aftermath of the 1979 Egyptian-Israeli peace treaty, families riving on both sides of the border built tunnels to transport illicit goods. Today, the tunnels are used to bring in weapons as well. (In order to end weapons smuggling, it will also be necessary to generate economic development, since many families depend on the smuggling of other goods for their livelihood.)

The MFO stationed in Sinai was created in 1981, when the ON refused to endorse the 1979 Egyptian-Israeli peace treaty and accept observer status under the terms of the treaty. It has some 2,000 troops from 11 countries (including two U.S. battalions) operating under U.S. civilian and military leadership, with its headquarters in Rome. Its mission is to prevent a massive remilitarization of the Sinai. The general calm on the Israeli-Egyptian border that has prevailed since the peace treaty means the MFO is likely to be open to adjusting its mission.

> # Washington should ensure that the Gaza withdrawl is complete and that Israeli security is not compromised in the wake of it.

The MFO is the most logical choice to supplement the work of Egyptian border patrols for a variety of reasons. It already has a mandate to work in the area and is familiar to all the parties involved. Senior Egyptian military officials, such as General Omar Suleiman, the head of Egyptian intelligence, say that they would welcome the MFO'S deployment on the Gaza border. Privately, both Israeli and Palestinian officials say the same. Most important, the MFO would be useful because the framework for it already exists, eliminating the need to embark on a lengthy process of creating a new force. (It took a year and a half to assemble the current MFO; every year, Egypt, Israel, and the United States each contribute $17 million for its upkeep.) Although simply diverting personnel from the MFO'S current mission may not suit the needs of the new task, its configuration could be quickly enhanced, and the parties involved could then determine if it should be legally considered an "additive" force or a separate "MFO II."

By sole virtue of its presence, the MFO would raise the diplomatic costs to the Egyptians of failing to secure the border with Gaza. It would at the same time ease the burden on them by monitoring the access of potential infiltrators and assisting on patrols. Perhaps with the help of the U.S. Army Corps of Engineers, an upgraded MFO could also provide technological assistance to help den with the tunnel problem; for example, seismic technology could detect new tunnels. Once Gaza has functioning air- and seaports, the MFO could also assist other forces (including private consultants) in providing security.

Alternative options are not attractive for a variety of reasons. Israel would not accept UN involvement, given the critical nature of the smuggling issue and Israel's adversarial interactions with the international body in recent decades. A NATO force—which must operate under the consensus of its 26 member states and risks becoming politicized—could not be established quickly enough and might not be nimble enough to adjust to changing realities on the ground; a NATO role may be viable at some point, but not in the short term. A U.S.-dominated force, meanwhile, is infeasible given that the U.S. military is already overburdened; with the MFO, the bulk of the troops already comes from other governments. What is required is Washington's leadership, not a massive number of U.S. troops.

To coordinate and implement effective border security, a consultative forum should be created for the PA, Israel, Egypt, and the MFO to discuss ongoing security concerns. Such a mechanism exists under the Egyptian-Israeli peace treaty, and it could be extended to cover issues relating to the Gaza border. The logical place for ongoing (and perhaps daily) consultation is Kerem Shalom, the border area adjacent to the southeast corner of Gaza (and the site of a new border crossing, designed to avert violence in the Rafah area). There is no reason why Egyptian, Israeli, Palestinian, and MFO officials should not have an "operations room" to discuss and iron out any problems that may arise. Such a move would facilitate security, ensure that affairs run smoothly, and allay suspicions on all sides. Without regular communication, misunderstandings are bound to occur.

OPEN AND SHUT?

IN ADDITION to security efforts, the United States should support programs to promote economic development in a post-withdrawal Gaza. Washington should work closely with the World Bank on infrastructure and construction projects, job creation, donor assistance, and trade facilitation through efforts such as upgrading the Karni crossing to ease travel conditions for Palestinians. Overseas Private Investment Corporation insurance could encourage further foreign investment.

Washington should do all it can to encourage Arab states to play a constructive role in the peace process. Some Arab states have begun to realize that violence in the West Bank hurts them by demonstrating their own impotence. Others, such as Egypt, Lebanon, Saudi Arabia, and Syria, are reeling from internal demonstrations in support of democratization, giving the United States a new source of leverage to get them behind the peace process. Tunisia, with its government under assault from Washington, recently invited Sharon to a conference in Tunis.

Arab states can first of all help delegitimize suicide bombing by making it clear that such "martyrdom" attacks are politically counterproductive, morally wrong, and only impede the creation of a Palestinian state. Such an imprimatur would make it easier for the PA to combat terror. Second, the Gulf states can help bolster the PA financially. Gulf governments have reaped tens of billions in extra oil profit over the past two years but have failed to meet even the minimal commitment of $55 million a month in aid to the PA. (World Bank officials say only $9 million a month has arrived.) In total, Arab League governments owe the PA $891.8 million. Arab governments must

demonstrate that their support for the Palestinians goes beyond mere rhetoric. Significant new funds would allow the PA to reliably deliver basic services, thereby weakening Hamas. Third, Arab states should put forward an "Arab road map" to demonstrate to the Israelis that steps toward peace will result in greater regional integration and not make Israel more vulnerable. The recent return of Egypt's and Jordan's ambassadors to Israel is a promising step. Washington can help by pushing for the reopening of quasi-diplomatic liaison offices throughout the Middle East (they were closed in 2000) and new multilateral talks on issues such as economic development.

By working with the PA, Israel, and Egypt, as well as its fellow members of "the Quartet"—the UN, the European Union, and Russia-the United States can at least somewhat reduce the likelihood that terrorist groups will ruin the Israeli disengagement from Gaza. Members of the Quartet must use their leverage over Iran, which backs the radical groups to varying degrees. Its support of terrorism should be raised during all consultations with Iran, including the current nuclear negotiations. Abbas recently sent an envoy to Beirut and Damascus urging them not to spoil the fragile cease-fire. Syria in particular—which is now feeling the heat for its occupation of Lebanon and for housing militants—must be forced to prevent rejectionist groups from operating from its soil to disrupt the calm.

Finally, the United States should put forward a new UN resolution ratifying Israel's Gaza withdrawal, guaranteeing that a full and complete disengagement wins not just the support of the parties, but also the support of the Security Council. Such a resolution would help guarantee that the terms of departure are upheld and would designate the PA as the authority in charge of the area from which Israel withdraws. The resolution should denounce further violence and make clear that, in the wake of the Israeli withdrawal, all militias must disband and turn their weapons in to the PA. If belligerents continue to menace Israel, Israel will have the right of self-defense in keeping with the UN Charter. Such a resolution could provide an incentive for Israel to move forward, so long as its security requirements are met.

Of course, nothing the United States—or anyone besides the Israelis and the Palestinians themselves—does can guarantee peace. But such measures can help move things forward in countless tangible ways. As the history of the conflict shows, windows of opportunity may be meaningful only when opened by the parties themselves, but unless others help keep them open, they can shut all too quickly.

DAVID MAKOVSKY is Senior Fellow and Director of the Project on the Middle East Peace Process at the Washington Institute for Near East Policy and a lecturer at Johns Hopkins University's Paul H. Nitze School of Advanced International Studies. This article is based on the forthcoming study *Engagement Through Disengagement: Gaza and the Potential for Middle East Peacemaking*, published by the Washington Institute.

Africa's unmended heart

KINSHASA—Congo's war, the bloodiest anywhere since 1945, is more or less over. The fear in the vast and shattered country is that it could restart

WHEN your correspondent tried to fly out of Kinshasa seven years ago, rebels attacked the airport and the flight was cancelled. Congo's great war was just starting and there were charred, petrol-soaked bodies on the capital's garbage-strewn streets. Now, millions of deaths later, the war appears to be over. The government has made peace with most of the main rebel groups. The half-dozen foreign armies that ravaged the country have withdrawn, and been replaced by 16,000 UN peacekeepers. Tribal militias still battle and pillage in the northeast, but the UN, after a long and shameful spell of inertia, is finally starting to protect the innocent by shooting or capturing the bad guys.

Arriving in Kinshasa today is less fretful than it was. By hiring a retired boxer as an escort, your correspondent passed through the airport without losing any possessions. (The gentleman in question is something of a local celebrity: on seeing him, customs agents smile, mime a left hook and forget to rifle through one's bags.) Like Congo itself, the airport is shabby, dysfunctional and packed with ill-paid, predatory officials. But at least the guns are quiet.

The question is, for how long? The war formally ended in 2003, when most of the warring parties came together to form a power-sharing transitional government. This was hardly an ideal solution. None of the new regime's members was elected, few are honest and some are mass murderers. But Congo's weary people were happy to accept gunmen in government as a temporary measure

to end the fighting. Under the terms of the peace deal, nationwide elections for a legitimate regime were to be held by June 30th.

They won't be. This is partly for logistical reasons. Congo has held no census since 1984, so no one knows how many voters there might be. (The electoral commission guesses 28m, out of a population roughly twice that number.) In a country with no experience of democracy, no functioning civil service and virtually no roads, organising a ballot takes time, even with the best political will in the world.

Such will is notably absent. After two years in power, the government only last month got round to approving a draft constitution laying the ground rules for the poll and defining the powers of whatever regime is elected. And that constitution still needs to be approved by a referendum. The electoral commission, meanwhile, has registered no voters and hired hardly any of the 40,000-60,000 electoral workers it thinks it will need, let alone taught them how to use the 9,000 voter-registration machines it finally ordered last month.

Realistically, no election can be held this year. The peace accord allows, under exceptional circumstances, for up to two six-month delays. So everything is on track, really, says the government. Many Congolese think this is self-serving baloney. They believe their unelected rulers are postponing the poll for as long as possible because they want another year to carry on stealing.

Rumours and stirrings

The leader of the main opposition party, Etienne Tshisekedi, says the government will cease to be legitimate on June 30th, and is calling for its members to stand down then. Probably most people in Kinshasa agree with him. Incendiary leaflets are circulating, and there are rumours of the mass distribution of machetes. In May, a general strike paralysed the town of Kananga and riots gripped the diamond city of Mbuji-Mayi. On June 5th, when one of the country's four vice-presidents attended a football match in Kinshasa, the 90,000-strong crowd chanted "Kill him".

It could blow over. But peace in Congo cannot be taken for granted. Half of Africa's modern wars have reignited within a decade of ending, typically because post-war regimes have not addressed the problems that caused them to flare up in the first place. Congo's transitional government has conspicuously failed to treat the malady that makes the country so war-prone: corruption.

The country first fell apart because its rulers stole the cement. The state was looted until the government could no longer control its own territory. Foreign invaders and domestic rebels filled the vacuum. Mending the devastation they wrought will require more statesmanship than the nation has ever known.

Congo is a big place—bigger, for instance, than all the states that voted for John Kerry in America's election last November put together.

It shares a border with nine other countries, and touches every sub-Saharan region: central, south, east and west. A stable Congo could be Africa's healthy heart. Arterial roads could be built through it; it has a huge trove of untapped minerals and enough hydroelectric potential to light up half the continent. Conversely, a return to chaos could be a continental heart attack.

Land of the surprising son

Optimists point out that the country's current president, 33-year-old Joseph Kabila, is probably its best since independence in 1960. Pessimists retort that that is not saying much. His best-known predecessor, Mobutu Sese Seko, who ruled from 1965 until 1997 and renamed the country Zaire, was a ruthless crook who fitted his palace with a nuclear shelter, hired Concorde for shopping trips and so gutted the treasury that inflation between October 1990 and December 1995 totalled 6.3 billion per cent. He lost control because he and his flunkies filched too much. Unpaid soldiers refused to defend the regime against a rebellion led by Laurent Kabila, Joseph's father, who seized power and restored the country's old name.

The senior Kabila turned out to be even worse than Mobutu. He was equally brutal and corrupt, but less intelligent. He had people executed while he was drunk and then forgot that he had done so. He ordered diamond dealers to sell their gems to a state-backed monopoly for near-worthless Congolese francs, and then wondered why the country's largest source of export revenues dried up. He provoked Rwanda, Congo's aggressive neighbour, by arming and supplying the perpetrators of the 1994 genocide, who were hiding in eastern Congo's forests. Mobutu had done the same, which was why the Rwandan army helped Kabila topple him. Now Rwanda backed a revolt to topple Kabila, too.

That was how the war started in August 1998. Zimbabwe, Angola and Namibia rushed to Kabila's de-

fence. Uganda sided with the Rwandans, but later fought them. Then it got really confusing. In all, six national armies and dozens of rebel groups and militias joined the fray. All sides plundered Congo's minerals. When Joseph Conrad spoke of "the vilest scramble for loot that ever disfigured the history of human conscience", he was describing Congo under Belgium's King Leopold II, but his phrase has not dated.

No one knows how many people died. A survey published in December by the International Rescue Committee (IRC), a pressure group, put the death toll at 3.8m between August 1998 and April 2004. The IRC calculated this by comparing death rates before, during and after the war and then extrapolating. Almost all the deaths—more than 98% in 2003-04—were from war-induced starvation or disease, rather than from bullets or spears. Clearly, the IRC's guess is open to challenge. But no one has come up with a better one.

Amid the mayhem, Laurent Kabila was shot dead by one of his bodyguards. His son Joseph took over, and quickly proved wiser than his father. Under western pressure, he signed a peace accord in South Africa in 2002. Unlike his father, who tore up treaties like old betting slips, he has honoured it, more or less.

He welcomed his former enemies into the transitional government that took office in June 2003. Two of his four vice-presidents are ex-rebels, and most armed factions have been given a ministry to mismanage. In theory, all the big rebel forces are to be merged into a single, slimmed-down national army. In practice, most remain under the control of their old commanders, and the process of demobilisation has been worryingly slow.

Most of Congo is now calm, although the east is much less so. A hot war continues in the north-eastern region of Ituri, where rival militias (backed by Rwanda and Uganda respectively) have killed at least 60,000 people in a quarrel over border trade and gold fields.

The other immediate threat comes from the Rwandan *génocidaires* who continue to lurk in the eastern Congolese rainforest. So long as they remain, Rwanda could invade again, as it has twice before, to try to eliminate them. Rwanda's president threatened to do so last year.

With help from the UN, both of these dangers are being tackled, says William Swing, the head of the UN mission. In Ituri, the blue helmets have started storming militia camps. Mr Swing says the UN's firepower has put the militiamen on the defensive and more than 12,000 have been persuaded to join a disarmament programme, leaving only a hard core of 1,500-2,000 at large. As for the genocide veterans, their main faction agreed on March 31st to lay down its weapons, go home and form a peaceful political party.

Not everyone is as optimistic as Mr Swing. By mid-May, the transit camps for surrendering Rwandan guerrillas were still empty. Last week a peacekeeper in Ituri was killed and two employees of the medical charity, Médecins sans Frontières, were abducted. And on June 2nd, Human Rights Watch, a pressure group, issued a fat report on continuing "widespread ethnic slaughter, executions, torture, rape and arbitrary arrest" in the region.

Even by the standards of war, some of the atrocities in eastern Congo are shocking. Zainabo Alfani, for example, was stopped by men in uniform on a road in Ituri last year. She and 13 other women were ordered to strip, to see if they had long vaginal lips, which the gunmen believed would have magical properties. The 13 others did not, and were killed on the spot. Zainabo did. The gunmen cut them off and then gang-raped her. Then they cooked and ate her two daughters in front of her. They also ate chunks of Zainabo's flesh. She escaped, but had contracted HIV. She told her story to the UN in February, and died in March.

The new "unified" Congolese army seems incapable of policing the east. When ordered to work together, its soldiers sometimes fight each other instead. So the heavy work is left to the UN.

Yet the show goes on

Nonetheless, the junior Mr Kabila's government is the only show in Congo, so donors have flocked to support it. In 2003, including debt relief, Congo was the largest recipient of French aid and the fourth-largest recipient of America's.

The donors' cash buys some influence. At the urging of the International Monetary Fund (IMF), Mr Kabila has restored economic sanity—inflation fell from 135% in 2001 to 4.4% in 2003, though it has since crept back up. As the violence eases, growth has rebounded, and is expected to hit 7% this year. But the average Congolese income, by the IMF's estimate, is still only $100 a year. That is an unbelievably low sum for a country where food is fairly easy to grow. The true figure must surely be a bit less wretched. But nobody really knows.

The country's recovery would be assisted if foreign investors, who mostly fled decades ago, were to return. Some are coming. Adastra, a London-listed mining firm, has signed a joint-venture deal with the government to turn two huge heaps of dirty-white dust into metal. The heaps are by-products of a moribund mine near the once-prosperous town of Kolwezi. Between them, they contain 1.7m tonnes of copper and 360,000 tonnes of cobalt. Tim Read, Adastra's boss, reckons the project has a net present value of $459m.

Freelance mining

Mr Read's firm is betting that Congo will not go back to war, and that its government—which has introduced a generally admired new mining law—will respect property rights, or at least those of big investors. The Congolese government is betting that Adastra will pay taxes and create jobs that are preferable to the local alternative, examples of which can be seen all around the big pile of dirt. Hordes of freelance miners scrabble for ore by hand, wash it in a river and sell it to local smugglers. Many are injured or killed when their tunnels collapse. Your correspondent saw a boy no bigger than his three-year-old son struggling under a half-sack of ore (he was probably older; malnutrition makes Congolese children quite short).

Adastra hopes to make a difference. But after the construction phase is over, the firm will employ only 700 people in a town of 250,000 where unemployment is practically universal. A shop assistant in Kolwezi makes $35 a month—less than those desperate freelance miners. The lowest-paid civil servant gets $2 a month. Congo needs many more Adastras, but they will come only when the country looks less daunting.

Congo's people have learnt to cope. In the absence of piped water, families in Kinshasa earn a living carrying buckets of it to building sites. A story from the southern town of Lubumbashi offers a snapshot of what works and what doesn't.

Veronique, an office worker, was separated from her daughter by the war. When peace broke out, she booked an aeroplane ticket for her (penniless) girl to rejoin her. But before the daughter could board the plane, she was detained. Her yellow fever vaccination card had been stamped by rebel health authorities, and so was invalid, the officials tut-tutted. Alas, she had no money for a bribe.

But Veronique was able to send her the equivalent of cash by mobile telephone. She bought $20 worth of telephone cards. These give you a code number which you key into your phone and thereby "recharge" it with pre-paid airtime. Veronique called the obstructive officials and gave them her code numbers to recharge their own mobile phones. It took only minutes to send her bribe across the country—faster than a bank transfer, which would in any case have been impossible, since there is no proper banking system.

That's Congo. Private cellphone networks and private airlines work because the landlines do not and the bush has eaten the roads. Public servants serve mostly to make life difficult for the public, in the hope of squeezing some cash out of them. Congo is a police state, but without the benefits. The police have unchecked powers, but provide little security. Your correspondent needed three separate permits to visit the railway station in Kinshasa, where he was stopped and questioned six times in 45 minutes. Yet he found that all the seats, windows and light fixtures had been stolen from the trains.

The government offers the usual excuses. Before a British audience in March, Jean-Pierre Bemba, one of the four vice-presidents, stressed that much of the blame for corruption rests with those (ie, western firms) who pay the bribes. But bribery, though rife, is less of a problem than simple theft.

For example, the central government is shelling out $8m a month to pay soldiers' salaries, partly in the hope that they will stop robbing civilians. But many soldiers are still not paid. In the east, where the problem is especially acute, observers assume that local commanders are snaffling their men's pay packets. Not so, says one of Mr Kabila's disgruntled advisers. In fact, he says, the money sent out east is intercepted and wired straight back to Kinshasa, where it is stolen by bigwigs. "It's called *Opération Retour*," he explains.

Supposing Congo weathers the likely storm of protest around June 30th and eventually holds elections, what then? It will depend on two things. First, who will win the presidency? Second, will the next president allow his powerful rivals enough of a stake in government to dissuade them from going back to war?

The most likely presidential contender is the incumbent, Joseph Kabila. In the absence of opinion polls, it is impossible to say how popular he is. But he is not personally blamed for any major atrocities, and he has all the apparatus of the state to bolster his cause.

Another possibility is Mr Bemba. As the vice-president in charge of the economy, he has colossal powers of patronage. But his patriotism is open to question: he used to lead a rebel army backed by Uganda. And his men are alleged to have eaten people. Mr Bemba dismisses the allegation, but it is widely believed and will not endear him to voters.

A third contender is Mr Tshisekedi, the opposition leader. His long and honourable opposition to Mobutu has won him much support in his home region and around Kinshasa. But he has no money to mount a nationwide campaign. It would boost his chances if he could organise big protests against the missed election deadline. Which gives the government another reason to crush any protests.

Whoever the next president is, he will have to placate several groups with contradictory desires. The men with guns must be given the right mixture of threats and inducements not to wreck the peace. People must feel they have enough stake in the new regime not to try to overthrow it. And donors must be shown enough progress that they do not walk off in disgust. Somehow, Congo must find a leader who will refute V.S. Naipaul's jibe. "It isn't that there's no right and wrong here," the novelist once wrote. "There's no right."

Sudan's Darfur: Is It Genocide?

"In causing civilian atrocities on such a massive scale, has the Sudanese government adopted a policy of cultural annihilation, or has it decided to crush a rebellion to protect its dominance?"

NELSON KASFIR

Haunted by the failure of the West to intervene in Rwanda while it was possible to save lives, some Western media and governments now insist that the civil war between the Sudanese government and Darfur guerrillas has resulted in genocide. The urgency of their concern is entirely justified; the label may not be.

Without doubt, the most frightening feature of the unfolding tragedy in Darfur is the scale and ferocity of armed attacks on civilians. Although civilians always form the majority of victims in guerrilla war, the numbers killed and displaced in Darfur—probably more than 2.5 million by December 2004—have been astounding for a war that began only two years ago. The wanton cruelty in these attacks, including massacres of unarmed villagers, sexual violence meted out to women, and methodical destruction of villages, including the poisoning of wells, far exceeds the brutality that characterized the episodic violence experienced previously by Darfur inhabitants.

Virtually all observers—humanitarian and human rights groups, UN missions and local participants—agree that most of the violence is being carried out by the Sudanese military in combination with local Arab ethnic militias, the so-called *Janjaweed* or "evil horsemen." Despite the difficulties in distinguishing between Arabs and Africans in Darfur, where intermarriage is common and almost everyone is a Muslim, most observers also conclude that these attacks are mainly against Africans.

The US government, so negligently reticent during the Rwandan massacres in 1994, has declared that the killings in Darfur amount to genocide. But others have not, most notably the United Nations commissioners reporting on the situation in Darfur to the UN Secretary General. Their January 2005 report (www.un.org/News/dh/sudan/com_inq_darfur.pdf) condemns the Sudanese government and the ethnic militias fighting with it for the enormous suffering they have caused, but argues that the violent attacks on civilians stem from counterinsurgency tactics.

The violence recorded and condemned in this report, however, is so disproportionate to the actions of the new and relatively inexperienced guerrilla groups in Darfur that it is hard to believe it is simply a tactical response to battlefield conditions. Furthermore, counterinsurgency tactics, however virulent, can only be instruments for the larger ends of war. Could any other purpose besides genocide be driving the Sudanese government to commit such carnage?

Perhaps. Since the government has only a precarious grip on national power, it is constantly mindful of threats to overthrow it. The regime depends on the support of wealthy political and economic interests that represent a small minority of the Sudanese people. Long before the present National Islamic Front (NIF) government took power in a military coup in 1989, national officials feared that their regime would not survive simultaneous rebellions arising in impoverished areas throughout the country. Thus, they may believe their survival depends on striking as hard as they can not only to destroy support for the insurgents in Darfur, but also to ensure that no other guerrilla groups take up arms elsewhere, especially in the northern part of the country.

The unprecedented and generous peace agreement the government made in January 2005 with a different set of rebels, the Sudan Peoples Liberation Movement/Army (SPLM/A), to end a 22-year guerrilla war in the south, poses puzzling questions for both the genocide and the regime-threat interpretations of the violence in Darfur. Why, if the government intends genocide, does it share power and wealth with some Africans? And why, if it is worried about survival, would the government encourage others to emulate the southerners by rebelling in hopes of compelling equally favorable negotiations?

Does it make a difference which of these views is correct when so many people are dying or displaced from their homes? Quarrels over explanations surely ought not get in the way of providing humanitarian assistance on a far greater scale than is occurring now. Even so, there is a history of repeated government attacks on civilians directly and through ethnic militias throughout the impoverished peripheral areas of the country, not just in Darfur.

If the problem is not only to stop this assault but to end future violence by the Sudanese government against its citizens, it is essential to understand better the motives of those who are prosecuting it so cruelly.

THE RISING TOLL

The rapid growth in casualties and the accompanying savagery in Darfur are startling. While extended wars and local conflicts, sometimes supported by national officials, have resulted in serious casualties and destruction several times during the past 15 years in Darfur, this guerrilla war is only two years old. The first attacks were initiated by a newly formed insurgent organization, the Sudan Liberation Movement/Army (SLM/A) in February 2003, followed a few weeks later by a second new group, the Justice and Equality Movement (JEM). The unprecedented scale of attacks on civilians in response has been compressed into an even shorter time. It started only after an SLA surprise attack on the airport at El Fasher, the capital of North Darfur State, destroyed seven military planes and killed about 100 soldiers in late April 2003. After the attacks, the Janjaweed was formed— mostly from members of previously existing nomadic tribal militias of Arab background—and armed by the government.

For centuries, Darfur's local villages and ethnic groups have been dependent on their own resources, both arms and customary mediation, to keep the peace. In the past 40 years, however, increasing tensions in the area have dramatically changed the nature of conflict and the methods for its resolution. The numbers of both conflicts and victims have grown rapidly. Reports of coordination of militias on the basis of Arab and African identities first appeared in conflicts in the late 1980s and the 1990s. These conflicts involved attacks that, while similar in character to the current crisis, were more limited in area and involved a few hundred or thousands of casualties, not the hundreds of thousands estimated in the current conflict.

Since October 2004, the Western press has badly understated the numbers of those killed in the latest conflict, invariably using an estimate of 70,000 deaths. This figure was mistakenly taken from an updated version of a World Health Organization (WHO) study of deaths from disease and malnutrition in camps organized for displaced persons; the report covered only the period between March and September 2004. After analyzing five studies of mortality in Darfur since February 2003, Dr. Jan Coeburgh, writing in the February 2005 *Parliamentary Brief*, estimated a range of 218,000 to 306,000 deaths through December 2004. And that number continues to rise, not only from new attacks, but also from disease and prolonged malnutrition. "This year," he added, "looks worse than last." As Dr. Coeburgh told BBC News in February, "the reality is that we just don't know the scale of the problem."

In addition, the January 2005 UN report on Darfur estimated that there are 1.65 million internally displaced persons (IDPs) living in 81 camps and safe areas, plus another 627,000 "conflict affected persons," and 203,000 refugees in Chad. This means that, out of Darfur's total population of 6 million, and in addition to the dead, approximately 2.5 million people have been profoundly harmed in this conflict.

To the devastation of lives must be added the destruction of communities. In her travels for Human Rights Watch through Darfur with the SLM/A in March and April 2004, Julie Flint observed that the "most striking thing . . . was a completely empty land—mile after mile of burned and abandoned villages." The UN commissioners in their report estimated that "600 villages and hamlets have been completely destroyed, while an additional 100 to 200 villages have been partially destroyed."

Even if we agree that these reports accurately estimate the alarming levels of death and destruction, it is important to determine which of the parties in this war have accounted for them. If these numbers were evenly distributed among the guerrillas, the government, and the Janjaweed militias, they would be just as tragic, but the case for either genocide or regime threat as the motivation for the violence would be harder to establish. It is clear, however, that the guerrillas have inflicted far less damage to civilians. The UN commissioners declare that "the vast majority of attacks on civilians in villages have been carried out by Government of Sudan armed forces and Janjaweed, either acting independently or jointly. Although attacks by rebel forces have also taken place, the Commission has found no evidence that these are widespread or that they have been systematically targeted against the civilian population."

Nor should these figures be divided between the Sudanese government and the Janjaweed. They are not operating independently. Outside observers consider the militias to be the tool of the government. As the UN commissioners point out, coordination of aerial bombing and militia attacks demonstrates close cooperation between the Janjaweed and the government. Musa Hilal, the sheikh of Um Jalloul (an Arab ethnic group in North Darfur State), reputedly one of the Janjaweed's organizers, frankly admitted to Human Rights Watch investigators on September 27, 2004, that "all the people in the field are led by top army commanders . . . [who] get their orders from the western command center, and from Khartoum." As one of several victims told a UN commissioner, "for us, these are one and the same." The evidence seems to establish clearly that in an extremely short time an extraordinarily large proportion of Darfur residents have been killed or driven from their homes by the government and its agents.

DISENTANGLING MOTIVES

Though essential to determine whether they have committed genocide, the motives explaining why the government and its militias have engaged in so much destructive behavior are difficult to establish. One problem in isolating the government's motives is that the Darfur crisis grows out of many conflicts at the local, regional, and national levels. These conflicts involve responses to diminished natural resources, to ethnic and cultural conflict, to negotiations and the peace agreement in southern Sudan, and to the relationship of the national government with impoverished and marginalized groups throughout the country. Consequently, both the government and the guerrillas enlist supporters who have their own motives for participating. To isolate the government's motives, it is important to identify the motives of the other contributors to the conflict.

Darfur, the westernmost region in northern Sudan, is the size of Texas. Even before this crisis began, its fragile semi-desert ecology could not easily support the people living there. A set of customary rules that evolved over centuries governs the sharing of water and land between nomadic herders and settled farmers. These rules have been deeply strained and increasingly violated because of advancing desertification and population growth. Average annual rainfall has declined over the past 50 years, while markets for peanuts and gum arabic, the main crops grown in Darfur, have shrunk over the past 20. As a result, living standards have fallen rapidly throughout the region. Since the government has never effectively policed Darfur, clusters of villages trained their young men as warriors to defend themselves from outside attack. Armed with spears, neither attackers nor defenders could cause many casualties. In the 1970s, however, rifles became widely available. All these factors reduced the ability of ethnic leaders to mediate ensuing disputes. Their capacities were also profoundly weakened when President Gaafar Nimeiri in the 1970s abandoned official recognition of customary administration in favor of centrally appointed local officials.

In addition to these essentially local conflicts, an increasing tendency to politicize cultural identities has occurred at the Darfur regional level over the past 20 years. The religious process of Islamization and the linguistic and cultural process of Arabization have proceeded unevenly. In Darfur, virtually all inhabitants are Muslims, while also holding different additional ethnic identities that are often multiple and fluid. Intermarriage and ethnic switching among local groups have been common throughout the area, even while the status hierarchy has firmly placed Arab above African for centuries. Long before the current civil war began, Africans in Darfur believed that the national government's policy of Islamization hid a policy of Arabization.

Both the idea that Arabic culture is a civilizing mission and the idea that African cultures retain valuable heritages have the potential to mobilize people throughout Sudan. Unfortunately, and entirely unnecessarily, these ideas are perceived as contradictory. Over the past two decades in Darfur, certain intellectuals have styled themselves the "Arab Gathering" to demand greater Arab representation in positions in Darfur state governments. This

has led to equivalent demands for African representation in national posts, most prominently in a *samizdat* publication called the Black Book, which was photocopied and surreptitiously handed out in mosques in 2000.

As the carnage has grown and people have had to choose sides, African and Arab identities have gained greater currency, although perhaps only temporarily. Thus, African students and notables, particularly from the Fur, Massaliet, and Zaghawa ethnic groups, formed the SLM/A and JEM, the most important guerrilla groups, while the loosely organized Janjaweed have been recruited primarily from Arab groups in Darfur. The fighting forces in the war have deepened cultural identities that in past years were remote from daily concerns. Nevertheless, there are Arabs fighting with the SLM/A and African ethnic groups that support the government. While the extent to which ordinary citizens have redefined themselves as Africans and Arabs remains unclear, the hardening of more inclusive identities has expanded perceptions of the stakes in the conflict from the local to the regional level.

Both the fighting in Sudan's south and its resolution have also influenced the motives of actors involved in the Darfur conflict. The Darfur guerrillas and the SPLM/A have had close connections. SPLA officers trained some SLA fighters as Darfur hostilities began, and the SLM/A's (and JEM's) basic platform is almost identical with that of the SPLM/A. Furthermore, the liberal terms of the Comprehensive Peace Agreement that ended the separate civil war in the south in January 2005 may have strengthened the prospects for rebellion in many areas of Sudan.

The impact of the peace agreement on other regions is unclear, although deeply contradictory. Its achievement has been argued to show that either peace pays or rebellion pays. Up to the date of the agreement, the duration of negotiations between the government and the SPLM/A coincided almost exactly with the period of fighting in Darfur. For his part, UN Secretary General Kofi Annan insisted hopefully at a November 2004 meeting of the Security Council in Nairobi that the peace agreement "would . . . serve as the basis and catalyst for the resolution of existing conflicts." In other words, the settlement in the south would lead to settlement elsewhere.

But Darfur notables and guerrillas were excluded from the negotiations between the SPLM/A and the government (mainly because the donors who organized them believed that it was better to avoid additional complications). The lesson that the SLA and JEM took from the success of the peace agreement was the opposite of Annan's—that rebellion pays. To be taken seriously as a negotiating partner, it is necessary to rebel first. No one knows what lesson the Sudanese government drew, as it negotiated with one group while fighting another. But Khartoum surely understands that if it negotiates an agreement with the Darfur guerrillas similar to the generous one it signed with the SPLM/A, it greatly increases the probability of several new rebellions.

CORE AND PERIPHERY

The relationship of Darfur to the national political economy also affects its civil war. Paradoxically, all the regional combatants are poor relative to those who control the economy from the center. As consciousness of this inequality has spread, the position of the national government has become more precarious.

Sudan presents a classic case of uneven development, which took root during Ottoman rule in the nineteenth century, deepened after 1898 during the Anglo-Egyptian Condominium, and intensified after 1956 during each postindependence government. For the last century, investment and development measures have been concentrated in the central area located at the convergence of the White and Blue Niles to the neglect of the rest of the country. Under British rule, nominally shared with Egypt, the Sudanese families ruling the two largest Islamic brotherhoods, the Khatmiyya and al-Ansar, were given special political and economic opportunities that they quickly translated into significant wealth. In addition, official capital investment was almost entirely devoted to the cultivation of cotton in Gezira, which eventually provided over half of Sudan's export earnings.

Over time, the concentration of wealth in this core area stimulated schools, jobs, and further investment among the peoples living in the region, particularly those identifying with three riverain ethnic groups: the Danagla, the Ja'aliyin, and the Shagiya. This wealth also created powerful economic interests that acted to protect the advantages of those living in the region. In particular, the profits from Gezira, intended originally as the engine of growth for the whole country, were blocked from investment into development projects in other areas. All other parts of Sudan in both the north and the south have progressed more slowly than the core in economic development, education, and infrastructure. In general, they provide less profitable opportunities for new investment. They have developed commercialized sectors more slowly and, as a result, have become even more peripheral.

The government is sending a message to potential guerrillas everywhere that if they rebel, civilians in their region will face atrocities.

Predictably, these growing economic disparities have fed into the construction of African and Arabic identities. Policies of Islamization and Arabization have helped to institutionalize the dominance of core economic interests and vice-versa. In the south, where fears of northern hegemony had existed since the nineteenth-century slave trade, civil war began soon after independence and continued, albeit with a significant interruption, until 2005. But in the north, even though Muslim and Arabic groups in peripheral areas did not share the wealth or power of those in the center, they did not threaten or organize rebellions against the dominance of the core until recently.

One of the important changes in the economy that has awakened political resentment in the periphery was the discovery of oil in the 1970s. The oil is located entirely in peripheral areas, including South Darfur State, but the oil revenues have been controlled exclusively by Khartoum. Oil has undoubtedly contributed to recent demands by political and guerrilla groups that wealth be shared, particularly when it is extracted from their own areas. The lesson of the government's concession in the peace agreement that it would split oil profits with the new southern government is not likely to be lost on other groups.

The common interest of core elites in political and economic dominance never meant that contending members of the leadership group agreed on policies or even on basic economic or political orientations. All Sudanese governments since independence have been riven by conflicts and frequently overthrown. New leaders often have treated their predecessors harshly. But these leaders have always emerged from the same core group. For all their disagreements, they have chosen to defend the economic and political interests of the core. And their hegemony has always depended on the absence of a challenge by groups living in the periphery.

This began to change in the 1980s. But the NIF, like its predecessors, has unhesitatingly used its formidable economic and political advantages throughout the periphery to appoint replacements for local officials who object to its policies and to disrupt local acts of defiance. Whenever it felt it might be losing control of a local population, it has formed and rewarded local ethnic militias to attack the groups represented by its opponents. The formation of these militias has resulted in splitting local populations on cultural rather than uniting them on economic grounds.

Darfur represents the latest example in which Khartoum has used its policy of Arabization in an effort to bolster or restore its hegemony. The groups from which the Janjaweed are recruited are just as marginalized as those the Janjaweed are attacking. Indeed, it has been argued that they have more in common with each other than either has with the groups that have long controlled the national government.

THE CRIMINAL ELEMENTS

Mass murder in Darfur raises the question of genocide. It does not answer it. Genocide is a complex crime requiring attention to each of its elements. The term and the concept were originally conceived and named by Raphael Lemkin to ensure that the Holocaust in Nazi Germany would never be repeated. When the United Nations made it a crime in 1948, it said that "genocide means . . . acts committed with intent to destroy, in whole or in part, a national, ethnical, racial or religious group. . . ." As one international tribunal characterized it, "the crime is horrific in its scope; its perpetrators identify entire human groups for extinction." The strong feelings it arouses can interfere with careful analysis.

Because genocide connotes immoral political activity, it has been applied to disparate events and spawned definitions that differ from the UN convention. The most important ambiguities in the UN definition relevant to the Darfur crisis concern the meanings of "in part" and a "group." The "in part" issue poses the question of how many people must be attacked before an event can be labeled a genocide. International case law makes clear that the "part" must be substantial, such as an attempt to eliminate all members of a group in a region or a country.

The "group" issue raises the problem of how permanent the cultural entities that are attacked must be before they qualify for protection from genocide. Does this category include only groups whose membership can be objectively determined by observers, or does it also include groups whose formation is based on subjective identification by its members? International tribunals have held that if parties on both sides of a conflict share objective traits such as language and religion, the subjective identification of the victims as a separate group can be the basis for establishing genocide. The classic example is the determination that genocide occurred in Rwanda even though Hutu and Tutsi share a language, territory, and various cultural practices.

Whether the Sudanese government has committed genocide in Darfur can be evaluated by considering each of the four elements of the UN definition—an attempt to destroy, a perpetrator, a group, and an intent. The evidence for three of these four elements supports the claim of genocide, though not each to the same degree.

First, an attempt to destroy has unquestionably occurred. The numbers killed, forced from their homes, and facing starvation constitute a substantial part of the regional population. Second, the balance of the evidence implicates Sudan's government as the perpetrator, acting both on its own and through its agents, the Janjaweed. Third, while it is difficult to distinguish Africans and Arabs as objective groups, since both are Muslims and speak Arabic, the polarization caused by war has heightened victims' identification with these groups. Although the evidence is not as clear, it seems likely that the victims perceive themselves to be attacked because they are Africans and thus can be assigned on this subjective basis as members of a protected group.

IS GENOCIDE THE INTENT?

The remaining element is intent: In causing civilian atrocities on such a massive scale, has the Sudanese government adopted a policy of cultural annihilation, or has it decided to crush a rebellion to protect its dominance? The available evidence can only provide inferences about the government's motive. Showing that others involved in the Darfur conflict have genocidal motives is not suffi-

cient to establish the national government's policy. For example, eyewitness accounts of atrocities indicate that members of the Janjaweed often have attacked Africans with genocidal intent. These attacks implicate the government, since members of the Janjaweed have acted as its agents. But more direct evidence is necessary to show that the government adopted genocide as its policy.

In their January 2005 report, the UN commissioners did not find genocidal intent by the government. However, the two arguments they make are not persuasive. They reject genocide because they found cases in which the attackers discriminated among members of the targeted group rather than attempting to exterminate all of them, and because the government allows victims driven out of their villages to live in IDP camps run by humanitarian organizations.

The first point is based on only a few examples without any suggestion that these incidents are representative. The second might be plausible if life were secure in the IDP camps, but it is not. The previously mentioned WHO report showed an extremely high death toll from disease and malnutrition in the camps. Death through starvation would still be genocide. Residents in the camps face frequent assaults when they venture outside to collect firewood and are sometimes attacked inside the camps. The government often forces the IDP camps to relocate. The NGO workers who staff these camps have also been harassed.

Yet the terms of the peace agreement that the government signed to end the war in the south are strikingly inconsistent with the presumption that it acts with genocidal intent in Darfur. Most southerners are Africans, but not Muslims, and are therefore even more plausible a target for a government motivated by genocide. In fact, an estimated 2 million southerners have been killed since civil war resumed in 1983.

Why, if the government intends genocide, does it share power and wealth with some Africans?

And yet the agreement has ended this civil war by giving southerners political control over the region and an equal share of oil revenues. The settlement also permits southerners to choose secession in a referendum that must be held after six years. It makes John Garang, the chairman of the SPLM/A, vice president in the national government and president of a new government for the southern region. Southern officials are to receive 30 percent of the positions in the central government. Garang has announced that he will even use his new position to negotiate an end to fighting in Darfur.

Why did the Sudan government agree to the peace agreement? No one believes it had a change of heart about the southerners. Instead, it appears to have negotiated in expectation of development aid and direct investment from Western public institutions and private companies. The prospects for local wealth through rapid expansion of oil exports are a strong attraction for national leaders. US firms have been prohibited from doing business with Sudan since the country was added to the US list of state sponsors of terrorism in 1993 because it gave safe haven to Islamic terrorist groups, including one that Osama bin Laden formed. China has taken advantage of Sudan's pariah status to invest heavily in its oil extraction. Oil profits already have allowed Khartoum to double its military budget since it began exporting oil in 1999. The Chinese helped Sudan build three new factories to produce weapons in the late 1990s.

The government believes that signing the peace agreement is sufficient to normalize its relationship with Washington and permit direct US investment. It remains to be seen whether Western countries, which made promises contingent on a successful agreement, will respond to internal public revulsion by introducing new demands to settle the war in Darfur first, thereby risking resumption of the civil war in the south. One telltale sign of the West's response is the prompt reopening of the World Bank's Khartoum office just after the peace agreement was signed. The World Bank had pulled out of Sudan several years ago when the government stopped making debt repayments.

Khartoum's commitment to honor the terms of the settlement cannot be taken for granted. Nonetheless, it is hard to argue that it has genocidal intentions toward Africans living in one area of the country when it has settled a civil war in another area on terms that bind it to work closely together with other Africans. While the evidence is not clear, the government's decision to sign the agreement seems just as consistent with a calculus of greater wealth to protect itself as with cultural annihilation of Africans.

OR IS IT REGIME SURVIVAL?

If holding on to power is its primary motive, why does the government persist in causing so much devastation to civilians in Darfur? The most likely reason is the threat the government faces if rebellions were to spread throughout the periphery. Since the Mahdist revolt against the Egyptians in the 1880s, the government has never faced insurgency throughout the north. To prevent the emergence of simultaneous rebellions, the government is sending a message to potential guerrillas everywhere that if they rebel, civilians in their region will face atrocities on a scale similar to those in Darfur. As John Ryle noted in the August 12, 2004, issue of *The New York Review of Books*, "The ruthlessness of the government's response to the Darfur insurgency is a sign of fear: any hint of weakness is liable to encourage other insurgencies...."

Aside from the long-running southern rebellion, there was little violent opposition from the periphery after independence until the 1980s. The first southern civil war, begun shortly after independence, was fought over polit-

ical control of the south. Southern rebels did not question that the core elite in the center would continue to rule the national state. This assumption became the basis for the 1972 Addis Ababa agreement, which ended the first civil war in the south.

When the SPLM/A began the second civil war after that peace agreement broke down in 1983, it proposed a radically different objective by calling for a "new" Sudan in which all peripheral areas would share power and wealth equitably with the center. Rebel leaders demanded an entirely new political and economic system in Sudan, not merely changes in relations between the center and the regions. This is why the "S" in SPLM/A stands for Sudan and not for Southern.

The national government was forced to take the SPLM/A's perspective seriously, because the southern guerrillas held most of the rural areas in the south and, for short periods, some areas in the north. The SPLM/A's ideas have spread to political activists in other peripheral areas. They form the ideology of the SLA and, to a lesser extent, that of the JEM. Leaders of both Darfur guerrilla organizations also argue that the government's policies discriminate in favor of peoples from one part of northern Sudan at the expense of those living everywhere else. They insist they are fighting for a change in Sudan, not for secession or for political autonomy on a cultural or racial basis.

Rebellion throughout peripheral areas in the north has been spreading for the past two decades. In the late 1980s revolts broke out in the Nuba Mountains in Southern Kordofan State and in Blue Nile State as a result of alliances formed with the SPLM/A. The government responded in the Nuba Mountains with attacks by local ethnic militias coordinated with Sudanese troops, just as it has now in Darfur. The devastation to civilians in Nuba areas was also called genocide by some outside observers. There are signs of revolt in other parts of Kordofan as well. Notables have recently demanded that the government share the profits from oil pumped from their areas. In addition, a new rebel group has emerged in western Kordofan, which borders Darfur.

Guerrilla outbreaks also occurred in the 1990s in eastern Sudan. The SPLM/A has had a military presence in this area for several years. In January 2004, the SLM/A signed an agreement with the Beja Congress, one of the groups involved in both political and guerrilla activity in the east. In an action in January 2005 suggesting the government's continued nervousness, police fired on peaceful demonstrators in Port Sudan following their presentation of a memorandum to the Red Sea State governor that demanded wealth and power sharing for the peoples of eastern Sudan. Nineteen protesters were killed and several more wounded.

All of these rebel groups are making the same demand: power sharing in a united Sudan. Paradoxically, the use of disproportionate violence by the government to quell each of them has led to new conflicts, greatly increasing its own insecurity.

IN SEARCH OF SECURITY

Explaining why governments engage in mass atrocities is important for identifying the remedy most likely to prevent their repetition. The recent history of Sudan demonstrates that the government has repeatedly engaged or been implicated in massive attacks on its citizens in region after region. Both genocide and threats to the regime's survival provide plausible motives to explain the Sudanese government's vicious behavior.

But they frame the issue differently. Genocide focuses attention on ending the violence in a specific place: Darfur. Threats to the regime's survival call for a political solution bringing peace to the entire country. Different frames mean different solutions. Intervening with enough external force could stop the killing and destruction in Darfur. And forcing the parties to develop new bases for sharing wealth and power through a national constitutional conference could bring lasting peace to the nation. Finding a solution that will not only stop the attacks in Darfur but also ensure they are not repeated elsewhere is clearly superior to ending the violence in Darfur alone. Neither solution is conceivable without sustained Western and African intervention.

If genocide were established and if international intervention were sure to follow, responding to attacks in one area might be considered the better solution, since intervention for other reasons would be less likely. The UN Genocide Convention does require intervention once a determination has been made. But the absence of effective involvement following the US announcement that genocide occurred in Darfur has stripped away the illusion that a mere declaration would lead to significant action.

Two major concessions by the Sudanese government provide a possible path forward. First, it has agreed grudgingly to cooperate with an admittedly undersized force of African Union peacekeepers in Darfur, financed and facilitated by the West. Expanding the peacekeepers to other parts of the periphery would provide an opportunity for serious negotiations involving all the parties. Second, it has responded, also grudgingly, to sustained Western and African pressure by accepting the peace agreement with the SPLM/A.

The premises underlying the agreement's new arrangements for the south are basically those that the guerrillas in Darfur and elsewhere in the north want for the whole country. The national government is not about to liquidate its hegemony willingly—especially not when it has China, with its considerable Sudanese oil stake, as its ally in the UN Security Council. To achieve a nationwide peace settlement, the Western powers would have to build aggressively on their commendable role in bringing about the peace agreement. If the Western media and public opinion could turn their attention from declaring genocide in one region of Sudan to bringing sustained pressure on Western governments to insist on all-party negotiations, security for civilians might have a chance.

NELSON KASFIR *is a professor of government at Dartmouth College. He has written extensively on Uganda and Sudan and is preparing a book on how guerrillas govern civilians.*

Blaming the victim:

Refugees and global security

The bulk of the world's refugees remain in the developing world. And the industrialized states, more worried after September 11, are taking new steps to keep them away.

by Gil Loescher

T HE ISSUES OF HUMAN SECURITY AND the security of states are intimately linked. For example, a greater respect for human rights, more equitable development, and the spread of democracy in war-torn places like Afghanistan, Kosovo, and the Democratic Republic of Congo would not only prevent and/or resolve the problems of refugee movements, they would also help establish a more stable and secure international order.

As a general rule, individuals and communities do not abandon their homes unless they are confronted with serious threats to their lives or liberty. Flight from one's country is the ultimate survival strategy, the one employed when all other coping mechanisms have been exhausted. Refugees serve both as an index of internal disorder and instability and as *prima facie* evidence of the violation of human rights and humanitarian standards. Perhaps no other issue provides such a clear and unassailable link between humanitarian concerns and legitimate international security issues.

Whether refugees find safe haven in the countries to which they flee depends in part on regional stability. This reality was brought home to me in recent months when I visited Turkey, Syria, and Kenya—countries that serve both as host states to refugees in their respective re-

gions and as transit countries for those seeking to migrate to Europe and North America. All three countries are located in extremely unstable regions from which some of the world's major refugee flows originate. The bulk of the refugees in these regions—Somalis, Sudanese, Iraqis, and Iranians—come from countries where conflict and persecution have persisted for years, making it unlikely that they will be able to return home any time soon.

The refugees I interviewed complained that their greatest concern was the poor security in these countries of first asylum. The human rights records of these countries are poor. Physical harassment, detention, and deportation to other countries where refugees risk greater persecution are commonplace. Police and security forces arbitrarily harass, detain, and arrest refugees. Corruption is rampant, especially among poorly paid border guards and police. Many refugees fear being attacked by agents from their home countries, often with the connivance of the authorities in the countries in which they have sought refuge.

This fear was given real meaning in Kenya, the host country for relatively large numbers of refugees from the Great Lakes region of Africa. In April, an assailant broke into a so-called secure resi-

dence established in Nairobi for refugees at particular risk. The assailant murdered two Rwandan refugee children, aged nine and 10, by slitting their throats. Their mother, a close relative of a former Rwandan president, was also seriously injured with multiple stab wounds. She and the children had been waiting in Kenya for 11 months for their resettlement application to be processed.

Refugees also face severe economic and social insecurity in these countries of first asylum; their freedom of movement is severely restricted; they cannot integrate with local populations; they are given inadequate or no assistance; they are refused permission to work. They live in limbo.

Refugees and local, national, and regional security

Refugees who flee persecution or violence may experience threats to their security on a daily basis, but when the connection between security and migration is highlighted by the media or politicians, it is rarely with reference to threats faced by refugees. Rather, the refugees themselves are usually characterized as posing a security threat to the receiving states and their citizens. This viewpoint is as likely to be expressed in the developing world as it is in the industrialized democracies. Most governments today perceive migration and refugee movements as a threat to their national interests.

For developing countries, displaced populations are both a consequence of conflict and a cause of continuing conflict and instability. Forced displacement can obstruct peace processes, undermine attempts at economic development, and exacerbate intercommunal tensions. Refugee flows also can be a source of regional conflict, causing instability in neighboring countries, triggering external intervention, and sometimes providing armed refugee groups with base camps from which to conduct insurgency, armed resistance, and terrorist activities.

In recent years, both new and long-established refugee populations have come to be viewed by local host governments as a threat to the internal order of the state, as well as a threat to regional, or even global, security. States perceive refugee groups as posing both direct and indirect security threats.

The direct security threat posed by the spillover of conflict and armed exiles is by far the strongest link between forced migration and state security. In many regions, such as the Balkans or the Great Lakes region of Africa, refugee exoduses have been deliberately provoked or engineered by one or more parties to the conflict with the specific objective of furthering their own political, military, or strategic objectives.

The use of refugee camps by combatants draws refugee communities directly into cross-border conflicts and accelerates conflicts and tensions within host countries. In recent years, many refugee camps in Africa, Asia, and Central America have housed not only those fleeing persecution and armed conflict, but also combatants and guerrilla forces who use the relative safety of the camps to launch violent campaigns of destabilization against their countries of origin.

In many host countries, governments prefer refugees to remain in camps and settlement areas close to the border of their home countries. Not only are the physical security and material safety of refugees not guaranteed in these remote areas, but the proximity of camps to countries of refugee origin also makes it easy for combatants to cross borders to engage in guerrilla warfare.

But indirect security threats posed by protracted refugee situations also impose important burdens on receiving states. Developing countries shoulder the social and economic strains of the world's vast majority of refugees and displaced persons. The long-term presence of refugees can exacerbate existing tensions and heighten intercommunal conflict, particularly when a state has ethnic rifts of its own, a vulnerable economic or social infrastructure, or hostile neighbors.

Generally, both the governments and the citizens of host countries view refugees negatively, associating them with problems of security, violence, and crime, and as a threat to social cohesion and employment. They are sometimes seen as posing a threat of insurgency or terrorism. In many regions, these negative perceptions have begun to generate a backlash against refugees—and especially, lately, against Islamic groups. Given the regionalization of conflict and the domestic instability caused by both new and protracted refugee situations, the indirect security threats posed by refugee flows, if left unaddressed, are likely to have serious consequences for regional and global security.

Western government responses

Western governments have failed to recognize these regional refugee situations and have not devoted sufficient resources, either financial or diplomatic, to long-standing refugee problems; long-term refugee needs induce "donor fatigue." For example, the refugees I recently visited have been adversely affected by recent cutbacks in donor funding to the U.N. High Commissioner for Refugees (UNHCR). This means that vital assistance programs in Turkey have been ended, and it also means that UNHCR has not been able to deploy the number of protection officers it needs at the dangerous and insecure refugee camps in Kenya. The deprivations and frustration endured by people living in insecure and precarious camps for long periods of time can generate various kinds of violence and instability. This is particularly the case in the Kenyan camps, where sexual abuse and violence against women and girls are common. In addition, banditry and armed robbery occur regularly within the camps.

Western governments have not developed effective policies to address the often deplorable conditions in regional host countries. The European Union, for example, administers development programs inside Somalia, but spends nearly nothing to secure better protection or improve the environment for the more than 250,000 Somalian refugees in neighboring Kenya. The involvement of the European Community Humanitarian Office in these regions is limited exclusively to projects in countries of origin and to non-refugee emergency projects in countries of first asylum.

International attention and assistance are in large part a reflection of politics, geo-strategic interests, and fickle international donor and media priorities. The U.S. and European governments have not fully examined the impact of their foreign and economic policies on refugee-producing regions. For example, more than a decade of economic and political isolation and sanctions against Iraq have made life so miserable and untenable for large sectors of the Iraqi population that hundreds of thousands have fled their homes to neighboring countries. Western governments have not developed a comprehensive policy to deal with these and other migration and refugee problems.

Ultimately, this is self-defeating. If refugees lack security in one country, they will try to move on to another, safer place. For example, most Iraqi refugees are unable to find either physical protection or material security in their host countries. Thus, insecurity, coupled with a lack of resettlement opportunities, has led large numbers of asylum seekers and migrants to move on to Western countries and to risk their lives by using illegal means of entry, including smuggling and trafficking organizations. Not surprisingly, Iraqi refugees now constitute the largest national group of asylum seekers in Europe.

The "securitization" of refugees

Although the bulk of the world's refugees remain in the developing world, the industrialized states feel increasingly threatened by an influx of refugees. Asylum-seekers are no longer limited to neighboring states; "jet age" refugees appear at the doorstep of distant nations. This comes, of course, on top of a steep rise in the number of undocumented immigrants to the West from the developing world and from Eastern Europe.

Throughout most of North America and Western Europe huge backlogs of asylum applications have built up, exposing the inability of the advanced industrialized states to establish administrative and judicial systems that can cope with the growing numbers of asylum applicants. In 1999, Britain, for example, had a backlog of some 100,000 unprocessed asylum applications.

Many governments find it virtually impossible to apprehend and deport those whose claims to refugee status have been rejected. At a time when governments are seeking to reduce public expenditure, they are simultaneously spending large sums on processing asylum applications and providing social welfare benefits to asylum seekers. There is a widespread belief that Western states have lost control of their borders and that refugees and immigrants pose a threat to national identities and economies of host societies.

Not surprisingly, Western worries about asylum seekers influence electoral politics. Across Western Europe, politicians allege that asylum seekers take away jobs, housing, and school places, dilute national homogeneity and culture, and exacerbate racial and ethnic tensions in local communities. As a result, political parties that invoke a popular xenophobia, campaigning on explicitly anti-refugee and anti-immigrant platforms, are now enjoying success in traditionally liberal democratic states like Denmark, the Netherlands, and France.

More recently, asylum seekers and refugees have come to be seen as direct threats to national security. Some communities of exiles are alleged to support terrorist activities and to be engaged in supporting armed conflicts in their countries of origin. In fact, the vast majority of immigrants are grateful for being given asylum and live by the rules.

> An angry, excluded world outside the West will inevitably turn to forms of extremism that will pose new security threats.

Prompted in part by the perceived threat to security, for two decades Western countries have been initiating a wide range of measures to curb the entry, admission, and entitlements of people claiming refugee status. Restrictive measures include extending border controls through stringent visa requirements, imposing sanctions against airlines and other carriers for transporting undocumented individuals, stationing immigration officials abroad, detaining asylum seekers before they reach national borders, negotiating agreements to send home those refused asylum, and threatening to withdraw financial and development aid if regional host countries will not take back those asylum seekers rejected in the West.

This trend has been exacerbated by the war against terrorism. In an effort to toughen immigration laws to prevent terrorists from entering their countries, North American and European governments have rushed through measures that may threaten the concept of the right of asylum. These include measures that allow for the indefinite detention of non-citizens suspected of terrorism, including asylum seekers and recognized refugees, without adequate rights of appeal. Since September 11, some governments have resorted to detention as a matter of first course, in some cases denying individuals their fundamental right to seek asylum and detaining them indefinitely until a deportation order can be executed. Such measures run foul of the U.N. Refugee Convention and accepted human rights standards.

In Europe, the gulf between the cultural background of contemporary refugee groups and that of Europeans causes special concern. Refugee groups may resist assimilation, and Western publics may be unwilling to tolerate aliens in their midst. These feelings, reinforced by racial and religious prejudices, pose difficult social and political problems for European governments. Xenophobic and racist attitudes are increasingly obvious among some segments of the population, and racist attacks have increased in every country that hosts immigrant minorities. Islamic groups in particular have been targeted, especially since September 11, 2001. The anti-immigrant, anti-refugee backlash is being exploited not only by the extreme right wing, but also by mainstream political parties throughout Europe. As a consequence, ethnic profiling and detention of members of Islamic groups and other minorities, including immigrants and asylum seekers, have increased dramatically.

Greater abuse?

The war on terrorism has given policymakers and law enforcement agencies a ready pretext to abuse the rights of refugees and other immigrants. Newly enacted measures to enhance internal law enforcement mechanisms to protect the state against terrorist threats can lead to an even greater deterioration of the rights of all citizens—and particularly the rights of refugees—leading to their increased vulnerability and exclusion. Indefinite detention, governmental restrictions on disclosure of evidence, the establishment of military tribunals with defined jurisdiction over non-citizens, and an array of possible new interior controls to deter potential terrorist abuse of the asylum systems have resulted in a tightening of visa systems around the world, making it even more difficult for refugees to escape persecution.

In many countries around the world, governments have seized on the rhetoric of anti-terrorism to steamroll domestic opposition. In this highly charged environment, politicians and the media are targeting refugees and immigrants as scapegoats for their countries' economic and social problems.

Politicians often exaggerate the various domestic threats associated with refugees in order to win short-term electoral gain. For example, in the run-up to the upcoming general elections in Kenya, President Daniel Arap-Moi and members of Kenya's parliament have stepped up the rhetoric, blaming refugees for depleting the nation's resources, degrading the environment, and endangering national security.

The way ahead

While the security threat associated with refugees is often exaggerated, history demonstrates that refugee movements are not only a humanitarian problem. They have a strong political and security dimension that can adversely affect domestic and international order. The management of migration, and particularly of refugees and displaced people, is not a side effect of political and economic instability and conflict, but an integral part of regional and international insecurity, and an integral part of conflict settlement and peace building within communities.

Forced displacement is a major factor in national and regional instability. Establishing effective responses to refugee needs should be a vital part of any broad model of security. The real challenges for policy-makers, practitioners, and researchers lie not so much within the humanitarian system itself, but in the wider policy-making world, including security, post-conflict development, the enforcement of human rights, and the development of civil societies.

Reacting to terrorist threats by placing unduly harsh restrictions on the free movement of people will simply lead to greater isolation and deprivation. An angry, excluded world outside the West will inevitably turn to forms of extremism that will pose new security threats. A failure by both the industrialized and developing countries to take action to stem the tide of poverty, violence, persecution, and the other conditions that create refugee flows will be costly in security terms.

In the realms of forced migration and state security, international and regional stability and justice coincide. Policy-makers need to build on this coincidence of factors to achieve the political will necessary to address both these issues more effectively and even-handedly. It is in the self-interest of states and coincides with their search for long-term global stability.

The international upheavals reverberating around the terrorist attacks of September 11 underline the important connections between refugee movements and international security. It is now impossible to overlook the strategic importance of the global refugee problem, and both national governments and international organizations need to make greater efforts at finding solutions to it. Western governments must recognize that the most obvious and logical solution lies in improving conditions in countries that produce refugees as well as in local host or transit countries in the regions of refugee origin.

Gil Loescher, Senior Fellow for Forced Migration and International Security at the International Institute for Strategic Studies in London, is the author of The UNHCR and World Politics: A Perilous Path *(2001).*

UNIT 4

Political Change in the Developing World

Unit Selections

Key Points to Consider

- What are the current trends in democracy throughout the world?

- What are the two views regarding the seriousness of Islam's threat to the West?

- How might democracy in the Middle East pose a challenge to western interests in the region?

- Can democracy be reconciled with Islam?

- What is the source of demands for reform in Saudi Arabia?

- What challenges does South Africa face?

- What are the implications of India's recent elections?

- What is the source of dissatisfaction with Latin American democracy?

- How do NGOs affect democracy?

Student Website

www.mhcls.com/online

Internet References

Further information regarding these websites may be found in this book's preface or online.

Latin American Network Information Center—LANIC
http://www.lanic.utexas.edu

ReliefWeb
http://www.reliefweb.int/w/rwb.nsf

World Trade Organization (WTO)
http://www.wto.org

simply to retain power. In some cases, leaders experimented with socialist development schemes that emphasized ideology and the role of party elites. The promise of rapid, equitable development proved elusive, and the collapse of the Soviet Union discredited this strategy. Other countries had the misfortune to come under the rule of tyrannical leaders concerned with enriching themselves and who brutally repressed anyone with the temerity to challenge their rule. Although there are a few notable exceptions, the developing world's experiences with democracy since independence have been very limited.

Democracy's "third wave" brought redemocratization to Latin America during the 1980s, after a period of authoritarian rule. The trend toward democracy also spread to some Asian countries, such as the Philippines and South Korea, and by 1990 it also began to be felt in sub-Saharan Africa. The results of this democratization trend have been mixed so far. A recent survey by Freedom House shows an increase in freedom around the world during 2003. Almost twice as many countries registered gains in freedom as those that showed a decline, but there are still many countries that lack effective democracy.

Latin America has been the developing world's most successful region in establishing democracy, but there is widespread dissatisfaction due to corruption, inequitable distribution of wealth, and threats to civil rights. Several heads of state in the region have been ousted, support for the left has increased substantially, and populist politics is on the rise.

Africa's experience with democracy has also been varied since the third wave of democratization swept over the continent beginning in 1990. Early efforts resulted in the ouster of many leaders, some of whom had held power for decades, and international pressure forced several countries to hold multiparty elections. Political systems in Africa range from consolidating democracies to states mired in conflict. In 2004, South Africa celebrated ten years of democracy with its third round of national elections. South Africa's successful democratic consolidation has occurred in the face of major challenges and stands in sharp contrast to the circumstances in other parts of the continent, especially next door in Zimbabwe where President Mugabe continues to rule through intimidation and manipulation.

Political change has begun in the Middle East but it will be a long-term challenge. Yasir Arafat's death and the election of Mahmoud Abbas as head of the Palestinian Authority have altered the dynamics of the search for peace. The assassination of former Lebanese Prime Minister Rafik Hariri prompted Syria to withdraw from Lebanon—creating political turmoil and posing a possible

Political change in the developing world has not necessarily produced democracy, in part because developing countries lack a democratic past. Colonial rule was authoritarian and the colonial powers failed to prepare their colonies adequately for democracy at independence. Even where there was an attempt to foster parliamentary government, the experiment frequently failed, largely due to the lack of a democratic tradition and a reliance on political expediency. Independence-era leaders frequently resorted to centralization of power and authoritarianism, either to pursue ambitious development programs or often challenge to Syrian President Bashar Assad's regime. Iraq's elections went more smoothly than many anticipated but the new government is dominated by Shiites and Kurds. To strengthen efforts at democratization and make progress against the insurgency the Sunni minority must be accommodated. Iran's recent elections saw the replacement of a moderate president with a hardliner but much of the public has become apathetic. In any event, bringing greater democracy to the region will have to involve Islamist elements in the process.

While there has been significant progress toward democratic reform around the world, there is no guarantee that these efforts will be sustained. Although there has been an increase in the percentage of the world's population living under democracy, nondemocratic regimes are still common. Furthermore, some semidemocracies have elections but lack civil and political rights. International efforts to promote democracy have often tended to focus on elections rather than the long-term requirements of democratic consolidation. More effective ways of promoting and sustaining democracy must be found to expand freedom further in the developing world.

NATIONAL INCOME
AND LIBERTY

Adrian Karatnycky

On balance, freedom registered upward trends in 2003, despite deadly, sporadic terrorism around the world and a year of significant political volatility. In all, 25 countries showed improvement, with 3 of the 25 making gains in political rights and civil liberties large enough to shift them from one political-status category to another: Argentina went from Partly Free to Free, while Burundi and Yemen went from Not Free to Partly Free.

The year also saw 13 countries suffer declines in freedom, with 5 of these dropping far enough to move them from one political-status category to another: Bolivia and Papua New Guinea slid from Free to Partly Free, while Azerbaijan, the Central African Republic, and Mauritania went from Partly Free to Not Free. The three status categories—Free, Partly Free, and Not Free—together represent Freedom House's broader-gauge metric for denoting a country's level of freedom. To capture subtler changes in political rights and civil liberties, we use a 1-to-7-point scale on which 1 represents the highest levels of freedom and 7 represents the most repressive practices. Thus 22 of the 25 countries that made freedom gains, and 8 of the 13 that suffered losses, did so in ways that the 7-point scale captured, even if the changes in either direction were not sufficient to warrant a change of category.

In sum, then, the changes in both political status and 7-point numerical ratings meant that 25 countries registered freedom gains, while 13 suffered setbacks—a ratio of almost 2 to 1.

What is the picture across the entire assemblage of the world's 192 sovereign states? As the year drew to a close, 88 countries were enjoying the political status of Free. In each, people lived amid a climate of open political competition, respect for civil liberties, vigorous independent media, and vibrant independent civic life. The share of the world's states made up by Free countries was 46 percent—a figure never exceeded at any time before 2002 in the entire history of the survey, which commenced publication in 1972.

Partly Free countries numbered 55, meaning that they accounted for less than a third (29 percent) of the world's states. Such countries typically safeguard certain basic political rights and civil liberties, but are beset by some combination of rampant corruption; disrespect for the rule of law; and religious, ethnic, or other communal strife. In some Partly Free states, a single party dominates politics behind a facade of limited pluralism.

There were 49 Not Free countries, enough to make up 25 percent of all the world's states. In Not Free countries, basic political rights and civil liberties are widely and systematically denied.

Significantly, the data compiled by Freedom House show that liberty is not the exclusive province of wealthy lands. Many poor and developing countries achieve a strong record of respect for political rights and civil liberties. In all, the survey data show that there are 38 Free countries with an annual Gross National Income per capita (GNIpc) of US$3,500 or less. Of these, 15 are places where yearly GNIpc is below $1,500. Equally important is the survey's finding that freedom levels are broadly similar when one compares middle-income countries with high-income countries (the former have yearly GNIpc of $1,500 to $6,000 with a median of $2,960, while the latter have GNIpc of $6,000 to $40,000 a year with a median of $19,750).

Argentina, the year's sole entrant into the ranks of Free countries, made the shift from Partly Free when successful elections in April marked the return to rule by a popularly elected president, Néstor Kirchner. The preceding elected president, Fernando de la Rúa, had been forced from office by violent protests in late 2000 and replaced by his main rival in a constitutional but nonelectoral process. Argentina's improved status was also a consequence of progress in fighting corruption and military and police impunity. At the same time, Bolivia fell from Free to Partly Free as a result of the resignation of its democratically elected president, Gonzalo Sánchez de Lozada, after violent street protests in October 2003 and a bloody police response;. Papua New Guinea, too, declined from Free to Partly Free amid widespread corruption and rampant crime.

Burundi and Yemen improved their rankings from Not Free to Partly Free. Yemen's progress reflected increased vibrancy in its political life. Burundi made progress as a result of incremental improvements in political rights that resulted from the integration into government of political groups that represent the majority Hutu population. The broadening of Burundi's government helped to improve interethnic relations and increase pluralism in a state emerging from a decade of devastating ethnic violence that has cost 200,000 lives.

FREEDOM IN THE WORLD

Freedom in the World is an evaluation of political rights and civil liberties in the world that Freedom House has provided on an annual basis for over 30 years. (Established in New York in 1941, Freedom House is a nonprofit organization that monitors political rights and civil liberties around the world.) The survey assesses a country's freedom by examining its record in these two areas: A country grants its citizens **political rights** when it permits them to form political parties that represent a significant range of voter choice and whose leaders can openly compete for and be elected to positions of power in government. A country upholds its citizens' **civil liberties** when it respects and protects their religious, ethnic, economic, linguistic, and other rights, including gender and family rights, personal freedoms, and freedoms of the press, belief, and association. The survey rates each country on a seven-point scale for both political rights and civil liberties (1 representing the most free and 7 the least free) and then divides the world into three broad categories: "Free" (countries whose ratings average 1.0–2.5); "Partly Free" (countries whose ratings average 3.0–5.0); and "Not Free" (countries whose ratings average 5.5–7.0).

The ratings, which are the product of a process that includes a team of 22 in-house and consultant writers and 17 senior academic scholars, are not merely assessments of the conduct of governments. Rather, they are intended to reflect the real-world rights and freedoms enjoyed by individuals as the result of actions by both state and nonstate actors. Thus, a country with a benign government facing violent forces (such as terrorist movements or insurgencies) hostile to an open society will be graded on the basis of the on-the-ground conditions that determine whether the population is able to exercise its freedoms. The survey enables scholars and policy makers both to assess the direction of global change annually and to examine trends in freedom over time and on a comparative basis across regions with different political and economic systems.

For more information about Freedom House's programs and publications please visit www.freedomhouse.org

Note: The findings in this essay and accompanying table reflect global events from 1 January through 31 November 2003.

Three countries saw their status decline from Partly Free to Not Free. The Central African Republic became Not Free after a March 2003 military coup ousted a civilian president and suspended the National Assembly. Mauritania entered the ranks of Not Free countries amid further erosion in political rights, including signs of pressure on the opposition that further reduce the chances for competitive electoral politics. And Azerbaijan entered the ranks of Not Free states after manifestly unfair presidential balloting and a massively fraudulent vote count in October 2003 resulted in Ilham Aliyev succeeding his ailing father as president of the Turkic-speaking, oil-rich former Soviet republic.

Regional Patterns

In Western and Central Europe, 24 of the states are rated Free. Turkey, which is included in this group of European states, is rated Partly Free—the only NATO member to belong to that category.

In the Americas and the Caribbean there are 23 Free Countries, 10 Partly Free, and 2 (Haiti and Cuba) Not Free.

In the Asia-Pacific region there are 17 Free countries, while there are 11 Partly Free and 11 Not Free states. Over the last decade, impressive economic-growth rates have accompanied political progress in South Korea, Taiwan, and Thailand.

In Eastern Europe and Central Asia there are today 12 Free countries, 8 Partly Free, and 7 Not Free. Dramatic progress in terms of rights has been registered primarily in Eastern Europe, however, where there are 12 Free and 3 Partly Free states. The significant progress that East European states have made has been confirmed and reinforced by their rapid integration into the security and economic structures of Europe and the Euro-Atlantic community. By contrast, among the 12 states (here we are leaving aside the three Baltic republics) that once formed the USSR, there is not one single Free country, and Not Free post-Soviet lands outnumber their Partly Free neighbors 7 to 5 in a year that saw Azerbaijan decline to Not Free status.

In sub-Saharan Africa, 11 countries are Free, 20 are Partly Free, and 17 are Not Free. Africa continues to register the greatest variations of any region year-to-year. It has seen significant instability, with steps forward in some countries often followed by rapid reversals.

In the Middle East and North Africa, Israel remains the sole Free country. There are 5 Partly Free and 12 Not Free states, an increase of one Partly Free state as a result of the aforementioned positive trends in Yemen. Africa, which suffered through three coups (one of which failed) was the year's most politically volatile region. Western Europe and the ex-communist countries of Eastern Europe and Central Asia registered the fewest changes, despite the dramatic end-of-year civic upheaval in Georgia.

Gains and Declines in Freedom

In addition to Argentina, which went from Partly Free to Free, and Yemen and Burundi, which raised their status from Not Free to Partly Free, 22 countries registered gains in freedom reflected in improved numerical scores that were significant but not large enough to warrant a change of category. Brief, alphabetically arranged, case summaries follow:

Improvements in Benin's democratic electoral processes, as signified by vigorously contested free and fair legislative elections, led to improvements in political rights.

The Atlantic island republic of Cape Verde, one of Africa's more vibrantly Free countries, registered improvements in women's rights as well as in public awareness of the legal protections that women now enjoy. The year also saw increased private-sector vibrancy and increased business opportunities. Taken together, these further improved civil liberties.

Chile's ongoing democratic consolidation saw further improvements in political rights as the military's once-overweening political influence waned.

In Congo (Brazzaville), political rights improved as a consequence of the signing of a cease-fire agreement on 17 March 2003. The agreement appears to be durable and has helped stabilize the country's fragile political environment.

Côte d'Ivoire's civil liberties improved after an internationally negotiated settlement to a civil war. A fragile government of national unity offered hope for an end to a period of extreme violence and strife.

Ghana's civil liberties consolidated and deepened in a year that saw increased openness in civic discourse and general improvements in respect for human rights and the rule of law.

Postwar and post-Saddam Iraq, while ravaged by terrorism and rampant crime, nevertheless saw a relaxation of the former Ba'athist state's controls on civic life, more open public and private discussion, and a range of newspapers and broadcast media, which today—despite limited controls by the U.S.-led occupation—are among the most diverse in the Arab world. All these factors on balance contributed to improvements in civil liberties.

Jordan's political rights made modest gains with the restoration of a national legislature with limited powers that was elected in a relatively open election.

The effects of Kenya's free and fair 2002 national elections continued to be felt in 2003, as official transparency and accountability improved while civic and political life showed increased vibrancy. All these factors contributed to incremental improvements in political rights and civil liberties.

Civil liberties in Madagascar improved as normalcy and calm returned to the country's civic, political, and associational life after violence that disrupted and destabilized the country following the bitterly contested December 2001 elections.

Malawi's democratic political rights improved as a consequence of the judicial nullification of a law that had eroded the rights of legislators and the parliamentary defeat of a controversial effort to lift term limits for the presidency.

Malaysia's civil liberties improved amid signs of greater resilience in academic freedom and improvements in personal autonomy.

Mali's democracy scored modest improvements in civil liberties as a result of further consolidation of democracy and incremental changes in public discourse.

Nigeria's civil liberties improved as a consequence of a degree of abatement in the intercommunal violence that had beset the country in 2002.

Paraguay showed gains regarding political rights via a free and fair April 2003 election that brought Oscar Nicanor Duarte to the presidency and helped to boost official transparency.

Political rights in the genocide-ravaged land of Rwanda ticked upward after multiparty presidential and legislative elections that led to freer political discourse in a setting of circumscribed political choice.

An improved security environment and increased pressures to punish those guilty of civil-war atrocities led to modest gains in Sierra Leone's civil liberties.

Sri Lanka's political life may have been shaken by the president's dismissal of three cabinet ministers and a temporary suspension of parliament. But its civil liberties improved as the result of a significant decline in violence and modest improvements in the rule of law resulting from an ongoing, though tenuous, ceasefire with the Tamil rebels.

Uganda made modest gains in political rights after a Constitutional Court ruling that removed restrictions on political party activity, potentially opening the door to multiparty politics in what has heretofore been a one-party state dominated by long-ruling president Yoweri Museveni.

The survey also records upward adjustments in the civil-liberties scores of the Pacific island states of Micronesia, Nauru, and Palau, respectively. The Freedom House team made these adjustments after refining its evaluation of the scope of freedoms enjoyed by trade unions, media organs, and nongovernmental organizations in these countries.

Categorical declines in freedom occurred in Bolivia and Papua New Guinea (from Free to Partly Free) and in Azerbaijan, the Central African Republic, and Mauritania (from Partly Free to Not Free). In addition, eight countries saw an erosion of political rights, civil liberties, or both.

Djibouti's political rights declined after unfair elections in which the incumbent government grossly exploited the advantages of office in order to win.

In the Dominican Republic, corruption scandals and a growing rejection of transparency by the government of President Hipolito Mejia resulted in the erosion of political rights.

Guinea-Bissau registered a decline in political rights as a September 2003 military coup toppled unpopular but democratically elected president Kumba Yala, who had been in office since early 2000. Paradoxically, the military eased some of the constraints on civil liberties by releasing some opposition members whom Yala had jailed. Nevertheless, on balance, the new dominance of the military in the country's political life represented a setback for freedom.

Nepal saw a decline in political rights as a result of the continued suspension of an elected parliament and the king's failure to schedule new national elections amid a violent Maoist insurgency.

Political rights in impoverished São Tomé and Principe declined after a brief coup led to the temporary displacement of that island country's government. The coup was quickly reversed as a result of the active and resolute engagement of Nigeria and its president, Olusegun Obasanjo. Still, the coup attempt traumatized political life in what had previously been an open and successful democracy.

The already-limited political rights enjoyed by the people of Swaziland took another hit from constitutional changes designed to entrench more deeply the institution of rule by royal decree.

Civil liberties declined in the United Arab Emirates amid serious impediments to equality of economic opportunity and the right to own property.

Vanuatu's political-rights score declined as a result of a technical reevaluation of the country's political life, not as a result of any specific changes that occurred in 2003.

Freedom, Wealth, and Poverty

The volatility of politics in poverty-riddled Africa—which experienced two military coups and a failed coup attempt—reinforces by counterexample the impression that prosperity correlates well with stability, democracy, and freedom.

This year's survey examines the relationship between income and levels of freedom in 192 countries.[1] To no one's surprise, the data confirm that the world's most prosperous states are as a group freer than the world's poorest countries. But the correlations also show that countries of middling wealth, including a broad array of developing nations, do nearly as well in terms of freedom as high-income countries. Moreover, our examination shows that a low level of economic development need not always condemn a society to an absence of freedom. Indeed, there is a large cohort of low-income to lower-middle-income countries that guarantee their inhabitants a broad range of political and civil liberties. As noted previously, 38 countries with a yearly GNIpc of $3,500 or less are rated Free.

As mentioned above, the biggest cluster of Free countries is found among high-income countries, defined as those with a GNIpc of more than $6,000 per year. In this group, whose median GNIpc is $19,570 per year, 38 countries are Free, 5 are Partly Free, and 5 are Not Free. The five Not Free countries in this group (Brunei, Oman, Qatar, Saudi Arabia, and the United Arab Emirates) draw the vast bulk of their wealth from natural energy resources and investment income derived from these resources. Two Partly Free countries (Bahrain and Kuwait) also fit this description. This means that among those upper-income countries which derive most of their wealth from enterprise and knowledge, 38 are Free and only 3 are Partly Free. Societies that are most successful in producing wealth are almost uniformly Free.

Among middle-income countries—meaning those with a GNIpc of $1,500 to $6,000 per year—35 are Free, 11 are Partly Free, and 7 are Not Free. With a median GNIpc of less than $3,000 per year, this set of 53 countries is a far from prosperous group. Yet 66 percent of middle-income countries are Free, a proportion that does not differ dramatically from that found among high-income countries, 79 percent of which are Free. At the same time, while the proportion of middle-income countries that are Not Free stands at 13 percent, the like figure for high-income states is only modestly lower at 10.5 percent. Moreover, the average numerical freedom score of a middle-income country is 2.7 on Freedom House's 1-to-7 (most- to least-free) scale, while 2.0 is the average score for a high-income country. Low-income countries, by contrast, receive an average numerical freedom score of 4.4.

Among the poorest countries, however, where GNIpc is less than $1,500 per year, freedom levels are significantly lower. This lowest-income cohort has 37 Not Free countries, 39 Partly Free countries, and only 15 Free countries. And among the poorest of the poor in this cohort (meaning the 29 countries with a yearly GNIpc of $300 or less) only three—Ghana, Mali, and São Tomé and Principe—are Free.

Among the world's 88 Free countries, 38 (or 43 percent) are high income, 35 (or 40 percent) are middle income, and 15 (or 17 percent) are low income. Of the 55 Partly Free countries, 5 (or 9 percent) are wealthy, 11 (or 20 percent) are middling, and 39 (or 71 percent) come from the poorest group. Among the world's 49 Not Free states, 5 (or 10 percent) are high-income, 7 (or 14 percent) are middle-income, and 37 (or 76 percent) are low-income.

Such data indicate that the lowest income levels correlate with significantly lower levels of freedom. But our look at incomes and freedom levels also reveals that low-income countries can establish strong democratic practices and respect for civil liberties rooted in the rule of law. Out of 128 countries with an annual GNIpc of $3,500 or less, 38 rate as Free in the survey.

Freedom's Tenure

This year, we have also taken a closer look at what the survey shows in terms of the tenure of freedom in the world's polities. While 88 countries stand rated as Free at the end of 2003, for the bulk of them, freedom—and in some cases sovereign statehood—is a recent arrival. Indeed, long-term uninterrupted freedom has been rare for most countries. Of the globe's 192 states, only 24 (or 12.5 percent) have been Free throughout the entire 31 years spanned so far by the Freedom House survey. An additional 20 of today's Free countries have had 15 to 30 years of uninterrupted freedom. Thus half the world's Free states have enjoyed high levels of freedom for fewer than 15 years.

Our time-series data show that over the last 31 years, a total of 112 countries have at some time or other been rated Free and experienced at least some period of democratic governance in an environment of broad respect for human rights. Significantly, of this group of 112 that have at any time in their history been Free, 88 are currently Free, while only 2 today are Not Free. This means that about 79 percent of all countries that have ever known high levels of freedom over the last 31 years are Free now. As importantly, when a country attains a high degree of freedom (or in other words, merits being ranked as Free in the survey), it rarely slides back into the kind of severe and systematic repression denoted by a Not Free ranking. Indeed, out of the 112 countries that have attained high levels of respect for both political rights and civil liberties, only 14 have ever seen their political-status rating lapse back to Not Free.[2]

Conversely, among the 109 states that have at any time in their history lived under the denial of most basic freedoms as indicated by a Not Free rating, 28 are today Free, 32 are Partly Free, and 49 are Not Free.

TABLE—INDEPENDENT COUNTRIES, FREEDOM IN THE WORLD 2004: COMPARATIVE MEASURES OF FREEDOM

COUNTRY	PR	CL	FREEDOM RATING	COUNTRY	PR	CL	FREEDOM RATING
Afghanistan	6	6	Not Free	Egypt	6	6	Not Free
Albania	3	3	Partly Free	El Salvador	2	3	Free
Algeria	6	5	Not Free	Equatorial Guinea	7	6	Not Free
Andorra	1	1	Free	Eritrea	7	6	Not Free
Angola	6	5	Not Free	Estonia	1	2	Free
Antigua & Barbuda	4	2	Partly Free	Ethiopia	5	5	Partly Free
Argentina	2▲	2▲	Free	Fiji	4	3	Partly Free
Armenia	4	4	Partly Free	Finland	1	1	Free
Australia	1	1	Free	France	1	1	Free
Austria	1	1	Free	Gabon	5	4	Partly Free
Azerbaijan	6▼	5▼	Not Free	The Gambia	4 ▲	4 ▲	Partly Free
Bahamas	1	1	Free	Georgia	4	4	Partly Free
Bahrain	5	5	Partly Free	Germany	1	1	Free
Bangladesh	4	4	Partly Free	Ghana	2	2 ▲	Free
Barbados	1	1	Free	Greece	1	2	Free
Belarus	6	6	Not Free	Grenada	1	2	Free
Belgium	1	1	Free	Guatemala	4	4	Partly Free
Belize	1	2	Free	Guinea	6	5	Not Free
Benin	2 ▲	2	Free	Guinea-Bissau	6▼	4▲	Partly Free
Bhutan	6	5	Not Free	Guyana	2	2	Free
Bolivia	3▼	3	Partly Free	Haiti	6	6	Not Free
Bosnia-Herzegovftia	4	4	Partly Free	Honduras	3	3	Partly Free
Botswana	2	2	Free	Hungary	1	2	Free
Brazil	2	3	Free	Iceland	1	1	Free
Brunei	6	5	Not Free	India	2	3	Free
Bulgaria	1	2	Free	Indonesia	3	4	Partly Free
Burkina Faso	4	4	Partly Free	Iran	6	6	Not Free
Burma	7	7	Not Free	Iraq	7	5 ▲	Not Free
Burundi	5 ▲	5	Partly Free	Ireland	1	1	Free
Cambodia	6	5	Not Free	Israel	1	3	Free
Cameroon	6	6	Not Free	Italy	1	1	Free
Canada	1	1	Free	Jamaica	2	3	Free
Cape Verde	1	1 ▲	Free	Japan	1	2	Free
Central African Republic	7▼	5	Not Free	Jordan	5 ▲	5	Partly Free
Chad	6	5	Not Free	Kazakhstan	6	5	Not Free
Chile	1 ▲	1	Free	Kenya	3 ▲	3 ▲	Partly Free
China (PRC)	7	6	Not Free	Kiribati	1	1	Free
Colombia	4	4	Partly Free	Korea, North	7	7	Not Free
Comoros	5	4	Partly Free	Korea, South	2	2	Free
Congo (Brazzaville)	5 ▲	4	Partly Free	Kuwait	4	5	Partly Free
Congo (Kinshasa)	6	6	Not Free	Kyrgyzstan	6	5	Not Free
Costa Rica	1	2	Free	Laos	7	6	Not Free
Côte d'Ivoire	6	5 ▲	Not Free	Latvia	1	2	Free
Croatia	2	2	Free	Lebanon	6	5	Not Free
Cuba	7	7	Not Free	Lesotho	2	3	Free
Cyprus (G)	1	1	Free	Liberia	6	6	Not Free
Czech Republic	1	2	Free	Libya	7	7	Not Free
Denmark	1	1	Free	Liechtenstein	1	1	Free
Djibouti	5▼	5	Partly Free	Lithuania	1	2	Free
Dominica	1	1	Free	Luxembourg	1	1	Free
Dominican Republic	3▼	2	Free	Macedonia	3	3	Partly Free
East Timor	3	3	Partly Free	Madagascar	3	3▲	Partly Free
				Malawi	3 ▲	4	Partly Free
Ecuador	3	3	Partly Free	Malaysia	5	4▲	Partly Free
				Maldives	6	5	Not Free

COUNTRY	PR	CL	FREEDOM RATING	COUNTRY	PR	CL	FREEDOM RATING
Mali	2	2▲	Free	South Africa	1	2	Free
Malta	1	1	Free	Spain	1	1	Free
Marshall Islands	1	1	Free	Sri Lanka	3	3▲	Partly Free
Mauritania	6▼	5	Not Free	Sudan	7	7	Not Free
Mauritius	1	2	Free	Suriname	1	2	Free
Mexico	2	2	Free	Swaziland	7▼	5	Not Free
Micronesia	1	1▲	Free	Sweden	1	1	Not Free
Moldova	3	4	Partly Free	Switzerland	1	1	Free
Monaco	2	1	Free	Syria	7	7	Not Free
Mongolia	2	2	Free	Taiwan (Rep.			
Morocco	5	5	Partly Free	of China)	2	2	Free
Mozambique	3	4	Partly Free	Tajikistan	6	5	Partly Free
Namibia	2	3	Free	Tanzania	4	3	Partly Free
Nauru	1	1▲	Free	Thailand	2	3	Free
Nepal	5▼	4	Partly Free	Togo	6	5	Not Free
Netherlands	1	1	Free	Tonga	5	3	Partly Free
New Zealand	1	1	Free	Trinidad &			
Nicaragua	3	3	Partly Free	Tobago	3	3	Partly Free
Niger	4	4	Partly Free	Tunisia	6	5	Not Free
Nigeria	4	4▲	Partly Free	Turkey	3▲	4	Partly Free
Norway	1	1	Free	Turkmenistan	7	7	Not Free
Oman	6	5	Not Free	Tuvalu	1	1	Free
Pakistan	6	5	Not Free	Uganda	5▲	4	Partly Free
Palau	1	1▲	Free	Ukraine	4	4	Partly Free
Panama	1	2	Free	United Arab			
Papua New				Emirates	6	6▼	Not Free
Guinea	3▼	3▼	Partly Free	United Kingdom[1]	1	1	Free
Paraguay	3▲	3	Partly Free	United States	1	1	Free
Peru	2	3	Free	Uruguay	1	1	Free
Philippines	2	3	Free	Uzbekistan	7	6	Not Free
Poland	1	2	Free	Vanuatu	2▼	2	Free
Portugal	1	1	Free	Venezuela	3	4	Partly Free
Qatar	6	6	Not Free	Vietnam	7	6	Not Free
Romania	2	2	Free	Yemen	5▲	5	Partly Free
Russia	5	5	Partly Free	Zambia	4	4	Partly Free
Rwanda	6▲	5	Not Free	Zimbabwe	6	6	Not Free
St. Kitts & Nevis	1	2	Free				
St. Lucia	2	1	Free				
St. Vincent &							
Grenadines	1	2	Free				
Samoa	2	2	Free				
San Marino	1	1	Free				
São Tomé							
& Príncipe	2▼	2	Free				
Saudi Arabia	7	7	Not Free				
Senegal	2	3	Free				
Serbia &							
Montenegro	3	2	Free				
Seychelles	3	3	Partly Free				
Sierra Leone	4	3▲	Partly Free				
Singapore	5	4	Partly Free				
Slovakia	1	2	Free				
Slovenia	1	1	Free				
Solomon Islands	3	3	Partly Free				
Somalia	6	7	Not Free				

PR and CL stand for Political Rights and Civil Liberties, respectively; 1 represents the most-free and 7 the least free-rating.

▲ ▼ up or down indicates an improvement or a worsening, respectively, in Political Rights or Civil Liberties since the last survey

[1]excluding Northern Ireland

The Freedom Rating is overall judgement based on survey results. See the box on p. 133 for more details on the survey.

Note: The ratings in this table reflect global events from 1 January through 31 November 2003.

The time-series record shows that 80 countries have not experienced broad-based freedom at any time during the 31-year record of the survey.

In all but the few worst cases, even highly repressive Not Free countries have been unable to rob their people completely and lastingly of all civil liberties and political rights. Out of 109 countries that have been rated Not Free in the history of the survey, only 13 (or 12 percent) have sustained in every year the high levels of political control and repression represented by the Not Free rating. These 13 countries are Burma, Chad, China, Congo (Kinshasa), Cuba, Equatorial Guinea, Iraq, Libya, North Korea, Rwanda, Saudi Arabia, Somalia, and Vietnam. This suggests that over time even the most repressive rulers find it hard to achieve the complete and continuous suppression of their citizens' desires for a broad array of political rights and civil liberties.

A year of significant momentum for freedom—despite the threats to liberty posed by widespread global terrorism—is encouraging. Equally heartening is the evidence that suggests political rights and civil liberties can thrive even in conditions of significant economic privation. We also can derive some comfort from evidence in this year's review of long-term data suggesting that countries and populations which have experienced a high degree of freedom are finding ways of protecting that freedom and preventing its complete reversal.

At the same time, the survey's findings show that while there is much promising news this year, important challenges remain. Many states—particularly the Partly Free countries—are finding it exceedingly difficult to make the transition to stable democratic rule rooted in the rule of law. Other states, a broad array of Not Free countries, have proved resistant to democratic change. Free societies, which account for 89 percent of the world's wealth, have the capacity to help solidify and deepen democracy where it is fragile. They also have the capacity and the resources significantly to assist indigenous movements that seek to bring democratic change to closed societies.

NOTES

1. Economic data are for 2001 and come from World Bank Development Indicators and www.internetworld-stats.com and are compared with Freedom in the World ratings of events through 30 November 2003.
2. These 14 "backsliders" are Argentina, Burkina Faso (formerly Upper Volta), Chile, the Gambia, Ghana, Grenada, Guatemala, Guyana, Lebanon, Maldives, Nigeria, Seychelles, Suriname, and Thailand.

Adrian Karatnycky is counselor and senior scholar at Freedom House and serves as the senior analyst of its annual Freedom in the World survey. More information on the survey is provided in the box on the following page, and on Freedom House's website at www.freedomhouse.org. For the rankings of individual countries for 2003, see the Table on pp. 90–91. In June 2004, the survey covering 2003 will be published in book form as Freedom in the World 2004 (Rowman & Littlefield). Aili Piano and Mark Rosenberg of the Freedom House staff gave invaluable assistance in the preparation of this article.

The Democratic Mosaic

By Martin Walker

The administration of President George W. Bush has been defined by the war on terrorism, its response to the appalling terrorist attacks of September 11, 2001. But it wants to be remembered for a grander and more positive strategy, as unveiled by the president at the National Endowment for Democracy in November 2003 and further elaborated in his State of the Union address this year. This "forward strategy of freedom in the greater Middle East" seeks to promote free elections, free markets, a free press, and free labor unions to advance democracy and opportunity in 22 Arab countries, stretching from Morocco on the Atlantic coast to Oman on the shores of the Indian Ocean. The inhabitants of those countries number some 300 million, speak diverse Arabic dialects that are often mutually incomprehensible, and have long endured violence, poverty, and arbitrary rule. The United States has little choice but to attempt this daunting challenge, said Bush: "As long as the Middle East remains a place of tyranny and despair and anger, it will continue to produce men and movements that threaten the safety of America and our friends."

The grandly ambitious project is inspired partly by the Helsinki treaties of 1975, which gave crucial breathing room to human rights groups in the old Soviet bloc, and partly by the success of American policies after 1945 that led to democratic governments in Japan and West Germany. To be sure, 59 years after victory in World War II, American forces remain deployed in those two countries, and the new strategy for the Middle East may similarly depend, in part, on a U.S. military presence.

But merely to prescribe democracy is not to settle the matter, because democracy comes in such a bewildering variety of forms. There are parliamentary monarchies without any written constitution (Britain), highly centralized presidential democracies (France), federal democracies (Germany), democracies with separated powers and a venerable constitution (United States), and democracies that seem to flourish despite an effective one-party system (Japan). There are new democracies (South Korea and Taiwan), and democracies that maintain most of their essential freedoms despite the strains of war and terrorism (Israel). Some democracies have survived and deepened despite poverty (Costa Rica), violent separatist movements

(modern Spain), recurrent wars (much of Europe), and deep ethnic divisions (Brazil). India's democracy has flourished despite all those challenges and the further complications of a debilitating caste system.

There are democracies so decentralized that the "central" government is almost impotent (Switzerland), and democracies so young and fragile that they exist only by means of a powerful and intrusive outside authority (Bosnia-Herzegovina). There are democracies restored from within (Spain and Portugal) and democracies born in the defeat of military dictators (Greece and Argentina); in Chile, a vigorous democratic movement eventually ended the military rule of General Augusto Pinochet, who had led a coup in 1973 to topple the elected government of Salvador Allende.

DEMOCRACY'S STUNNING ADVANCE HAS BYPASSED THE ISLAMIC WORLD.

Democracy, however defined, has scored some stunning advances since Allende's fall. According to Freedom House, which for 30 years has published an annual survey of political rights around the world, democracy's reach has grown ever more extensive. In 1972, the year of its first survey, Freedom House rated 43 countries as "free," 38 as "partly free," and 69 as "not free." The 2004 Freedom House survey rates 88 states as free, 55 as partly free, and 49 as not free. So the number of free countries has more than doubled over the past 30 years, the number of partly free states has grown by 17, and the number of repressive (i.e., not free) states has declined by 20. (The absolute number of states has grown over the same period.)

Democracy has proved so diverse over the past half-century that it confounds easy definition. It's a strikingly robust plant, capable of almost infinite variety. But in the Islamic world, democracy struggles on unfriendly soil. The Freedom House survey of the 47 nations with an Islamic majority found only nine electoral democracies, none of them in the Middle East. But even the electoral democracies often lack fundamental rights. Of states with an Islamic majority, Freedom House ranks only two, Senegal and Mali, as free. Why should this be? India's example suggests that the influence of colonialism is not an adequate explanation. Nor is poverty, which, in any case, is not an issue in the oil-rich states. The explanation must lie elsewhere.

Most political theory about the key components of democracy focuses on three important preconditions: the role of certain key state institutions, the strength of civil society, and socioeconomic and cultural structure. The key institutions include elections, in some form, with a secret ballot; reasonably free speech and media; and the rule of law, as administered by a tolerably independent judiciary to protect the rights of minorities. The rule of law is critical. (Without it, Thomas Jefferson's somber definition of a democracy as "nothing more than mob rule, where fifty-one percent of the people may take away the rights of the other forty-nine," might well discredit the enterprise.) It should extend to all citizens, and cover commercial as well as criminal matters; otherwise, property rights and the sanctity of contract are at risk. But the rule of law can take many forms. The countries of the European Union, for example, manage to function with fundamentally different legal systems. Most Continental nations prefer variants of the French system, in which a state-employed magistrate acts as investigator and as prosecutor before a judicial panel. The British retain trial by jury and an adversarial system in which the Crown presents the prosecution and the defense then tries to refute it.

But such distinctions between the legal forms of Western democracy are mere details by comparison with the gulf that separates Islamic law, sharia, from Western concepts of law. Although democracy can function with a state-established religion, as in Britain or Israel, the question of whether it can emerge in the shadow of sharia remains open. The difficulty is less the hudud, the stern code of punishment for fornication (flogging), theft (amputation), and adultery (stoning), than it is sharia's fundamental objection to any separation of church and state. Nor can there be much freedom of individual conscience when the penalty for converting from Islam to another religion is death. This is not to say necessarily that democracy cannot prosper under sharia, but finding an accommodation will be difficult, and is unlikely to be peaceful. It took centuries of war and dispute—and eventually the Reformation—for medieval Europe to resolve a similar clash of prerogatives between the canon law of the Roman Catholic Church and the secular law of earthly sovereigns.

The importance of civil society in the emergence of democracy has long been recognized. "Among the laws that rule human society," Alexis de Tocqueville suggested in *Democracy in America*, "there is one that seems to be more precise and clear than all others. If men are to remain civilized or to become so, the art of associating together must grow and improve in the same ration." Samuel Huntington, in his seminal *Political Order in Changing Societies* (1969), saw the insufficient development of this art as explaining the problems of "the modernizing countries of Asia, Africa and Latin America, where the political community is fragmented against itself, and where political in-

---

Freedom in the Middle East

While there has been steady progress toward greater freedom around the world in recent decades, the Middle East still lags behind. Freedom House, a nonpartisan research organization that annually surveys the status of freedom in 192 countries, reports that only one of the 18 countries it groups in the Middle East and North Africa is rated "free," and that is Israel. Worldwide, 88 countries are rated free. The good news is that Yemen, once a refuge for Osama bin Laden, has moved from "not free" to "partly free." Six Middle Eastern countries are now partly tree. According to Freedom House, the presence or absence of elections is not decisive in rating a country. In partly free countries, "political rights and civil liberties are more limited [and] corruption, dominant ruling parties, or, in some cases, ethnic or religious strife are often the norm." Eleven countries in the region (and 49 worldwide) are considered not free.

---

stitutions have little power, less majesty and no resiliency, where in many cases governments simply do not govern." Huntington discerned in the countries being destabilized by rapid change "a lack of civic morale and public spirit capable of giving meaning and direction to the public interest," and concluded that "the primary problem of politics is the lag in the development of political institutions behind social and economic change."

DEMOCRACY CAN FUNCTION WITH A STATE-ESTABLISHED RELIGION; THE QUESTION OF WHETHER IT CAN EMERGE IN THE SHADOW OF SHARIA REMAINS OPEN.

To give life to those political institutions, a civil society is needed, in the form, for example, of sports and hobby clubs, labor unions, cafés, and other nongovernmental and political entities within which people can gather and argue and cooperate outside state structures. All of these—and an increasingly independent news media spurred by satellite TV and the Internet, charitable bodies, and women's groups—exist throughout most of the Arab world. Not all of them are organized through the mosques, and many thrive despite political repression, the customary restraints upon a public role for women, and the competing tug of tribal tradition. In countries that are making significant steps toward representative government, such as Morocco, Jordan, Oman, Qatar, and Kuwait, civil society is blossoming fast. Those five countries, all monarchies, have sovereigns who seem prepared to enlarge the political space for their subjects. The prospects for "the art of associating together" in these states are promising, in part because long-established royal dynasties with their own religious credentials do not seem intimidated by the Islamist clerics.

Civil society is inextricably linked with socioeconomic structure, but the economic circumstances of successful democracies are widely divergent. India is an obvious example of democracy unimpeded by poverty, as is Costa Rica, with a long and exemplary record of representative government in Latin America. In the most populous countries of the Arab world, wealth is actually distributed more equitably than in the United States.

Economists measure income distribution in a state by means of the Gini index (named for Corrado Gini, the Italian statistician who devised it). The lower the index, the more evenly income is distributed in a country; the higher the index, the greater the share of wealth owned by the rich. So a fully egalitarian society would have a Gini figure of 0, and a society in which the richest person owned everything would have a figure of 100. The table gives the Gini figures for selected countries, with gross domestic product (GDP) shown in purchasing power parity. It's important to note, however, that figures for the Arab world are notoriously unreliable, and that, for the oil-rich states, a Gini index is almost meaningless because of the extraordinarily high proportion of foreign workers.

Income disparities are a crude indicator, concealing both regional differences (a low income in New York City can be relatively high in Mississippi) and many social subtleties. But the figures suggest that democracy can flourish in countries with sharp disparities of income, and survive even in countries such as Brazil, where the disparities tend toward the acute. If reasonably even levels of income distribution are a useful predictor, then many Arab countries are in promising shape.

LONG-ESTABLISHED ROYAL DYNASTIES WITH THEIR OWN RELIGIOUS CREDENTIALS SEEM LESS INTIMIDATED BY THE ISLAMIST CLERICS.

Incomes may not be a helpful indicator, however, in analyzing a particularly distinctive characteristic of democracies—the middle class, which plays a stabilizing political role. The middle class is hard to define because income is only one factor in its measurement; social origin, education, career, and lifestyle all contribute to the making of a middle class. Nonetheless, there are common features. Members of the middle class have homes and savings. They make some provision for their old age. They invest in the education of their children. Thus, they have a stake in a stable future, and that provides a strong personal incentive for them to be politically active—to ensure that schools are good, that the financial system will handle their savings honestly, that police will safeguard their property, that courts will be honest, and that the government will not tax them too highly or waste their savings through inflation. They need a free press to tell them what the government and courts are doing, and freedom of speech and assembly and elections to organize their opposition if the government lets them down. In short, though it may be simplistic to say that a middle class, by definition, will demand the kinds of institutions that help sustain democracy, such institutions and a socially active and politically engaged middle class will mutually reinforce each other.

Wealth and Inequality

Country	Gini index	Per capita GDP (U.S.$)
Japan	24.9	25,130
Sweden	25.0	24,180
Yemen	33.4	790
Egypt	34.4	3,520
Britain	36.0	24,160
Jordan	36.4	3,870
Morocco	39.5	3,600
China	40.3	4,020
United States	40.8	34,320
Russia	45.6	7,100
Mexico	53.1	8,430

Parts of the Arab world may enjoy less income inequality than the United States. A low Gini index connotes low levels of income inequality.

The middle class is growing fast in most Arab countries, although it's growing most quickly in the state bureaucracies. But no doubt as a consequence of the subservient role of women, the Arab middle class is not growing nearly quickly enough to cope with the stunningly high birthrates that give the region such a high proportion of young people under the age of 25. According to the United Nations Department of Economic and Social Affairs, the median age in Egypt and Algeria is now 20; in Lebanon it's 18, and in Iraq it's 17. On average, annual population growth remains about three percent in many Arab countries, compared with two percent globally.

The role of women in the Arab world points to a deeper issue: the degree to which democracy depends on culture. The long stability of Britain and the United States, the first countries to produce a mass middle class, is telling. Some political theorists suggest that the tradition of juries and common law, property rights, elected parliaments, a free press, and largely free trade, along with the low taxes permitted by a happy geography that precluded the need for a vast standing army, endowed the English-speaking world with a special predisposition to democracy. The theory is beguiling, but it turns ominous when used to suggest that some peoples and cultures are inherently antipathetic to democracy—as has been said at various times of Germans, Japanese, Indians, Africans, and Russians, and as is now being said of the Islamic world in general.

The debate on democracy's potential in the Middle East will continue, even as democracy's green shoots are evident in Oman's elections, Qatar's new constitution (which gives women the right to vote), and Jordan's and Morocco's significant steps toward representative government. But these potential democracies remain works in timid progress, proceeding under two baleful shadows. The first is the example of Iran, where a democratically elected parliament and president have been unable to establish their authority over the ayatollahs of the Guardian Council, who control the judicial system, the Pasdaran

Revolutionary Guard, and the domestic security agencies, and who are deeply suspicious of democracy. As Ayatollah Ruholla Khomeini wrote in 1977, "The real threat to Islam does not come from the Shah, but from the idea of imposing on Muslim lands the Western system of democracy, which is a form of prostitution." The second shadow is the nagging fear that a democratic election in most states of the Arab world is likely to be won by the well-organized Islamists. The army intervened in Algeria to prevent the Islamic Salvation Front from taking office after it won the elections of 1992. That triggered an insurgency in which more than 100,000 people have since died.

Still, it's not entirely clear that the separation of religion and state, a concept Islam finds difficult to embrace, is a prerequisite for democracy. The British have functioned tolerably well with an established Church of England for nearly five centuries; Germany's Christian Democratic and Christian Social Union coalitions have provided impeccably democratic government; and France's proud republican tradition of laicism has not spared the nation political anguish over the right of Muslim women to wear headscarves in school. But there's little left in modern European politics of the religious passions that unleashed war, massacres, and persecution in the 16th and 17th centuries.

Islam, at least in the Arab world, has yet to undergo its Reformation, and those Islamic states that have produced a more relaxed religious form have their own difficulties. Indonesia is a tremulous democracy, rent by ethnic as well as religious tensions, with the army constantly poised to intervene again. Malaysia, economically the most dynamic of Islamic countries, has seen Islamist extremist groups win power in two states—one of which they lost in recent elections—after years of well-funded Wahhabi proselytizing. Turkey, where a moderate Islamic party has now come peacefully to power by election, remains the

most promising example of the way in which Islam and democracy might prosper together. Since the reforms of Kemal Atatürk, Turkey has had 80 years of secular rule, 50 years of NATO membership, and now the lure of joining the European Union to strengthen its democratic commitment.

Turkey, of course, is a constant reminder that there's little in history or political theory to suggest that Islamic nations cannot become democracies. Indeed, the constitutional monarchy and parliamentary system that ruled independent Iraq from 1932 to 1958 produced the freest press, the most vibrant civil society, and the most impressive levels of health and education in the Arab world during that period. Yet Iraq was a clouded democracy: The elected prime minister, Nuri Said, was an authoritarian figure, susceptible to British influence, who routinely suspended parliaments when they proved hostile. At least the latest efforts at democratization in the Arab world take place under happier circumstances, without the looming presence of the Cold War.

President Bush's new "forward strategy of freedom" will need a great deal of international support, both political and financial, if it is to succeed, and a patient world will have to persuade a highly skeptical Arab public that the United States is resolved to achieve a fair peace settlement between Israel and the Palestinians. Ultimately, however, as the president made clear in January, his strategy rests on an act of faith: "It is mistaken, and condescending, to assume that whole cultures and great religions are incompatible with liberty and self-government. I believe that God has planted in every human heart the desire to live in freedom. And even when that desire is crushed by tyranny for decades, it will rise again."

MARTIN WALKER *is editor in chief of United Press International and a former public policy fellow at the Woodrow Wilson Center.*

Keep the Faith

Islamists and democracy

MARINA OTTAWAY

POLITICAL CHANGE is coming to the Middle East. Countries that slumbered for decades with no sign of political evolution are finally awakening. The Egyptian National Assembly has approved a constitutional amendment to allow direct, competitive presidential elections—a first in that country's history. Syrian troops left Lebanon when mass demonstrations signaled the people's discontent with their lack of power under the status quo. Iraq elected a government (albeit as a result of U.S. occupation rather than a renewal of Iraqi politics). Political competition has become possible in Palestine and will undoubtedly increase ahead of this summer's elections for the Legislative Council. Even Saudi Arabia has held elections—for powerless local councils, to be sure, and without the participation of women, but elections nevertheless. Only a few Arab countries—most notably Tunisia, Libya, Syria, and the United Arab Emirates—appear immune to the stirrings of change.

There are many reasons for this budding political activism, including the sheer magnitude of socioeconomic transformation in recent decades, which leaves static governments increasingly out of sync with their rapidly changing societies; satellite television, which has deprived Arab governments of their traditional monopoly over information; and the assassination of former Lebanese Prime Minister Rafiq Hariri and the death of Yasir Arafat. But the Bush administration certainly contributed to this political ferment, in ways both intentional and not. By announcing his determination to bring democracy to the Arab world, the president triggered a lively debate about democracy in the region's press—a debate that started by questioning Bush's credibility and denouncing democracy promotion as a thinly veiled justification for the Iraq war but that soon admitted that Bush's hypocrisy did not excuse the absence of democracy in the Middle East.

Most Arab governments, faced with the Bush administration's rhetorical onslaught, thought it prudent to respond by introducing some mild reforms that would not threaten their power, hoping to avoid real pressure to change. Liberal reformers are now grudgingly admitting that perhaps U.S. pressure is not a bad thing. The just-published 2004 U.N. Arab Human Development Report, for example, concludes that even policies as fundamentally flawed as the intervention in Iraq and as hypocritical as Washington's criticism of the autocracies it supported for decades might, in the end, encourage democracy. Even the region's Islamist movements are debating the merits of representative government.

A major obstacle to the continuation of this process, however, is the fact that liberal reformers, upon whom many Americans place their hopes for democratization, are mostly intellectuals without organized constituencies. There is no equivalent of the broad-based Civic Forum that made the transition to democracy in the former Czechoslovakia such a success. In Egypt, for example, the *kifaya* (enough) movement, which has attracted a lot of attention by holding demonstrations in defiance of a government ban, appears incapable of mustering more than a few hundred people to each of its rallies. Throughout the region, it is the Islamists, not the democrats, who have the organized constituencies. That's why religious Shia parties won the largest share of the vote in Iraq, and that's why free elections would give significant power to religious parties in other countries as well. In Morocco, Jordan, and Algeria, for example, religious parties that have been allowed to register and compete in the elections have shown that they have a substantial following.

This means that, for the United States to further encourage democratization, it will have to cultivate the Islamists. Of course, liberal organizations also deserve support. But, unless Islamist movements become convinced that it is in their interests to advocate democracy in their countries and participate in a democratic process—even if the process does not produce an Islamist state—democratic transitions will be extremely difficult. Despite the mutual dislike and suspicion between Washington and Islamists, the United States must recognize the legitimacy of their participation and defend their rights as much as it does those of liberals.

ISLAMIST ORGANIZATIONS—that is, organizations that appeal to the religious values and social conservatism of the Arab public in their call for political reform—are the key to democratization in the Arab world. They have considerable support, as measured by the votes they receive when they are allowed to participate in elections, the turnout at their demonstrations, and the audiences attracted by radical preachers during sermons at mosques. They are also well-organized, maintaining strong networks of educational, health, and charitable programs.

Clearly, such participation and popularity suggests these organizations would be expected to play—and ought to play—a central role in Arab democracy. Indeed, there is little alternative: It would take considerable repression to prevent such groups from participating in a democratic political process, and that repression would in turn undermine the possibility of democracy itself. Furthermore, the post-cold war experience in Eastern Europe showed that a crucial element for the success of democratic transitions is the willingness of old, undemocratic parties to reinvent themselves as new organizations capable of becoming part of the new politics. The success of old communist parties in transforming themselves into new democratic actors was crucial in establishing political competition in Eastern Europe. New parties appear by the dozen in all democratic transitions around the globe, but they rarely survive. One need only look at Iraq to see that this dynamic also holds true in the Middle East today. The former exile movements and many religious parties are all represented in parliament, whereas most of the roughly 200 parties formed after Saddam's fall hardly received any votes, despite the considerable help they received from the U.S. government and democracy-promotion organizations.

The Bush administration's answer to the challenge of political Islam has been to promote moderate Islamist organizations and moderate interpretations of the religion. The Muslim World Outreach Policy Coordinating Committee—set up in July 2004 by the National Security Council to improve communication with Islamic organizations and better the image of the United States in the Muslim world—is an important part of this strategy. But trying to promote moderate interpretations of Islam is probably futile and certainly risky. Remember that, when the Iranians took over the embassy in Tehran, conventional wisdom was that the Shia were the most dangerous Muslims, whereas Wahhabis were thought to be conservative socially but rather apolitical. Obviously, that view has changed in recent years. Nevertheless, the idea has taken hold in some parts of the administration that Sufism represents a benign, moderate form of Islam that should be encouraged. It is doubtful that

most people making the argument understand the roots of Sufism and why it has spread in some regions but not in others. It is even more doubtful that they have given much thought to the resentment that favoring one sect over another could cause in the Muslim world. (Imagine the reaction in this country if secular France, concerned about radical evangelical groups in the United States, decided to support moderate churches.)

The challenge that Islamist organizations pose to democracy cannot be met by befriending moderate but marginally important groups. It can only be met by dealing with the mainstream, powerful organizations that will determine the future of Middle East politics. A good place to start is with the Egyptian Muslim Brotherhood, the first and still the most influential modern Islamist movement. Much of the radical thinking about the moral corruption of Arab states and the need for dramatic change has come from the Muslim Brotherhood. Beginning in the 1920s, Muslim Brothers were the first to denounce what they saw as the moral decadence to which Arab societies had sunk as a result of abandoning the precepts of Islam, and they were the first to call for the building of Islamic states. Egyptian Muslim Brothers, often acting as teachers, have been at the root of radical Islam's growth in the Middle East. Even the militancy and missionary zeal manifested in recent decades by Wahhabis is attributed by scholars and analysts to the influence of the Muslim Brothers. But the Egyptian Muslim Brotherhood long ago renounced violence and is seeking to become a legal political party. Denied the right to do so—Egypt bans parties based on religion—Muslim Brothers have still sought to peacefully participate in Egyptian politics since the early '80s, running for office on the tickets of other political parties or as independents. Other Islamist organizations, such as Muslim Brothers in other countries, are also debating whether to embrace the democratic process.

Still, any opening toward Islamist groups raises the vexing problem of their commitment to nonviolence and to democracy. There are no simple answers to these issues. Saying that the United States will deal only with organizations that have renounced violence and thus do not have an armed wing is tempting but unrealistic. If Washington had taken such a position in Iraq, there would have been no elections—after all, the Kurdish parties and the Shia religious parties that won the elections all have armed militias. The Kurdistan Democratic Party and the Patriotic Union of Kurdistan control the *peshmerga*, which includes as many as 80,000 fighters. The two major parties in the Shia coalition, Dawa and the Supreme Council for the Islamic Revolution in Iraq, probably have an additional 10,000, if not more. If Washington wants elections in Lebanon this month, as it says it does, it has to accept that Hezbollah will run without disarming. If it wants elections in Palestine in July, it has to accept that Hamas will still have an armed wing. It also has no choice but to accept that there can be no certainty about any party's long-term commitment to democracy. Uncertainty is an inherent part of any transition.

ISLAMIST PARTIES ALREADY participate in some Arab countries, and they are not proving less—or more—democratic than other organizations. But there is ample evidence that participation in an electoral process forces any party, regardless of ideology, to moderate its position if it wants to attract voters in large numbers and avoid a backlash. In Turkey, the Islamist party that now governs the country and aspires to lead it into the European Union started with radical propensities in the mid-'90s. It won enough votes to form the government, but it also frightened the secular army enough to seek a court's disbarment of the ruling party in 1997. That experience spurred the rise of the moderates within the Islamist movement, leading to their victory, as the reorganized Justice and Development Party, in 2002.

Talking to the Muslim Brotherhood and other mainstream Islamist organizations should be a central, ongoing task for American diplomats in the Middle East. We need to understand these organizations as well as we possibly can. It is only when we understand them better that we can decide whether they have a legitimate role to play in a democracy or whether their ultimate goals are dangerous. Moreover, such engagement would strengthen the hand of the more moderate, democratic Islamists who favor reform. It would do more to restore the tarnished image of the United States in the Arab world than any public diplomacy initiative launched so far. Of course, engagement won't guarantee the success of democracy in the Middle East, but not engaging will guarantee its failure.

Marina Ottaway is a senior associate at the Carnegie Endowment for International Peace.

Voices within Islam:

Four Perspectives on Tolerance and Diversity

BAHMAN BAKTIARI AND AUGUSTUS RICHARD NORTON

When Muslim intellectuals interact with non-Muslims, they frequently find themselves in debates about Islam and its compatibility with democracy, or under what circumstances Islam supports political violence. These issues, along with the themes of renewal, tolerance, and dissent in Islam, formed the basis for a series of meetings we have held recently with prominent Muslim thinkers. In the following pages, we present a selection of the views of these intellectuals and religious figures.

The thoughts of Gamal al-Banna that are excerpted only hint at the breadth of his incredibly prolific writings, which are available only in Arabic and span 60 years. Al-Banna is the brother of Hassan al-Banna, the founder of the Muslim Brotherhood. Gamal, however, is usually considered a critic of the brotherhood for its conservative understanding of Islam. (In his 1946 book, *The New Democracy*, he enjoined the brotherhood "not to believe in faith, but to believe in human beings.") The book that may be most indicative of his work is *Islam Is Not Religion and State but Religion and Society* (2003), in which he argues that Islam does not offer a specific model for contemporary political systems and that the appropriate focus for activism should be at the level of society, not politics. Much of his work generates debate, including a recent volume, *The Veil*, which criticizes those activists who wish to repress women in order to symbolize their faith.

Gamal al-Banna has lent his powerful voice in support of both secular and Islamically oriented activists in his native Egypt, and he is revered for his open mind and his brave voice. If one were to sum up his work, it would be his insistence that "in Islam thinking is essential." He emphasizes consistently that the "Koran and the prophet accept entirely the concept of freedom of thought, welcome diversity of creed, respect the opinion of others and leave the matter of judging to God on Judgment Day."

While discussions about democracy are important among Islamic thinkers, many of the debates between leading Muslim intellectuals are concerned with how Muslims should understand and interpret their religion.

This can be clearly seen in the extended essay by the Syrian engineer Muhammad Shahrour.

Shahrour is famous for his best-selling first book, *The Book and the Koran: A Contemporary Reading*, which is especially popular with the educated middle class in the Arab world. Published in 1990, it is a large and often difficult work, but the substance of the argument is captured by Shahrour's insistence that the Koran should be read as though it were just revealed by the prophet Mohammed, not through the filter of centuries of interpretive dust. Like many contemporary Muslims who are thinking seriously about their religion, Shahrour was not trained as a scholar of religion; he earned his doctoral degree in soil engineering at University College in Dublin. He continues to work and write in Damascus, and sometimes appears on satellite television and speaks often in the Arab world.

The Iranian reformist thinker Mohsen Kadivar is an important and courageous voice of reform in a country that is locked in an intense struggle between reformers who want to make the system more responsive to the will of the people and conservatives determined to hold on to power through their rigid interpretation of Islam.

Kadivar, who is a *mujtahid* (a cleric qualified to interpret religious law), comes to this ideological battlefield supplied with one of the key weapons in the Islamic Republic: the language of religion. He writes of democracy, but he does not demand the overthrow of the Islamic state and its replacement with a secular form of government. In 1999, however, he was indicted and sentenced to prison for "disseminating lies, defaming Islam, and disturbing public opinion." Released after 18 months, Kadivar has been even more determined to present his views, and his prison conviction has only increased his popularity.

Kadivar is sometimes compared to Martin Luther or John Calvin, the clerics who transformed Roman Catholicism. His opinions and writings receive significant attention both among lay intellectuals and young clerics in seminaries. He has a doctorate in Islamic philosophy and theology, and achieved the clerical certification to perform *ijtihad* (independent interpretation) in 1997. Kadivar is in a unique intellectual position to influence the future of Muslim thinking on important issues such as human rights, tolerance, democratic governance, and relations with nonbelievers.

In contrast to Kadivar, Ayatollah Mohammad Boujnourdi is a leading member of the conservative clerical establishment in Iran today. He lived in Najaf, Iraq, prior to the 1979 revolution, and was a close confidant of Ayatollah Ruhollah Khomeini, advising the Iranian leader on a range of issues. In 1984, Ayatollah Khomeini appointed Boujnourdi as the head of the Supreme Judicial Council, a body charged with drafting legislation.

Boujnourdi describes himself as a "pragmatic man," and has criticized Iranian hard-liners for adopting ex-

tremist positions. He is known among the Iranian reformists as an "enlightened" conservative because he agrees with the reformists that the Islamic Republic has at times resorted to unnecessary force, alienating the population. Hence, in contrast to other senior conservative personalities, Boujnourdi represents a part of the Iranian clerical establishment that has engaged the reformists in discussions on democratization and human rights.

None of these four contrasting thinkers writes customarily in English, so the work reproduced here offers a unique glimpse of their ideas rather than an interpretation of their views by Western scholars or journalists. Obviously, there are many important voices and many perspectives, so what is offered is a small yet indicative taste of some of the key debates about Islam and pluralism that are under way today in the Muslim world.

Radicalism Emerges from Tyranny

GAMAL AL-BANNA

THE ADVENT OF tyrannical military rule precipitated the rise of fanatical groups that made violence and direct action a methodology. The mentor of some of the Islamic movements was Sayyid Qutb, who considered jihad a means of establishing and legitimizing "divine judgment" (*hakimiyya ilahiyya*) in place of all human law. The hard line adopted by the Islamic movements did not simply derive from a distorted interpretation of Islam. There are psychological and political factors as well, and the tyranny of the ruling military junta played an important role.

Torture in [President Gamal Abdul] Nasser's detention camps in the 1950s and 1960s led the mostly young detainees to believe that any government that follows such practices is not a Muslim government, but an apostate (*kafira*) government. It was thus in Nasser's prisons that the seeds of the accusation of apostasy were planted,

and it was the charge of apostasy that served as a rationale for jihad by the young Islamists.

The members of the Muslim Brotherhood in detention, who were wiser and more resilient than their young confederates, attempted to refute these ideas but to no avail. A book bearing the title *Preachers Not Judges* was released, but the damage had been done. The savage torture they had been subjected to rendered them impervious to appeasement. The first to be released from prison among them, Shukry Mustafa, founded the Forum of Heresy and Migration, which proceeded to abduct one of Al Azhar's fine clerics and went on to execute him when the government refused to negotiate for his release.

A vicious circle resulted: terrorism by the state was met with violence from the organizations provoking more terrorism by the state, leading to more violence.

A Call for Reformation

MUHAMMAD SHAHRUR

MANY WESTERN ANALYSTS, in their attempts to conceptualize Islamist groups that practice violence and terrorism, resort to the terminology of "fundamentalist movements." But Islam—in contrast to Christianity, where fundamentalism is a clearly stated doctrine—has no fundamentalist tradition. Accordingly, any rhetoric about violence and terrorism among Islamic fundamentalists refers only to armed political movements and not to ritualistic, legislative, or ethical Islam itself.

These political movements reflect important divergences between Islamic and Western experience. Historically speaking, in Muslim societies the rulers have made religion subservient to political authority. The legitimacy of political authority was based on identifying obedience to officeholders with obedience to God and the prophet. This was further augmented by the persistence of the doctrine of predestination from the first century of Islam. The idea that fate defines a person's life found many justifications in the Koran, such as "Nothing will afflict us except that which God ordered" and in popular proverbs, including "What is written in your destiny must be seen by your eyes." Obeying God and the prophet was, of course, separable from obeying rulers. But this distinction was completely overlooked by most of the *ulema* (religious scholars), who were already caught up in service to despotic authorities.

In 1924 the new Turkish ruler, Mustafa Kemal Ataturk, disestablished the caliphate, the nominal head of the community of all Muslims. With the caliphate gone, the legitimacy of despotism disappeared. Paradoxically, an alternative legitimacy has not since emerged in the minds of most people. And this lies at the heart of the bizarre combination of ruling regimes that we find in the Arab and Muslim world today. Rulers have found themselves with few choices except to complement their weak or even missing legitimacy by returning to religious sources and creating state offices such as grand mufti and sheikh al-Islam to gain an aura of legitimacy. We can see this in the Iranian revolution, where a despotic authority claiming revolutionary legitimacy exists today under the slogan of the "Guardianship of the Jurisconsult." This supreme group can block any legislation passed by the parliament.

Besides the failure to develop secular sources of state legitimacy, another difference between the West and the Islamic world has been the existence in the latter of Sharia. Sharia refers to the verses that inform rulings and judgments on a range of issues, including social and familial relations, personal affairs, punishments for crime, and financial and commercial transactions. In practice, these issues are inseparable from the work of the state. Unlike required religious practices (such as prayer, fasting, and pilgrimage), religious rules for marriage, divorce, inheritance, wills, education, adoption, selling and buying, and lending and credit cannot be separated from the scope of political power. This situation is unique to the Muslim world.

A third difference is the persistence in Muslim countries of an archaic worldview. While Europe managed to rid itself of its biblical worldview, timeworn interpretations of the Koran, sayings of the prophet, and causes of revelation still predominate in the Arab world despite the availability of enlightened critiques. Occasionally, we even hear *fatwas* (religious opinions) from here and there excommunicating those who suggest that the earth is round. The Muslim world's endemic crisis derives from stultifying, unenlightened interpretation and legal decisions that are inspired by the dead hand of anachronistic thinking.

This crisis is exacerbated by the division of the Islamic world into isolated schools of law, each with its own texts, jurisprudence, and scholars building up a gigantic yet unharmonious and contradictory heritage that can neither be accepted as is nor reformed. Meanwhile, fundamentalist Islamic movements of various orientations seek to impose their own interpretations, and try to supplant existing schools of law with their own, often narrow views.

THE MODERNIST FAILURE

Liberal movements in the Muslim world adopted the European model and hence rejected Islamic jurisprudence and its legislation. These movements did not discard Islam as monotheism, or as a divinely ordained message. They did not reject its value system or ultimate ideals. The liberals called for separation of religion and state, never targeting rites (such as prayer, fasting, good treatment of parents, and avoidance of cheating in markets) but only the Sharia and jurisprudence. However, the liberals were never able to find a context for success within Islamic Arabic culture because they lacked adequate philosophical and theoretical tools. They failed to make even minimal adaptations to the Islamic and Arab ethos. As a result, the liberals remained Westernized and outside the people's culture.

The story of the Marxist movements was even worse. They started from an absolute and assumed history—a kind of determinism akin to that of predestination nurtured by despotic Muslim dynasties. Believing in the deterministic development of societies from one historical stage to another and progressing ultimately to the communist stage, Marxists in the Muslim world became prisoners of dogmatism. Moreover, their inclination toward atheism led them to deconstruct religion itself, rather than the structures of religious despotic authority that did repress people's minds and chain their wills. Their attempt to justify their anti-religious stance by arguments critical of the feudal despotic institution was akin to drying up a water source because the water wheel is dirty.

The Arab world has also seen nationalistic movements that dogmatized science, progressivism, and modernity. These subordinated freedom to the slogans of Arab unity, socialism, and progressivism. Nationalists developed their own rhetoric to delegitimize nonconformists. They used the vocabulary of "reactionary/agent/traitor" as opposed to the "infidel/atheist/polytheist" and the "capitalist/imperialist/enemy of the people" terminology used by Islamists and communists, respectively. But they shared an antagonism to pluralism.

The 1967 Arab-Israeli war exposed the failure of all of these movements. In the ruins of defeat, it became clear that projects for modernity in the Arab world had betrayed their original promises. Nationalistic ideologies especially appeared as romantic and idealistic formulations that lacked any concrete theory of state and society and failed to fulfill the need for justice or to reconcile competing interests. As a result, oppressive police regimes emerged.

THE ISLAMIST RESPONSE

It is with this history in mind that we can better understand the Islamist movements—that is, the groups of Muslims who want to reestablish an Islamic state. These movements became noticeable in the wake of the 1967 war, spawned by what is usually referred to as an Islamic revival or resurgence. Yet, with their reliance on traditional Islamic literature, the Islamists were unable to provide any creative ideas on how the state and society should deal with the new variables introduced by the twentieth century.

The Islamist movements had no insights appropriate for an age of dramatic scientific and information progress—not to mention new political concepts and innovations that have spread worldwide, such as civil society, freedom of speech, elections, constitutionalism, plebiscites, women's economic rights, and elected political offices. Historical Islamic literature was silent on constitutional jurisprudence that clearly defines the rights and prerogatives of the ruler, how he is selected, and the duration of his rule.

Similarly, Islamic historical thought had developed no notion of individual freedom as it is understood today. Freedom of speech itself enjoys only minimal importance in Islam. Traditionally, individuals did not enjoy any genuine rights to speech in Islam; they were to follow the way of their monarchs and jurists. Religious judgments and not plebiscites are the foundation of classical Islamic jurisprudence.

Furthermore, when it comes to principles and methods of political action, it is clear that members of Islamist movements are deeply committed to their religion but suffer from extreme naïveté. This naïveté is part of a more general political fragility and awkwardness among Islamists that help to explain their turn to terrorism and violence.

Fatally, Islamist movements overlap politics with excommunication, Islam with belief, piety with rites, and jihad with armed violence. Rather than propagating religion by good examples, as advised in the Koran, they do so by the sword. Moreover, they distort Koranic texts with serious consequences—such as when they identify killing (*qatl*) with fighting (*qital*)—and consider these harsh distortions as basic to Islam. These errors stem from the lack of a contemporary genuine Islamic theory on state and society that puts jihad, piety, and debate in their proper place. Misunderstanding and ambiguity compounded with a vigorous religious zeal can easily lead to incidents of bigoted armed violence—including the kind of killing out of ignorance that was seen in the 9-11 attacks on the United States.

Another reason why Islamic political movements have resorted to violence is itself political. Their violence is a fundamental element of the spiral of violence and counterviolence caused by the Muslim world's deeply entrenched incumbent authoritarian regimes. These regimes' weak legitimacy entices them to engage violently with the opposition. Their uncompromising response helps create a mood for violence that aggravates the already violent tendencies inherent within Islamist movements. Poverty, unemployment, the unequal distribution of wealth, class privileges, and ignorance further fuel popular backing for radical Islamists.

Finally, Islamists have been pushed toward violence by the failure of contemporary modernization movements. Islamists appeared to fill the intellectual and cultural vacuum from which Arab Muslims have suffered. But the Islamist movements resorted to experiences and wisdom drawn from the distant past. Official religious institutions, handicapped by conservatism, exacerbated the problem by also fixating on the past and grounding this view inside the "Arab mind." As a result, the response to political violence was based on a model of jihad drawn from the past.

REINTERPRETING ISLAM

The prospects for Islamist movements are gloomy unless they can articulate a contemporary Islamic theory on state and society that provides for freedom of thought, political opposition, the transition of power, plebiscites, parliamentarianism, freedom of faith, and individual and collective human rights—especially those of women. For this to happen there needs to be a complete transformation of the concept of legal reasoning in Islamic terms. Unless this transformation occurs, the threat of the rise of fundamentalist powers inimical to the structure of civil society and its institutions will remain.

Let me emphasize this point: the basic texts of Islamic law are the same texts that a movement like the Afghan Taliban used. We need to adopt modern methods of interpretation. But the obstacles are large: when the great Egyptian thinker Sheikh Muhammad Abdu proposed a reinterpretation late in the nineteenth century he was subjected to public defamation by the traditionalists, who condemned him as a Mason and a Western agent.

The process of Islamizing reality is ultimately a sociological process. A civilized society produces a civilized Islam and a Bedouin society produces a Bedouin Islam. Perhaps the most debilitating event in the Muslim world has been the rise of political movements whose agenda is to pull Muslim societies backward in history under the tempting slogan of applying Sharia. The latter serves as a hollow label; underneath it lurk all sorts of ruling private interests that lie at the real core of policy making in Muslim countries. Remember that Afghanistan under the Taliban became the world's largest producer of narcotics. Muslims are no different from other people in their susceptibility to corruption once they achieve unchecked power.

The Islamists have emerged because the modernity projects in the Arab world betrayed their promise, creating a pressing need for an alternative. The inherited traditional culture was more than ready to offer that alternative. But this alleged revival did not go beyond rites and worships as understood by people of tradition—for example, prayer, fasting and pilgrimage, spending considerable time in mosques, men growing beards, and the adoption of the female veil. (The veil in particular served as a political symbol and slogan. Warring factions mobilized around it, and ruling powers accepted and encouraged it as long as it diverted people from their real grievances and problems.)

What really would have had an important effect on the rulers were doctrines on constitutional jurisprudence, checking power, and ensuring governmental accountability to the people. But these concepts were not found in the inherited traditional culture. Thus, by emphasizing tradition, the rulers benefited from the poverty of the Islamic tradition regarding these issues. They were aided by official religious institutions that supported the spread of the Islamic heritage by funding religious education and the publication of millions and millions of tracts.

WHOSE SHARIA?

Islamic traditions that govern social transactions and personal matters do offer a semblance of diversity in legal and jurisprudent schools of thought. But this provides a superficial kind of legitimacy, reflecting an artificial richness of ideas and disagreement among religious scholars. Examples would include debates over when the month of fasting, Ramadan, begins. Or under what circumstances interest on bank loans is permitted. Or how Sharia is to be applied in cases of theft. (Is the amputation of a thief's hand literally or only metaphorically required? For that matter, if amputation is permitted, what is the "hand"?)

Some would argue that the mere call for applying Sharia is antithetical to religious, political, and cultural pluralism. To consider this argument, we must first be clear about what we mean by Sharia. If we mean the divine revelation of the Koran and the prophetic tradition, then this argument would imply that God and his prophet are opposed to pluralism within the confines of the Muslim state. But this is impossible, of course. God declared it clearly and uncompromisingly: "No compulsion in religion, righteousness is already differentiated from falsehood" and also, "If God wills it, he could have made all of those on earth believers: Would you force the people to be believers?"

So it is a question of defining the conceptual boundaries of the term "Sharia." By what criteria should "law" beyond the Koran be accepted? What about the books of the *Hadith* (narrations of the life of the prophet Mohammed), assembled and written by jurists? In fact, many of the texts now considered part of Sharia are historical words, the products of human labor. These texts, moreover, were formulated according to legal proofs and reasoning—also a historical human product. If the objective is to project all that is mentioned in these texts onto our current world, then we will have a Taliban in every Arab and Muslim country, albeit with various local versions. This prospect poses a threat to pluralism and civil society. It also would represent for society at large a return to a past life—not of the prophet's companions, but of the medieval ages.

It would be completely different if we comprehended Sharia as a general guiding umbrella of the rulings, injunctions, and principles mentioned in the Koran and Sunna (the body of customs and practices based on the prophet's words and deeds) that should be projected forward over time and space. Only in this sense can we posit a Sharia that is not in conflict with civil society and pluralism. We only need to do what our predecessors did when they first read the Koran and Sunna in the light of their reality and time. We too should have our own reading of the Koran and Sunna so that they can provide us with new fundamentals of jurisprudence and legislation. These fundamentals should stem from the following bases:

• Supreme ideals (ethics and value systems). These were subject to accumulation from the days of Noah until the prophet Mohammed. They include upholding the ultimate universal human values, such as respecting parents, not committing suicide, keeping promises, and engaging in honest trade.

• Rites and rituals. These are the centerpiece of belief. But prayers, fasting, and pilgrimage were subject to change and diversity. Prayer is found in all religions, and fasting among Muslims, Christians and Jews. Accordingly, the state should accept diversity in rituals and rites, and the existence of many houses of worship (mosques, churches, and synagogues) should be acceptable to civil society. (Indeed, this is already the case in most Muslim countries.)

• A Sharia subject to development in understanding and application (except for monotheism and rituals).

THE PATH TO SALVATION

We have looked at the development and ideologies of the Islamist movements, but a last point needs to be addressed: what is their goal? The answer is not as simple as one might assume. Is the goal the pursuit of power by example? I will assume the truthfulness of the raised slogans about justice, equality, *shura* (consultation), fighting corruption, and ensuring security. But these slogans require a mechanism to achieve them and gradualist programs at the core of that mechanism. Obviously, Islam's original heritage and traditions are deficient when it comes to these mechanisms and programs.

Fundamentalism essentially starts from sacralizing tradition and subscribing to it literally, regardless of its contradictions or inconsistencies. Accordingly, it will find itself obliged to force people to go back to the past with all its details and leave aside the present with all its novelties in order to apply unchangeable traditional texts under the pretext of respecting constants.

If the Islamists' goal is to participate in power and not to monopolize it, then with whom would they cooperate in ruling? With the nationalists or the liberals? Or would they transform themselves into a new official religious institution whose function is to legitimize a new system within which it would be an active partner? The Islamist movements' choice—to monopolize power or to share it and leave the door open to all coming movements, whether religious or not—will determine whether the cycle of destructive violence will resume yet again, with merely a changing of places and roles.

An experiment taking place in Lebanon features an Islamist movement, Hezbollah, that is undoubtedly fundamentalist in terms of its foundations and ideology. Yet it is trying to prove that terrorism is not an option, and that violence was employed only against occupation, colonialism, and subjugation. Hezbollah is unabashedly attempting to entrench its position more and more inside the politics and culture of its society. However, it cannot be foretold to what extent this movement can cooperate with nationalist powers and other political movements. Nor is it clear how it will engage with external powers that cannot tolerate collective action led by a religious fundamen-

talist movement. (There is also the compelling question of whether Hezbollah will deviate from Iran, its major patron and supporter.)

The central concern for the Arab Muslim world is the need to appreciate the urgent necessity of a second contemporary reading of the Koran and Sunna, guided by the imperatives of the world today. This process should be freed from the perspectives of early thinkers, with due and deep respect for all of them, because we need a current reading. The exercise of self-conscious and critical reason is the only safeguard against terrorism and violence. This process is of course arduous and still remote, and the hopes built around it are imbued with idealism. Nevertheless, for good or bad, I see no other way to salvation.

Translated from the Arabic by
Ashraf N. El-Sherif, Boston University

Freedom of Thought and Religion

MOHSEN KADIVAR

TO UNDERSTAND THE PLACE of tolerance in Islam, we need to examine what we mean by freedom of thought in Islam. I argue that freedom of thought and faith is not only beneficial to Islam and to Muslims, but that it is also mandated by fundamental religious rules.

Islam is one of the three great religions but it is frequently thought to be a religion that does not accept diversity of viewpoints. In historic Islam the text of the Koran, the traditions of the holy prophet Mohammed, the behavior of the authorities of religion, and consensus among Muslim scholars are considered to be permanent precepts, beyond time and space. Thus they are regarded as divine and not subject to criticism. While proponents of this approach believe in religious rationality, referred to as "wisdom," this rationality is thought to exist beyond the human mind.

According to Islam, Muslims are free to openly practice their religion, express their religious beliefs, practice their rituals alone or in groups, and teach religion to their children. They have the right to criticize all other religions and to ensure the supremacy of Islam. Nobody has the right to force a Muslim to leave his religion under duress or to prevent him from practicing the religious ceremonies. There is a consensus in this and there are no differences in this area.

Yet a Muslim is not allowed to change his religion to become, for example, a Christian or a Buddhist or become an atheist. A Muslim who for any reason leaves his religion, or in other words becomes an apostate, would be severely punished. The child of a Muslim who has chosen to become a Muslim after maturity and then renounces Islam is subject to execution, even if he repents. His wife would be

separated from him without divorce, and his property expropriated and divided among Muslim heirs. Also, a youth with one Muslim parent is not free to choose another religion other than Islam after maturity. If she or he does not become a Muslim the charge of apostasy would apply, although she or he would first be asked to repent. If the apostasy continues the person would be sentenced to death or to life imprisonment with forced labor.

There are several "traditions" that are frequently cited as justification for these punishments. Sunni Muslims refer to the tradition of the prophet that states: "Kill any one who changes his religion." Shiite Muslims refer to a tradition from their sixth imam, Jaafar Sadeq, that also reportedly makes death a penalty for anyone who leaves Islam. In the history of Islamic thought, few Muslim thinkers have dared to question these traditions. Why have Muslim thinkers shied away from analyzing them? How can a religion that wants its followers to research and accept a religious faith with the help of reasoning and analysis argue for killing a Muslim should he or she decide to follow another faith that is as rational and accepting of its followers?

"DURESS IS NOT PERMISSIBLE"

The Koran has a verse that states: "Duress is not permissible in religion, as the path has become clear from falsehood to light, therefore anyone that takes the idols as tyranny and starts to have faith in God, has truly found a support that is never separated from him. . . ."

This verse means that we as Muslims cannot deny that God has prohibited us from imposing faith on anyone, since forced faith and tyranny are not valid. The disapproval of force in this verse equals accepting freedom in religion and its requirements are freedom in both matters: freedom in bringing religion and freedom in leaving it.

How can a religion that denies the freedom of religion and thought expect to be freely chosen and when those who choose may have their freedom taken from them? If people are free to think seriously about religion, it is irrational to argue that they must choose Islam. If they are free, then the result cannot be determined beforehand. If they have no choice but to accept it then they are not free. What is the difference if an individual has been born in a Muslim family and has matured in an Islamic society and therefore is a Muslim and if someone has been born in a Christian family, has matured in a Christian society, and as a result is a Christian? Good and operative ideas are the choices for conscious individuals.

As stated in the Koran, "We send the book [the Koran] righteously to you for the people, therefore anyone who finds the right path has done so to his own benefit and anyone who deviates has done so to his own loss and you are not their guard." The Koran has revealed the right of people to choose their faith, and people in this world are free to go by it or to ignore it. It is not in this world but in the other world [that is, at Judgment Day] that one is to be evaluated and awarded.

Unfortunately, the subjects of freedom of religion and thought in Islam have not been studied in the context of how individual Muslims perceive their faith. Like any idea, people choose their religion, or choose to abandon it for another idea or faith. We live in an age of rational thinking. People do not see a conflict between reason and faith. Faith is strengthened by reason and principle, not by coercion and pressure. That which is created with force and pressure is only a superficial idea and no more than that.

I believe that all ideas and religions found in human societies do not all enjoy the same validity and justification, and there is no doubt that some proponents of Islam find their religious faith superior to others. If non-Muslims or skeptical Muslims do not accept our reasoning, we do not have an obligation to impose our version of truth on them. Force and terror in the name of religion would undermine religion itself. When a person sees a benefit in a religion, such as well-being and spiritual peace, he or she will not let go of it. Change comes when people are convinced, not when they are forced.

Restricting thought and ideas is not the solution to our problems, and as Muslims we cannot ignore the fact that in today's world our ideas have to exist with other ideas, even if we disagree with them.

Translated from the Persian by Bahman Baktiari

Islam and Tolerance

Mohammad Boujnourdi

INTOLERANCE IS ON the increase in the world today, causing violence, religious persecution, and even genocide. Sometimes it is racial and ethnic, sometimes it is religious and ideological, sometimes it is political and social. In every situation it is evil and painful. How can we solve the problem of intolerance? How can we assert our own beliefs and positions without being intolerant of others? How can we bring tolerance to the world today? I would like to discuss some of these issues from an Islamic point of view.

Given the Muslim view of God as rule-giver, tolerance in Islam is understood to be the undeserved and capricious generosity of a ruler toward the ruled. Epistemologically, tolerance is defined according to the regulations of the Sharia and the normatively interpretative example of the prophet Mohammed and the first Muslim community. Theologically, Muslims view everything in light of the destiny of Islam to rule the world and, therefore, they are committed to what they believe is God's will.

What is tolerance? Literally the word "tolerance" means "to bear." As a concept it means respect, acceptance, and appreciation of the rich diversity of world's cultures, forms of expression and ways of being human. In Arabic it is called *tasamuh*. There are also other words that give similar meanings, such as *hilm* (forbearance) or *afw* (pardon, forgiveness) or *safh* (overlooking, disregarding). In the Persian and Urdu languages, we use the word *rawadari*, which comes from *rawa*, meaning "acceptable or bearable" and *dashtan*, meaning "to hold." Thus it means to hold something acceptable or bearable.

A RELIGIOUS DUTY

Tolerance is a basic principle of Islam. It is a religious moral duty. It does not mean "concession, condescension, or indulgence." It does not mean lack of principles, or lack of seriousness about one's principles. Sometimes it is said, "People are tolerant of things that they do not care about." But this is not the case in Islam. Tolerance according to Islam does not mean that we believe that all religions are the same. It does not mean that we do not believe in the superiority of Islam over other faiths and ideologies. It does not mean that we do not present the message of Islam and do not wish others to become Muslims.

But is it not true that Islam grants Jews and Christians living within Muslim-ruled nations a special status as *dhimmis* [Arabic for "protected people"]? This concept of *dhimmi* began in 628 AD, when the prophet Mohammed defeated a Jewish tribe that lived at the oasis of Khaybar and made with members of the tribe a treaty known as the *dhimma*. This treaty allowed the Jews to continue cultivating the oasis as long as they gave half of their produce. This agreement has served as a model for Muslims ever since.

Some Western scholars point to the taxing of non-Muslims (*jizya*) as an example of discrimination. But it is important to note that the prophet was not attempting to make the taxes a form of indirect pressure on non-Muslims. He commanded that the total amount of taxes be proportionate with the economic capability of non-Muslims. *Jizya* was not enforced on them as a kind of "punishment" because they refused to convert to Islam, nor to humiliate them. Quite the contrary, it was meant to enhance their feelings of citizenship, since it was clear that *jizya* was paid to cover the expenses of protecting non-Muslims against outside attacks. As citizens they had the right to share in their societies' protection. Moreover, the poor among them did not have to pay the *jizya* and had the right, like Muslims, to be supported by the money collected through *zakat* (alms giving). In short, they did have citizenship.

We can thus say that throughout history Muslims have been very tolerant people. We must emphasize this virtue among Muslims and in the world today. Tolerance is needed among our communities. Muslims must foster tolerance through deliberate policies and efforts. Our centers should be multiethnic. We should teach our children respect for one another. We should not generalize about other races and cultures. We should have more exchange visits and meetings with others. Even marriages should be encouraged among Muslims of different ethnic groups.

With non-Muslims we should have dialogue and good relations, but we cannot accept things that are contrary to our religion. We should inform non-Muslims what is acceptable to us and what is not. With more information, I am sure respect and more cooperation will develop.

Translated from the Persian by Bahman Baktiari

BAHMAN BAKTIARI *is director of the international affairs program at the University of Maine, and* AUGUSTUS RICHARD NORTON *is a professor of international relations and anthropology at Boston University.*

First Steps: The Afghan Elections

"For [Hamid] Karzai, winning a nationwide plebiscite made him the first elected leader in Afghan history and legitimized his government. . . . But his electoral victory will prove hollow unless he succeeds in using this window of opportunity to permanently change the dynamic of Afghan politics."

THOMAS J. BARFIELD

The October 9, 2004, Afghan presidential election proved skeptics wrong. That day saw a reasonably fair electoral process unmarred by the violence that had been threatened by the Taliban, relatively high participation by women, and balloting that produced a clear first-round majority for interim President Hamid Karzai, who took 56 percent of the vote.

What produced a successful election in a country that still lacked security and had a leader who was often derided as only the "mayor of Kabul"? A key reason was the overwhelming support for the election by ordinary Afghan voters who went to the polls in astonishingly high numbers. They were keen to move away from war and saw the emerging political process as a way to bring about more security in their lives. And despite the regional and ethnic divisions in the country, no faction (other than the Taliban) saw any advantage in boycotting or disrupting the election. Each wanted to stake a claim in a new political forum.

For Karzai, winning a nationwide plebiscite made him the first elected leader in Afghan history and legitimized his government. The election demonstrated broad political support that went well beyond his own ethnic group, the Pashtuns. But it was also clear that the vote for Karzai signified more an expression of hope for the future than approval of his past performance. In particular, the Afghan people expressed a desire that he carry out wider-ranging reforms and bring more political and economic stability to the country.

The election alone could not address Afghanistan's many serious problems, including the continued power of regional armed militias, lack of security, slow reconstruction, corrupt administration, and record opium production. Karzai has moved on some of these issues with a major reshuffling of his cabinet that seeks to exclude the old militia leaders and replace them with technocrats. But

his electoral victory will prove hollow unless he succeeds in using this window of opportunity to permanently change the dynamic of Afghan politics.

THE TALIBAN'S RISE AND FALL

The roots of the current political situation in Afghanistan lie in the civil war that began with the Soviet withdrawal in 1989 and ended with the expulsion of the Taliban in 2001. When the Soviet-backed government collapsed in 1992, its forces splintered and joined factions of the mujahideen resistance. Ethnic, personal, and regional ties proved more important than ideology in this reshuffling. The Pashtuns allied mainly with Gulbuddin Hekmatyar's Hizb-i-Islami (Islamic Party), which was dominant in the south and east. The Sunni Persian speakers (generically labeled Tajiks) joined with the Jamiat-i-Islami Afghanistan (Islamic Society of Afghanistan) and came to control the west and northeast. Jamiat's political leader was Burhanuddin Rabbani, but the group's significant military power lay under the command of Ahmad Shah Masud in the Panjshir Valley and Ismail Khan in Herat. The smaller number of Hazaras—Persian-speaking Shiites from the Hazarajat region in central Afghanistan—united under the banner of the Hizb-i-Wahdat Islami (Islamic Unity Party), while the Turkish-speaking Uzbeks became an autonomous faction led by their former communist militia general, Abdul Rashid Dostum.

From 1992 until 1996, these factions fought one another in a bewildering series of alliances and betrayals that produced no decisive outcome. What political credit the mujahideen leaders had gained by defeating the Soviets was squandered in their relentless pursuit of personal power. Kabul was left in ruins while other parts of the country, particularly in the Pashtun south, teetered on the verge of anarchy in the grip of locally abusive militia leaders.

The Taliban emerged in 1994 in the southern city of Kandahar as a reaction to this stalemate. Under the leadership of Mullah Omar, the Taliban promised to bring strict order by implementing a severe Islamic regime. Although founded as a religious movement, its membership was predominantly Pashtun. With Pakistan's help the Taliban came to dominate the southern and eastern regions of the country at the expense of Hekmatyar, whose militia they defeated and incorporated into their own. After a series of advances and defeats in 1995, they succeeded in seizing Kabul from Masud and Herat from Khan in 1996. After some setbacks the next year, the Taliban took control of northern Afghanistan from Dostum in 1998 and followed this by conquering Hazarajat.

The only force left to stand against the Taliban was the umbrella Northern Alliance led by Masud. He, however, had been pushed back into his northeastern mountain redoubts. Here Masud was pressed not only by Taliban forces but also by units composed of "Afghan Arabs," the Taliban's foreign jihadist allies led by Osama bin Laden. These foreigners provided the Taliban with money and shock troops to be used in the Afghan civil war in exchange for training bases and freedom to run their own affairs.

To the outside world the Taliban government was most notable for its severe Salafi Islamist regime. The Taliban banned all forms of entertainment, compelled religious observances, restricted women to their homes, destroyed art treasures deemed idolatrous, and enforced penalties such as amputation for theft and stoning for adultery. The regime never received the international diplomatic recognition it craved, and its harboring of bin Laden and Al Qaeda put it at odds with both the Bill Clinton and George W. Bush administrations. Internally, the Taliban's excesses and lack of administrative capacity alienated large sections of the population but without threatening their hold on power.

As tensions between the United States and bin Laden rose over his responsibility for the attacks on American embassies in East Africa in 1998 and the USS *Cole* in 2000, the Taliban found themselves in a difficult position. They were unable to control the actions of their Arab allies that threatened to bring retaliation but were unwilling to expel them. On September 9, 2001, Al Qaeda assassins posing as journalists killed Masud in a suicide attack, throwing the Northern Alliance into disarray. Two days later Al Qaeda operatives attacked New York and Washington, provoking the United States to seek the overthrow of the Taliban and the destruction of Al Qaeda forces in Afghanistan.

The ensuing American war in Afghanistan against the Taliban was remarkably brief. Lacking the ability to get its own troops into the country quickly, the United States sought out the remains of Masud's Northern Alliance (now known as the United Front) to act as surrogates on the ground. With the help of American airpower the Taliban were pushed out of Mazar-i-sharif by opposition forces in November 2001. This signal defeat in the north caused local militia leaders in non-Pashtun regions to desert the Taliban. Within days the west fell into the hands of Khan's Tajik forces and the Taliban evacuated Kabul in panic as Panjshiri troops moved south. Fearing the creation of a new government without their participation, Pashtun leaders also deserted the Taliban en masse. This left Al Qaeda exposed to attack and forced Taliban loyalists to flee to Pakistan. Kandahar itself fell in early December, although Mullah Omar escaped the city by motorcycle.

DEMOCRACY, READY OR NOT

As agreed to by all the Afghan anti-Taliban factions then negotiating in Bonn, Germany, a Pashtun royalist, Hamid Karzai, was chosen as head of a 30-member interim power-sharing government. The Karzai administration was supported by a contingent of international troops under a UN mandate. Unlike Iraq, the Afghans retained national sovereignty from the beginning, and the Bonn Accord set out a timeline for the creation of a constitution and future elections.

The interim government faced a number of problems. It lacked a national police force and army, was dependent on the cooperation of local military leaders who had returned to power after the American invasion, and had little administrative capacity. The promised number of international troops who could have secured regions outside of Kabul never materialized; as a result, the so-called warlords, far from being marginalized, grew in power at the expense of the Karzai government. While the international community did an excellent job in providing humanitarian relief, the more difficult work of repairing the country's infrastructure and reviving the economy lagged. The one growth area was opium production. Although the Taliban were not able to provoke a national insurgency, most of Afghanistan remained insecure.

At the same time, millions of refugee Afghans in Iran and Pakistan thought the situation improved enough to return home. The political process moved forward too. A *loya jirga* (national assembly) ratified Karzai's appointment as president and approved his cabinet in June 2002. This assembly consisted of elected representatives from all parts of the country, but they were joined in Kabul by a set of powerful unelected regional leaders and old mujahideen commanders whom Karzai invited to participate. While the attendance of these figures may have recognized political realities in the country, their presence was seen as interference in what was supposed to be a popularly elected assembly. A second loya jirga was held in December 2003 to debate a constitution that created a parliament within a strong presidential system. The constitution was ratified in January 2004 and elections for both the new parliament and the presidency were scheduled to take place in March 2004.

While most representatives to the loya jirga were elected, the operation of the constitutional assembly itself more resembled a local tribal or village council; participants sought consensus or acceptable compromises rather than attempting to win issues through majority voting. In such an assembly discontent is traditionally reflected by walking out and refusing to participate. If enough people reject the process, the assembly is seen as a failure and cannot make binding decisions. This strategy was clearly in evidence at the constitutional loya jirga, which saw many walkouts and the subsequent coaxing of dissidents to return.

Also, although different sections of the constitution were debated, the loya jirga did not vote on them individually. Rather, after changes deemed necessary were made behind closed doors, the assembly was presented with the final document and adopted it by acclamation rather than by a counting of votes. While parties could be seen as winners and losers in this process, no faction faced the humiliation of a public loss and the final unanimous vote bound everyone.

By contrast, the proposed elections marked a major shift in the dynamics of Afghan politics. They would produce public winners and losers, clearly identify the electoral strength (or weakness) of individual candidates and parties, and require the mobilization of popular support. In addition, this would mark the first time the Afghan people would have the opportunity to determine their head of government, although a few elections for parliamentary seats had taken place in the 1960s. The timing of the elections raised immediate concerns. Critics argued that they should be postponed for at least a year, if not longer, on the grounds that the country was not prepared to carry them out.

THE DRIVE FOR ELECTIONS

The criticisms had strong foundation. The security situation in particular seemed to have little possibility of solution. The Taliban threat was taken very seriously, particularly after a number of incidents in which insurgents assassinated workers seeking to register voters. Because attempts to rebuild the Afghan national army had moved so slowly, it was unlikely to constitute an effective force to secure the registration or election process. While the international community supported the holding of elections, it was unwilling to commit many more security forces to the process. European Union representatives declared that the security situation was so uncertain that they would not even provide the usual countrywide teams of election observers, but rather restrict their efforts to observing polling stations in Kabul and a few other cities.

The lack of formal political parties had a greater impact on planning for the parliamentary than for the presidential election. Afghanistan never had a true political party tradition, and factions tended to be based on region or ethnicity rather than ideology. King Zahir Shah had banned political parties from participating in the parliament created by the 1964 constitution. And those groups that had previously seized power on the basis of strong ideologies, such as the socialist Peoples Democratic Party of Afghanistan or the Islamist Taliban, never allowed any opposition after they took control. Holding elections too soon, it was feared, would give an advantage to the existing regional power holders, most of whom represented old mujahideen factions and regional warlords who would attempt to rig any elections or at least maintain veto power over who could run. As the planned spring date approached, the registration of voters in rural areas fell so short of targets that it appeared it would be impossible to hold elections as scheduled in any event.

The large turnout of voters indicated that the Afghan people were genuinely and enthusiastically motivated to see the election succeed.

Despite these difficulties, the Karzai government gave a high priority to ensuring that the elections took place. Although outsiders saw the timing as set primarily by the Bush administration, keen to cite some progress before the American presidential election in the fall of 2004, Karzai's ministers insisted they had their own reasons for conducting the balloting on schedule. All previous Afghan governments that had promised elections had postponed them indefinitely as a strategy to maintain power. This government, many cabinet ministers claimed, needed to break this pattern to prove that it had popular support and legitimacy that went beyond the loya jirga process.

Winning a national election would give the Karzai government both a mandate to change the status quo and an electoral power base. It was also seen as a way to undermine the legitimacy of the warlords and weaken their grip on institutions of national power. Thus, although the constitution appeared to give Karzai the power to put off elections if he chose, he was determined to hold them—a posture that went against the usual criticism that he was more prone to equivocation than decisive action.

THE CAMPAIGN SEASON

When it became clear that registration of voters would not be even close to complete in March 2004, the Afghan government began a series of postponements—first to June, then August, and finally October 2004. October was the last realistic deadline before snows would begin to block the more mountainous regions of the country. The government also decided during the summer to separate the presidential election from the parliamentary vote. The presidential election would be held first as announced and the parliamentary balloting would be postponed until the spring of 2005.

Since the vote for president used the same ballot countrywide, this was much easier to carry out than the more complex parliamentary elections that would require the recruitment and listing of candidates at the local level. It also postponed the problem of dealing with local power holders who might seek to interfere with the election process. They would have relatively little impact on a national presidential election, but infighting was likely to emerge about the choice of local candidates for parliamentary seats.

The registration process was stepped up to meet the October deadline and proved surprisingly effective. By September about 10.5 million voters had registered, 41 percent of them female (although the percentage of women was far less in Kandahar and neighboring ethnically Pashtun provinces). Refugee Afghan voters in Pakistan and Iran were also allowed to register and participate. There was concern about multiple registrations in some areas where the number of voters appeared to be greater than the population. In the absence of a proper census, the significance of such irregularities was difficult to judge. The figures may have included registrants from the region along the Pakistan border where the definition of legal residence is fluid among tribes split by the contested Durand Line, demarcated by the British in 1893.

The election postponements did nothing, however, to resolve the potential security problems. Even as election day approached it remained an open question as to whether the Taliban and their allies would be able to intimidate people into staying away from the polls or launch terrorist attacks against poll workers or voters during the election itself. The country braced itself for trouble.

Karzai has argued persuasively that the rise of the drug trade is a greater source of instability to the Afghan state than the Taliban.

A total of 18 candidates chose to run for the presidency and qualified for the ballot. Of these only the Pashtun Karzai and three others—Yunus Qanuni, an ethnic Tajik from the Panjshir Valley; Hajji Mohammad Mohaqiq, an ethnic Shia Hazara; and Abdul Rashid Dostum, the ethnic Uzbek from northwestern Afghanistan—were expected to draw more than a few percent of the votes.

Given the difficulty of carrying out any campaigning, the electoral base of each of these candidates was expected to come primarily from his own ethnic group. But, because no ethnic group constitutes a majority of Afghanistan's population, a candidate would need to garner a significant number of votes from outside his own faction to win a first-round victory. In addition, the two largest ethnic groups (Pashtuns, 40 percent; Tajiks, 30 percent) were much more fractured in their potential political leanings than the two smaller groups that had candidates

running (Hazaras, 15 percent; Uzbeks, 10 percent), so bloc voting could not be taken for granted.

Because of transportation difficulties and security concerns, no real election campaigning took place other than through the mass media. Karzai's attempt to campaign outside of Kabul ended when a rocket was fired at his helicopter when it was landing in Khost and his security advisers afterward refused to let him travel outside of Kabul, much to his disgruntlement. The other candidates, lacking any logistical support, generally found it impossible to campaign outside of their respective regions.

Although not technically a campaign move, Karzai's removal of Ismail Khan as the governor of Herat in September undoubtedly raised the interim president's profile as a national leader who was not merely the mayor of Kabul. Since removal of warlords ranked high on the agenda of many voters, the successful dismissal of this powerful regional strongman without sparking widespread warfare worked in Karzai's favor.

THE VOTE AND ITS OUTCOME

The nationwide voting that took place on October 9 was conducted through 4,900 polling centers with 22,000 polling stations. These operated in all districts of Afghanistan's 34 provinces. An additional 2,800 polling stations served refugees in Iran and Pakistan. The turnout of slightly more than 8 million voters was greater than expected and constituted 70 percent of all registered voters. The balloting was carried out peacefully and in generally good order. No attacks were reported on either voters or poll workers.

While there were some complaints of double voting and indelible ink that was not indelible, the election process went surprisingly smoothly. Early protests lodged by a number of losing candidates on the day of the election were quickly withdrawn in part because they appeared to run strongly against public opinion that was happy with the process.

Karzai's success in garnering 56 percent of the vote meant that no runoff election was required. The second-tier candidates trailed far behind: Qanuni with 16 percent of the vote, Mohaqiq with 12 percent, and Dostum with 10 percent. While most of the press focused on Karzai's majority in the national vote, the regional breakdown was equally significant. Karzai not only drew overwhelming support from predominantly Pashtun areas of the country (90 percent-plus) and from expatriate voters in Pakistan (80 percent), he also won majorities in some mixed or non-Pashtun areas, and he won a plurality of votes in two northern provinces (Kunduz and Balkh). In most provinces won by strong regional candidates Karzai came in second.

The bottom-tier candidates were interesting for their lack of support. Ahmad Shah Ahmadzai, whose candidacy many thought would split the Pashtun vote with his more Islamist platform and allies, did very poorly. In fact, he was outpolled by the only woman candidate, Masuda

Jalal, no doubt an embarrassment for him. The Pashtuns basically decided on Karzai; complaints that he was too connected with foreigners or was soft on Islamic issues carried little weight.

Before the election major concern was voiced about whether women would be willing or able to vote. In accordance with Afghan tradition, particularly strong in rural areas, elections officials created separate polling stations for men and women. Overall, women's participation in the election was quite high, with around 40 percent of the total vote outside of Kabul. The major exception was the Pashtun provinces in the south. Women accounted for only 25 percent of the vote in Kandahar and less than half that in neighboring Uruzghan, Helmand, and Zabul, provinces where Taliban influence is still strong. By contrast, the Pashtun provinces in the east all had participation of women at 40 percent or above. Women's turnout in such conservative eastern provinces as Paktia, Patika, and Khost approached parity with men's and was among the highest in the country. Such results indicate a sharp regional difference in Pashtun areas on women's participation in politics, a split that may have consequences for the parliamentary elections since the constitution guarantees women approximately a quarter of seats in the *Wolesi Jirga* (House of the People).

The large turnout of voters indicated that the Afghan people were genuinely and enthusiastically motivated to see the election succeed. It was the Taliban, not Karzai's rival candidates, who were the losers. The insurgents proved unable to interfere with the process despite the many threats they made, provoking fears of voter intimidation and terrorist incidents. While extra security was put in place and undoubtedly had some impact, it would not have been possible to have such a smooth election if the Afghan people had not been so solidly in support of it. Many Afghans see this, combined with the failure of "Taliban-lite" candidates to garner significant support anywhere, as evidence that the Taliban are becoming a spent force politically.

Although Karzai's first-round victory increased the legitimacy of his government and gave him an opportunity to reorganize its administration, it was not necessarily a ringing endorsement of him or his record in office. Most Afghans wanted and expected much more than his government has been able to deliver. While considerable effort was made during 2004 to speed reconstruction and other foreign aid, progress in this area remains slow and has not involved enough Afghans at the local level. Nor have people been happy with the pace of disarming and reducing the power of local military commanders, who still have a disproportionate role in government.

EAT FIRST, THEN PRAY

But looking at their alternatives, particularly at the old and discredited mujahideen leaders who had embroiled the country in civil war, it was likely that most Afghans

who supported Karzai voted their hopes and not their fears. He has been given a mandate to work for changes and will be judged on his ability to bring them about. As an Afghan friend told me, "You Americans pray before beginning a meal, we Afghans pray after it is eaten." Thus, while most American politicians pile on benefits before an election and relax afterward, Karzai must do the opposite. He must use his vote of confidence to deliver more now that he has won the election.

Karzai has moved quickly to name a new cabinet. The new ministers are mostly technocrats drawn from all the country's ethnic groups. Some of the powerful warlords have been dismissed, including the Panjshiri defense minister, Mohammed Fahim, although the recent appointment of Ismail Khan as minister of power and water shows that the process is far from complete. On the other hand, since Karzai had earlier removed Khan from his power base as Herat's governor, this appointment may have been designed to attract the attention of other regional leaders thinking about their future. In a similar vein the current foreign minister, Abdullah Abdullah, kept his post despite his public support of Qanuni in the election. By keeping a least one key member of the Panjshiri faction within his government, Karzai was reaching out to the opposition. Qanuni himself refused to consider a ministerial post, announcing that he would head an opposition party in the upcoming parliamentary elections.

But new appointments in themselves will mean little without broader improvements. The central government needs to be overhauled to reduce corruption and favoritism, a legacy of the "division of spoils" and "balance of power" mentality that guided the appointments made by the provisional government after the Bonn Accord.

THE OPIUM WARS

Beyond the organization of government, Karzai will also need to eliminate the power of armed commanders in the provinces. If they are not removed it will be difficult to hold free and fair local parliamentary elections this year. These elections will be much more complex because of the number of participants involved and because they will potentially pit local rivals against one another. It is imperative that leaders whose power still derives from the barrel of a gun be removed or at least forced to compete for power in an electoral process. The international community has given too little support for this vital effort, and the national army and police forces are still not strong enough. While disarmament is a difficult task, it is not insurmountable because most Afghan commanders have direct control of only a small number of troops. The bulk of their subordinates' forces consist of men attached to their own commanders who themselves have no fixed loyalty to those above them.

Unfortunately, a dramatic increase in opium production has compounded the problem. Running a militia requires money, and the drug trade has given many militia leaders a new lease on life. Karzai has argued persuasively that the rise of the drug trade is a greater source of instability to the Afghan state than the Taliban. Tackling this problem is politically difficult since, in the absence of economic improvements, small farmers have grown increasingly dependent on poppy production.

Because the international community focuses primarily on policies of crop destruction, the burden of dealing with this problem falls on the segment of the population that has the fewest resources. Here the failure to bring about economic reconstruction in provincial areas has immediate political impact. A heavy-handed opium eradication program will alienate the population at a time when the government most needs its cooperation. And the more that opposition to the government is funded through illegal sources, the harder it will be to make reforms even with the support of the Afghan public.

Although the presidential election represented a positive step forward, Afghanistan is still a difficult place to run. Karzai certainly does not need to be reminded of this fact, but the international community does. The election results are all to the good, but they are a means to an end, not an end in themselves.

THOMAS J. BARFIELD *is chairman of the department of anthropology at Boston University.*

The Syrian Dilemma

THE RETREAT FROM LEBANON THREATENS THE VERY SURVIVAL OF THE BAATHIST REGIME.

David Hirst

Damascus

ONE ARAB NATION WITH AN ETERNAL MISSION: BAATH PARTY, SMASHER OF ARTIFICIAL FRONTIERS. Till not so long ago, this was the slogan emblazoned across a triumphal archway under which travelers passed at the Lebanese-Syrian frontier. It was a relic of that turbulent, post-independence era when revolutionary nationalist movements, bent on restoring the "Arab nation" to its former greatness, took power in various countries. No country was more central to this than Syria, the "beating heart" of Arabism, and no movement more than Syria's progeny, the Baath, or Renaissance, Party.

Since the 1950s Syria has embarked on countless, ultimately abortive unionist projects with other Arab states. None of them, oddly, involved the country that was closest to it—not at least until, in 1976, its army crossed the most "artificial" of its colonially drawn borders and, in what it portrayed as its pan-Arab duty, sought to rescue Lebanon from the civil war into which it had fallen. Thus began the overlordship, the far-reaching political, economic and institutional penetration of one Arab country by another, in which everyone—America, Israel, the Arabs—eventually acquiesced. Yet still the frontier post, odious symbol of Arab fragmentation, remained obstinately there; and eventually the decaying archway that supposedly heralded its disappearance disappeared itself, giving way to a new complex of immigration buildings and a duty-free emporium adorned not with unionist slogans but with ads for Dunkin' Donuts.

When, soon after the February assassination of former Lebanese prime minister Rafik Hariri, widely assumed to have been carried out by Syrian agents, I crossed that frontier, it had become more than just a standing reproach to Arabism; it was symbolic not merely of the Arabs' failure to unite but of the tearing asunder of the little degree to which they had. Normally teeming, it was almost deserted. Lebanese were uneasy about going to Syria. Syrians were positively fearful of going to Lebanon, where they have been insulted and assaulted, their residences attacked, a reported thirty of them murdered.

If assassinations sometimes accelerate history, Hariri's brutal, spectacular, but popularly unifying demise was surely one of them. At a stroke it unleashed, in a great and very public torrent, all the anti-Syrian sentiments that had been surreptitiously building over the years.

Ever-growing street demonstrations, unprecedented in modern Arab history, culminated in one on March 14 that drew perhaps a million people, a quarter of the population, to Beirut's Martyrs' Square. Not just the numbers were impressive; so was their composition. In this multi-confessional country, it was, if anything, a triumph over confessionalism. The people by and large stood in one trench, their Syrian-controlled rulers in another; that, not confessional antagonism, was now the fault line principally defining the course of events. True, one sect, the Shiites, was heavily underrepresented, and the Shiite resistance movement Hezbollah had earlier staged a huge—yet smaller, more regimented, essentially single-sect and tactically motivated—"pro-Syrian" rally of its own. No less true, however, the Sunni Muslim community now threw its full weight behind the hitherto mainly Christian and Druse opposition, the significance of that being that it was traditionally Sunnis, not Shiites, who chiefly stood for Lebanon's pan-Arab nationalist identity and looked to Syria to sustain them. But at bottom it was Lebanon's silent majority—of all classes, sects and stations—who had their say on March 14. And at bottom what they said was: Give us our independence, dignity and freedom back again.

For the Baathists it was surely the death throes of One Arab Nation. Here they were, its historic standard-bearers, being reviled and driven back across that "artificial" frontier from the one Arab state where they had had the means and opportunity, in their fashion, to implement it. But history may one day judge it to have marked the birth of something new. As Lebanese columnist Samir Kassir put it: "The Arab nationalist cause has shrunk into the single aim of getting rid of the regimes of terrorism and coups, and regaining the people's freedom as a prelude to the new Arab renaissance. It buries the lie that despotic systems can be the shield of nationalism. Beirut has become the 'beating heart' of a new Arab nationalism."

In its basic impulses, this was indeed a strictly Arab, and inter-Arab, affair. But where does it fit into the great debate about the degree to which America is contributing to the winds of change in the region? Certainly, at least, George W. Bush could rejoice at this timely convergence of "people power"—massive, authentic, homegrown—with his global crusade for "freedom and democracy." So could his Administration's neoconservative hawks, for whom, soulmates of Israel's Likud, the pan-Arab nationalism of the Baath is the very antithesis of Zionism and its inherent drive to keep the "Arab Nation" fragmented, weak and doomed, in the end, to make peace with Israel on Likudnik terms. The neocons have long targeted Syria as a prime candidate in their grand design for regime change throughout the region, an objective that Congress's latest "Lebanon and Syria Liberation" bill endorses in all but name. And compared with that other candidate for regime change, Iran, Syria is a temptingly "low-hanging fruit," as some in Washington put it, and probably harvestable by merely political, not military, means. No wonder Bush so smartly joined the Lebanese opposition in almost daily and peremptory demands for full and immediate withdrawal of Syrian troops and intelligence services.

This sudden and overwhelming confluence of the local and the international, the spontaneous and the long-envisaged, has shaken the Baathist regime, so much so that some in Damascus now feel that its end is only a matter of time. "Total defeat in Lebanon," said leading dissident Michel Kilo, "will mean defeat at home."

To be sure, Syria is not defeated yet. Even as its forces redeploy and some withdraw altogether, it is still sustaining Lebanon's president, Emile Lahoud, and his puppet administration without them. So far Hezbollah, now agonizingly torn between its pan-Arab, jihadist imperatives and increasingly irreconcilable Lebanese ones, remains potently, if very uncomfortably, at Syria's service. And soon after Syrian secret police departed Beirut, car bombs began to go off in Christian neighborhoods. Were these, Lebanese asked, Syria's opening shots in the manufacture of a scenario long hinted at? Namely, that if the world pushes Syria to leave Lebanon, the world will soon come begging it to return as Lebanon, sliding back into civil war, begins to look like another Iraq, another paradise for militants and terrorists of all kinds.

That remains to be seen. But even without such desperate expedients, Syria's extraordinary resolve to keep its faltering grip on Lebanon and the brutally coercive methods it has used are already evidence enough of how vitally important it deems Lebanon to be. "Along with the command economy and the apparatus of repression," said Louai Hussein, a Syrian commentator, "control of Lebanon was one of three main pillars on which [the late] President Hafez al-Assad built his power and prestige." In fact, Syria's rulers always instinctively strive for greater regional influence than the resources of Syria alone can command. They exploit their regional "cards" in a continuous quest to advance their interests—which now boil down to securing their mere survival in the new, US-dominated Middle East order. Iraq is such a card, hence the repeated recriminations over what Syria is, or perhaps isn't, doing to help the anti-American insurgency there. Palestine is another, hence persistent American charges that Syria is "unhelpful" to the peace

process, or Israeli ones that Palestinian suicide bombers get their orders from Damascus.

In a long-eroding regional hand, Lebanon, and its complete and exclusive hegemony there, is Syria's only remaining trump. It is Syria's front line, its arena of proxy war, its substitute for the military confrontation with Israel that—given its vast military inferiority—it could never risk directly from its own territory. Hezbollah is the formidable instrument of this proxy war; quiescent at the moment, it is ready and waiting to offer what, in some great showdown, Iran or Syria might require of it: its jihadist zeal, its guerrilla prowess and, according to Israel, the thousands of upgraded long-range missiles it could rain down on Israeli cities.

Economically, Lebanon is Syria's milch cow, such a cornucopia of extortion, racketeering and diversion of public funds that the distribution of the spoils—authoritatively put at about a billion dollars a year—among the Baathist oligarchy is said to be a factor in the stability of the regime. Lebanon is also the place where up to a million ordinary Syrians, facing at least 20 percent (and rising) unemployment in their own country, find illicit, low-paid work, or did so until, after Hariri's murder, they started fleeing in sizable, if unknown, numbers.

Perhaps even more dangerous to Syrian Baathism than the loss of this priceless Lebanese asset would be the potential domino effect inside Syria of the Lebanese "people power" that chiefly brought it about. First there were elections in Iraq and Palestine, which, however flawed, showed Syrians the shaming fact that Arabs enjoy more electoral choice if they are occupied than if they are sovereign. Then came this huge, unscheduled outbreak of popular self-assertion in a country where an Arab sister-state, not an Israeli or American occupier, is in charge.

The Baathist order has lost all legitimacy, sunk as it is in the most cancerous corruption and abuse of law and human rights.

And could any Syrians fail to grasp that what the Lebanese were rising up against was not (despite some ugly, chauvinist side effects) Syria itself but the extension, on Lebanese soil, of what they themselves more drastically endure at home? That is to say, the oppressions of a once-revolutionary new order that has long since betrayed its three great founding principles, pan-Arab unity, freedom and socialism. Like the now-defunct Soviet-style, single-party "people's republics" on which it was largely modeled, this Baathist order has lost all true legitimacy, sunk as it is in the most cancerous corruption, minority sectarian rule, intellectual and technical backwardness, bureaucratic ossification, abuse of law and human rights, imperviousness to dissent or criticism.

Syrians also know that those who brought the Syrian presence in Lebanon to its disastrous pass are the same people—the so-called "old guard," shadowy power centers in the army and intelligence services—who have brought fear to their own lives these past forty years, as well as blocked all reform and democratization. It is no surprise that Syria's dissident intelligentsia identify entirely with Lebanon's democratic uprising; call, like it, for full Syrian withdrawal; and use the international publicity the uprising has brought to dramatize their own campaign for human rights and civil liberties, a campaign whose most visible form is small snap demonstrations, outside courts or prisons, whenever opportunity arises.

But the Syrians aren't going to rise up like the Lebanese—not yet, anyway. Long repressed, they don't have the organized opposition or the strong residue of democratic traditions that the "Syrianization" of Lebanon never snuffed out. And the barrier of fear, always much higher than in Lebanon, remains strongly in place. "Hariri's murder," said a dissident who had no doubt about its authorship, "was a savage warning to us as much as to the Lebanese." Yet weak though it may be, and still confined largely to the intelligentsia, small political groups and human rights activists, the opposition is certainly gaining ground on a regime that is in little better shape itself, rattled and insecure as it is behind the despot's characteristic facade of lofty self-confidence, loyalist street demonstrations and the portrayal of obvious reverses as great achievements in the onward people's march.

Syrians find it hard to imagine that with Lebanon and all the domestic, regional and international pressures it has unleashed, President Bashar al-Assad doesn't realize he must do something—and do it decisively—to guard his regime, or even his country, against the gathering perils. But in this aptly dubbed "dictatorship without a dictator," has he the means, or the will? Whereas his late father, Hafez, was absolute master of what he had built, Bashar often seems more like its prisoner, forever torn between two alternative courses, reform or reaction, liberalization or repression, reaching out to the people as his source of authority or falling back on his old guard. Thus, on coming to power a few years ago, he initiated the "Damascus Spring," only to rein it in when, timid though it was, he thought it was going too far.

Such alternatives now confront him more starkly than ever. He can either make a clean break with Lebanon, purge the old guard, open wide the doors to domestic reform and appease America and the world; or he can cling to Lebanon by any means, bow to the old guard, revert to full-scale repression and defy the world. What he will probably try to do is essentially what he always has done: make no clear choice, temporize, hope that something turns up. But with his authority steadily fraying, both within his apparatus and in the country at large, how long can it be before someone, somewhere, decides it is time to rescue the regime—or overthrow it? These are the kinds of questions now being asked by Syrians, whose yearning for change is tempered only by fear of the way—liable to be tumultuous at best, civil war at worst—it might come about, and what would come after.

Not the least of the great imponderables is what America's role and objectives might be. The homily that a typical liberal, secular-modernist dissident might address to President Bush would go something like this:

> In principle we like your "freedom and democracy" and think that what you've been doing in the Middle East has, by accident or design, given a push in that direction. But the bad things your country does still so far outweigh the potentially good that the last thing reformists like us need is to be identified with you, especially if you or Israel physically attack us. For we know that whatever you do it is Israel's wishes, not ours, that concern you. And we fear that you really do take seriously your Israeli friend Natan Sharansky—the right-wing fanatic who inspires your speeches—and his preposterous theory that only when Arabs are democratic will they be ready for peace with Israel. No, we want democracy because it will serve our national interest far better than a despotic regime whose nationalism is just a cover to suppress democracy. And so long as your policies remain what they are, our national interest will be to oppose them. But in any case, if one day we do have free elections here, it won't be the likes of us who win them but—thanks largely to you—the Islamists.

David Hirst, a longtime Middle East correspondent for the Guardian, *is the author of* The Gun and the Olive Branch: The Roots of Violence in the Middle East (*Nation Books*).

Africa's Democratization:

A Work in Progress

> "Real, sustained efforts are being made across the continent to deepen democracy and reap the benefits of accountable governance. The success of these efforts has been mixed, but it is far too soon to write them off as failures."

JENNIFER WIDNER

In February 2005, Nigerian commentator Mobolaji Sanusi referred to political developments in Africa as "demo-razy." The target of his concern was the tendency of many of the continent's political leaders to set aside demands for serious competition and accountability and exploit elections by rigging ballots, changing constitutions to allow for longer terms, and ignoring other rules when convenient. Over the past year several donor countries have expressed similar anxiety about the behavior of political elites in Nigeria, Kenya, Zimbabwe, Togo, Uganda, and many other countries, often taking to task reformers they had previously championed.

This observation raises at least two important questions. The first is whether democratization has stalled in Africa, or whether the patterns to which Sanusi draws our attention simply reflect the last stands of the old-style authoritarian politicians, part and parcel of struggles to consolidate democratic reform. Certainly systematic regional backsliding is not all that evident, although movement toward more open systems has not accelerated either. Roughly 46 percent of sub-Saharan countries rank "four" or below on the Freedom House political liberties index (in which "one" indicates the highest standards of political rights and "seven" human rights standards on a par with North Korea). Almost 19 percent of sub-Saharan countries rank "one" or "two" on this index. And during the past year four countries received better scores than they had the previous year, compared to three that received worse scores, including Zimbabwe. Since 2000, the overall ratings have changed relatively little. In 52 per-

cent of countries, there was no change in the existing level of respect for political rights, while there was some improvement in 26 percent and some decline in 22 percent.

The second question is whether those governments that respect rights more now than in the past and hold regular elections perform any better than their predecessors. Much of the internal and external pressure for stronger observance of human rights and regular, competitive elections had its roots in the quest for accountable government. People in the street joined policy makers in reasoning that if citizens could vote against poorly performing politicians at the ballot box, political elites would have a stronger incentive to attend to the needs of ordinary citizens, choose policies that would generate economic growth, and reduce corruption. Greater respect for individual rights would also make it possible for information to circulate more freely, making corruption riskier, encouraging deeper discussion of policy issues, and drawing attention to problems, such as impending hunger, that left unaddressed would cause death or injury. Systematic information about state performance is hard to come by, but we do know that perceptions of corruption remain high in countries that have held elections and rank favorably on the Freedom House index of political and civil rights. Do these data suggest that our theories about the relationship between democracy and accountability are wrong, or does it just take time for the anticipated effects to materialize?

If the evidence suggests it is premature to answer these questions definitively, that in itself should give no cause for despair. Real, sustained efforts are being made across the continent to deepen democracy and reap the benefits of accountable governance. The success of these efforts has been mixed, but it is far too soon to write them off as failures. And, importantly, public participation in them is growing.

TROUBLED ELECTIONS

Pushing an authoritarian government to engage in political liberalization and hold multiparty elections is a difficult challenge. It requires not only an underlying sense of grievance or anger, but also people who are willing to assume the costs and risks of organizing others and relative weakness in the incumbents' ability to suppress or divert pressure for reform. Yet the tasks of consolidation, of building stable democracy, are if anything even greater, and although they may ease with time, they never go away. Most countries in sub-Saharan Africa not immersed in civil war have entered a difficult period: weak reform movements seek to maintain pressure on governments to adhere to promises, and responsive leaders are learning that to behave accountably they have to improve government capacity.

Sub-Saharan Africa is in an ambiguous transition period, with signs of hope as well as grounds for pessimism.

In any part of the world, incumbents and their entourages will try to hold on to power. How far they go toward that end is partly a function of popular and elite consensus on the rules and the bounds of appropriate behavior; it is also a function of the ability of civic groups and opposition parties to police and halt excesses. In sub-Saharan Africa, as in many places, we continue to observe problems with political norms, but overt violence at the polls as an intimidation tactic and tampering with ballot boxes are less common than they used to be. Africa, like other regions, suffers ever more sophisticated efforts to tip the electoral balance toward incumbents or toward the head of state's handpicked successor as consensus gels on norms against some of the most outrageous tactics.

Zimbabwe's elections at the end of March 2005 anchored the negative end of the spectrum during the past year. Although the balloting was less overtly violent than at other times in the country's history, President Robert Mugabe sought to hold on to power by various means. He gained the right to appoint people of his choice to fill 30 of the 150 seats in the parliament. The government designed electoral boundaries to generate more safe seats for the ruling party. In the two years prior to the election it arrested opposition leaders on several occasions and introduced a bill to ban human rights groups and other NGOs that received any foreign funding. Youth groups attached to the ruling party engaged in political harassment, and there were reports of threats to deny food to those who voted against Mugabe. Journalists worked under increasingly tight restrictions after the 2002 Access to Information and Protection of Privacy Act gave the minister of information broad powers to confer or deny licenses for reporters and publications. When the elections were held, an estimated 10 percent of registered voters were turned away without being able to cast ballots.

Opposition groups inside and outside Zimbabwe tried to bring pressure on Mugabe to relent and to abide by the norms that govern free and fair elections. The country developed a spirited underground opposition, Zvankwana-Sokwanele ("Enough!"), which signaled its presence in different parts of the country with the spray-painted Bob Marley slogan, "Get up, stand up!" Demonstrations continued— although, during the lead-up to the vote, the police cracked down, breaking up most peaceful protests. The archbishop of Bulawayo, Pius Ncube, observed that the contest had been rigged before it started and urged Zimbabweans to follow the example of protesters in Ukraine. Mugabe responded to the archbishop by calling him a "halfwit."

The reaction of neighbors was mixed. Prior to the election an attempted visit by a South African labor organization elicited a demoralizing denunciation from Mugabe. Civic groups assembled on the South Africa-Zimbabwe border in mid-March for a demonstration, concert, and overnight vigil in support of free and fair elections. The aim was to convince a regional group, the Southern African Development Community (SADCC), to bring pressure on the Mugabe government for an open process. Later in March, another delegation from South Africa, incorporating the Landless Peoples' Movement, Jubilee South Africa, and other groups, visited and concluded that elections would not be fair, citing a lack of voter education, inadequate poll monitoring, and a nonpublic voters' roll. The governments of other countries in the region were less helpful to the cause of democracy. For example, an SADCC observer team pronounced the elections free and fair despite the obvious problems in the run-up to voting day and the fact that many voters were denied access to the polls.

Elections went off more smoothly in several other countries in the past year, suggesting that know-how, norms, and accountability mechanisms may have improved in a broader regional context. In Ghana, John Kufuor won a second term as president in December 2004 with 53 percent of the vote and a turnout of 83 percent. Although the opposition objected to news of a coup plot—news it believed was designed to swing popular opinion in favor of the incumbent—the balloting was generally considered free and fair. Coup-prone Niger also conducted relatively free and fair elections in December 2004, returning President Mamadou Tanja to power for a second term. Ballots were helicoptered to insecure areas in the north, where security held for the duration of the election. Turnout was fairly low throughout the country, but the results were close in a hotly contested race. Coup master Francois Bozize ran for the presidency in the Central African Republic in an election that went off smoothly, although Bozize's candidacy met with disapproval in many quarters. In Cameroon's October 2004 elections, the picture was less happy than in Ghana and Niger, but fragmentation of the opposition was a greater cause for concern than outright manipulation of the vote that returned long-time president Paul Biya to office.

THE LIMITS OF TERM LIMITS

Elsewhere political leaders have continued to employ another favored tactic for preserving their rule: trying to overturn constitutional limitations on terms of office. Earlier, leaders in Zambia, Malawi, Zimbabwe, Guinea, Togo, and Namibia had attempted to secure constitutional amendments to permit themselves to run for third terms. In Zambia and Malawi, public protest quashed these efforts, while in Zimbabwe, Guinea, Togo, and Namibia the bids to lift term limits succeeded.

This past year saw several more efforts of this type, including some from unexpected quarters. In April 2004, Namibia's president, Sam Nujoma, sought to amend the constitution yet again, this time to ensure that he could run for a fourth term. (He retired instead, ceding the presidency this spring to Hifikepunye Pohamba, who was elected in November 2004.) In May 2004, the Burkina Faso legislature voted to amend the constitution to allow the head of state, Blaise Compaore, to remain in office for another term. Opposition parties attracted several thousand people to a protest demonstration in Ouagadougou, the capital, but some of the minor opposition parties allied with the government to pass the rule change. Gabon's parliament in July passed amendments allowing President Omar Bongo to run for office indefinitely. Bongo had already ruled for 36 years. In October 2004, Uganda's Yoweri Museveni, architect of the policies that helped move his country from civil war to donor darling, pushed to remove constitutional limits on the number of terms a president may serve. Although his intentions were not clear, this gesture would allow him to stand for election in 2006. Donor countries, international NGOs such as Transparency International, and civic groups within Uganda objected to the proposed constitutional alterations, and Museveni may still change his mind.

In the Ivory Coast, too, the ruling regime has tried to manipulate the constitution to secure its grip on power. In 2000, political leaders engineered the notorious Article 35, which restricted eligibility to run for president to candidates whose parents could both prove Ivorian citizenship. The constitutional amendment was a transparent ploy to eliminate Alassane Ouattara, the strongest opposition candidate, from the election. It also sent a signal that citizens of mixed parentage, especially northerners, would no longer have the same rights as others, and sub-

sequent policy decisions amplified that signal. Civil war then split a country long known as a bulwark of stability in the region. Under strong international pressure, 2004 saw the Ivorian government take obviously reluctant steps to repeal the measure, a requirement for peace. In December 2004 it began the process of amending the constitution once again, opening up eligibility and removing the restrictions imposed in 2000. In another maladroit move, however, the government then insisted that the measure be subject to a national referendum. Although this step was consistent with standard amendment procedures in many constitutions, it was difficult to argue that the existing charter had any power in the midst of civil war, and it was obviously going to be difficult to carry out a referendum process in insecure parts of the territory. A more measured approach would have been to accept the change as a requirement of peace negotiations and call for subsequent review of the constitution after the war's end.

Coups have become a less common means of securing power than in the past, thanks to concerted international efforts to cut off aid to countries whose leaders take office with the force of guns, and thanks to increasing domestic opposition. Togo tested this proposition on the death of President Gnassingbé Eyadéma on February 5, 2005. The country's military sought to install Faure Gnassingbé, Eyadéma's son, as president and forced a retroactive constitutional amendment through the legislature to make the move legal. Opposition parties immediately called for a two-day "stay at home" protest. A week later, demonstrators in Lome, Togo's capital, rejected the nonconstitutional transfer of power, and security forces killed three in the crowd. As Togolese protesters kept up the pressure domestically, people in neighboring countries took to the streets in support, and West African governments moved quickly to impose sanctions, despite a government promise to hold elections within 60 days. The sanctions set by the Economic Community of West African States included a recall of ambassadors, a travel ban on the Togolese leadership, and an arms embargo intended to help stem the importation of arms by the ruling party, which had already started to distribute weapons to its paramilitary supporters. The African Union followed suit. While the final outcome remains unclear, the strong regional and international response holds out the hope of curbing this attempt to evade the demands of democracy and accountability.

This quick inventory suggests that, although egregious efforts to circumscribe popular participation in politics are still occasionally evident and are likely to continue sporadically for some time to come, there are also growing signs of active domestic constituencies for accountable government. Demonstrations by civic groups are rarely if ever as large as those in Ukraine, Lebanon, and other countries outside Africa where citizens have tried to resist the usurpation of space for public dialogue, but they continue even in countries as traditionally inhospitable to such action as Togo. In some instances, such as the protests at the border of Zimbabwe, organizers have become increasingly creative in finding ways to communicate in the midst of oppression. With respect to democracy, the glass is half full.

CONSTITUTIONAL CONVERSATIONS

At least as interesting as electoral conduct as an indicator of changing norms and patterns of behavior is the apparent belief among churchgoers, lawyers, politicians, students, and many ordinary Africans that diverse national communities must hold continuing conversations about the allocation of power and the fundamental rules of the game that will structure political life in the future. Since the mid-1970s, African countries have engaged in constitution-drafting exercises more than 90 times. Although some of these efforts were smokescreens for power grabs, many more were serious, participatory ventures. What is different about the tone now, compared to earlier years, is that the process of reform is increasingly negotiated with civic groups and churches, instead of directed from the top, and nongovernmental organizations as well as opposition parties have persisted in monitoring deliberations and in lobbying for inclusion of important provisions.

Accountable government is not something achieved by the "simple" creation of a competitive electoral process.

The past year saw protests designed to move constitution-making processes forward in several countries. The most notable was popular action to try to shake up stalled deliberations in Kenya. In 1997, Kenyan opposition groups, religious leaders, and civic organizations launched a drive for constitutional change under the umbrella of the National Convention Executive Council. Pro-reform demonstrations in July 1997 were violently repressed and nine people died, while many others were injured. In 1998, the ruling party and opposition groups in the legislature hammered out a deal that set up a constitutional review commission and elaborated a five-part process. District forums and other bodies would nominate representatives to a national consultative conference. The government would then create a 26- or 27-member constitutional commission, with representatives of political parties, civic groups, religious communities, and women. The commission would sponsor hearings across the country and gather information. The 629-member national conference would weigh this information, and its recommendations would then go to a parliamentary committee for further review and presentation to the legislature, the body empowered by the existing constitution to adopt and ratify reforms.

Church-led protests kept the issue in the public eye through 1999, especially after the president, Daniel arap Moi, expressed his desire to keep the process within the national assembly—a procedure consistent with past practice but increasingly out of step with the more participatory processes employed across the globe. At one point civic and religious groups, which organized under the banner of Ufungamano, held their own conference and developed a draft. This coalition eventually won permission to nominate members to the constitutional commission. Its work informed deliberations at several stages. Other watchdog groups formed and continued to put pressure on the government to act through February and March of this year.

The process slowed at several points, partly as a result of internal disputes and misaligned incentives, and partly because of meddling by the Moi government. Mwai Kibaki's election as president in 2002 raised hopes of progress, but reform proposals remained mired in the parliament. Some of the disputes were substantive—for example, over the degree to which the president should share power with a prime minister. Others had roots in personal animosities and bids for influence. The president, then ill, had little inclination or ability to move the process forward. Rioting broke out in July 2004 as people grew increasingly impatient. Foreign diplomats and UN Secretary General Kofi Annan weighed in, reminding Kibaki to deliver on his campaign pledge to reform the constitution.

Popular action to push for constitutional dialogues has occurred in other countries as well. In Nigeria, President Olusegon Obasanjo's government has initiated a constitutional review process, arguing that the existing 1999 constitution was tainted by the influence of the military. Popular pressure helped put constitutional review on the agenda. The Electoral Reform Network, a coalition of 85 NGOs, has lobbied for renewed attention to the subject. The power to revise, adopt, and ratify a constitution lies with the legislature, but, like Kenya, Nigeria has broadened participation. The president created a National Political Reform Conference with 400 members, including delegates from each of the states, although preparation of draft language is in the hands of a committee of seven distinguished Nigerians.

In Swaziland, protesters in March 2005 expressed dissatisfaction with a protracted constitutional reform process and demanded the unbanning of political parties, an end to arbitrary arrest, and a number of other changes. The government had launched a reform initiative in 1996 but had not brought the deliberations to a conclusion. Mauritius, too, has revisited the design of its electoral system. In Zambia, NGOs have called for a new constitution. Although the Zambian government professes to be open to this possibility, it de-registered one of the groups promoting revision, the Southern African Center for Constructive Resolution of Disputes. Mozambique's Frelimo party has embarked on an ad hoc effort to recommend changes in that country's constitution, and September 2004 saw renewed discussion of fundamental charter changes in Angola.

DANGERS AHEAD

Deep political change does not happen overnight. Sub-Saharan Africa is in an ambiguous transition period, with signs of hope as well as grounds for pessimism. Currently there are two main dangers.

The first ground for caution is the possibility that donor aid agencies, eager to show they have used their money to leverage change, will decide they have failed, prematurely, and pack their bags. That would be disastrous. Solid domestic and regional pressure groups are beginning to gain a toehold and initiate action in several countries. Usually they are still too weak to bring misbehaving officials to heel. Concerted regional and donor action is important for forcing leaders to respond to domestic demands. The case of Togo (though the story there is not yet over) shows what concerted internal and external pressure can achieve. The case of Zimbabwe shows what happens when regional powers avert their eyes.

It would be unfortunate if a short-term quest for energy returned the world to the days in which Africa was nothing more than a source of primary commodities.

Accountable government is not something achieved by the "simple" creation of a competitive electoral process. It is the joint effect of several conditions, not all of which are yet present. Voters cannot send clear signals by voting out the bad guys unless they care about politics, know who is responsible for policies or practices they dislike, and can sort out performance from the many other attributes of a candidate they may find attractive or unattractive. Eliminating coercion in the polling process and giving candidates a level playing field are both important, but they are not all that matters. For most ordinary people in most countries, it is hard to acquire all the information necessary to cast a ballot thoughtfully. Instead, people tend to take cues from members of the "interested public" whom they respect—public intellectuals they trust or like, or organizations that represent people like them, kin groups for example. These people pay closer attention to politics and policy and their positions offer cues to those with less time, motivation, or capacity to make judgments. A challenge for many African countries is to create an environment in which nonsectarian groups that perform these functions can acquire information, communicate freely, and establish ties with rural majorities. Young political elites in some countries already play these roles, but in other places this group is small and constrained.

It is relatively easy for voters to punish leaders for corruption. Corruption is usually observable in the spending behavior of the entourage around a politician. By contrast, the responsibility of particular leaders for popular or unpopular policies or for good or bad management often remains unclear, complicating the quest for accountability. In a globalizing world, all countries are exposed to shocks and fluctuations that come from outside their borders, making it difficult to identify whether a policy failed or whether conditions changed in ways that would make it hard for any leader to navigate effectively.

At the same time, Africa faces a number of more unique problems in identifying responsibility. One comes from the perception that donors call the shots on development strategy. In some measure, this perception is valid. Governments are often beholden to carry out policies that bilateral aid agencies or the World Bank and IMF request, in return for grants and loans, even though many leaders have felt little compunction about abandoning such agreements. Changes in aid policy that give more ownership to governments for designing strategies should help alleviate this problem. Another challenge in holding leaders accountable is the increasing favor shown to semi-presidential or premier-presidential systems, such as in Mali, Madagascar, and Mozambique, which divide authority between prime ministers and presidents. Often it is not evident in these systems where the locus of authority lies. The desire to check the power of a strong president is understandable, but semi-presidentialism may not yield the results anticipated because of the difficulty voters have in locating responsibility for problems and improvements.

Finally, accountability is partly a function of the government's ability to respond. Since the 1970s, donors have often neglected capacity building, believing that money spent on training public officials, building facilities, or working closely with ministries is money in the hands of the corrupt or the abusive. Although the reluctance to work directly with governments is understandable, it is increasingly inappropriate in countries where civic groups can criticize openly and where the ability to monitor the use of resources has improved. A government cannot respond effectively if it lacks personnel able to do so or if it cannot manage its fuel supply well enough to send judges or other officials throughout the country.

RESISTING TEMPTATION

Strengthening the causal relationship between democracy and accountability will depend on external events too—and here is where the second danger arises. The next decade is likely to see competition between China and the West to buy Africa's oil. Already we have seen signs in Sudan that competition for oil has made some countries, including China, unwilling to press a government to behave responsibly toward its citizens. Willingness to keep quiet about abuses while paying cash on the barrelhead gives some purchasers an advantage over others. Ultimately this willingness to trade without regard for who gets the money or how it is used may undermine international and regional coalitions for good government and help generate high levels of conflict on the African continent. It would be unfortunate if a short-term quest for energy returned the world to the days in which Africa was nothing more than a source of primary commodities. Africans and their friends abroad should resist the temptation to trade recent tentative advances in pursuit of better governance for money and a renewal of the exploitative relationships of the past.

JENNIFER WIDNER is a professor of politics and international affairs at Princeton University and author of *Building the Rule of Law* (Norton, 2001).

The New South Africa, a Decade Later

ANTOINETTE HANDLEY

"What has changed has been unexpected: the politics of the country have stabilized with astonishing speed . . ., [the ANC government] has implemented a conservative macroeconomic policy; and an epidemic has emerged as the single greatest threat to stability and prosperity."

Any traveller returning from South Africa since its first democratic elections in April 1994 is inevitably asked "What has changed?" And the paradox is that everything has changed and nothing has changed; . . . stability has depended on the illusion among whites that nothing has really changed and among blacks that everything has changed."—*Shula Marks*

The paradox at the heart of the South African miracle has been sustained by an illusion. The real test will come, the historian Shula Marks suggests, when whites realize that everything has changed and blacks realize that nothing has changed.

Few anticipated that white South Africans would voluntarily hand over political power to the African National Congress (ANC), the liberation movement that they for decades had vilified as "terrorists" and "communists," or that the leaders of this movement would prove remarkably moderate custodians of the country's economy. And yet for the most part that is what has occurred. Ten years ago the ANC won the country's first genuinely democratic elections and has governed since. This is a good time to assess what is new about the new South Africa.

What has changed has been unexpected: the politics of the country have stabilized with astonishing speed, moving from violent conflict to quotidian electoral politics in a few short years; the ANC government, despite its radical heritage and its electoral alliance with the South African Communist Party, has implemented a conservative macroeconomic policy; and an epidemic has emerged as the single greatest threat to stability and prosperity.

SINCE THE MORNING AFTER

When he came to office in 1994, President Nelson Mandela of the ANC faced a daunting range of challenges in three broad areas: stabilizing the political and social order; addressing issues of socioeconomic justice while maintaining the fiscal health of the country; and restoring growth, employment, and productivity to a long moribund economy.

A decade into the politics of the new South Africa, it is easy to forget how profoundly uncertain the political outlook was leading up to the first nationwide democratic elections in April 1994. Political violence had peaked in the early 1990s, averaging more than 3,000 deaths a year as political negotiations progressed and the governing National Party, the ANC, and other parties moved toward agreement on an interim constitution and elections. The threat by Mangosuthu Buthelezi, leader of the Inkatha Freedom Party, to boycott the country's first national democratic elections further raised political tensions and seemed sure to blight the birth of the new South Africa.

Only a week before the polling, Buthelezi decided to participate. The party's name, symbol, and its leader's photograph were hastily glued on to millions of ballots. Despite disorganization at many polling stations and threats of further violence, South Africans from every formerly defined racial category and all classes cast their ballots on April 26, 1994. The ANC won overwhelmingly, yet was gracious in its victory. Mandela headed up a "government of national unity" that initially drew potential dissidents into governing. The head of the National Party, F. W. De Klerk, was made vice president and Buthelezi was given a seat in the cabinet. With a clear eye to reassuring whites who continued to dominate the economy, the National Party (which later, in the spirit of the new South Africa, renamed itself the New National Party) was granted control of portfolios including finance, mineral and energy affairs, and agriculture.

Black South Africans woke up to a new president, to the decisive defeat of apartheid, and to a democracy that many had thought they would never see. White South Africans woke up in the same houses and drove their children to the same schools before heading to their usual jobs, while their homes were cleaned by the same maids. This seemingly banal reality—over a million women, overwhelmingly African, work in white households as maids—captures the cold fact that unequal access to jobs and education and consequent extreme levels of income inequality were perhaps the defining characteristic of apartheid. (A telling statistic: in 1994, while white South Africans could expect to live into their 70s, the average life expectancy for black South Africans was around 55 years.)

Since its founding in 1910, South Africa had been built on racial discrimination. After 1994, South Africans set about trying to bring some normalcy to the country's near pathological politics. De Klerk and Mandela quarreled in public, but for 14 months the government of national unity provided reassurance to the formerly powerful white minority, and political violence dropped to one-fourth its previous level. (The National Party, citing its inability to meaningfully shape government policy, withdrew from the national unity government in June 1996.)

> Black South Africans, who make up 75 percent of the population, are generally considered to own between 2 percent and 7 percent of overall market capitalization in terms of stock shares.

However, not all of South Africa's problems could be resolved through political magnanimity. After 1994 the security of the average South African was more routinely threatened by violent crime than by political violence. A surge in reported crime from the early 1990s continued after 1994. Whether this surge represented a rise in actual incidents of crime or just improved reporting rates was unclear.

The new ANC government poured significant resources into policing and embarked on much needed reforms that were intended to improve the quality of policing and detective work and to increase public trust in the police. These efforts may have produced some success. After topping out in the late 1990s, overall crime rates have stabilized and violent crimes may even have declined. Still, crime remains at horrifyingly high levels: there are, for example, 45 murders per 100,000 people, a figure almost 10 times higher than the US average. Most South Africans (white and Indian South Africans in particular) continue to feel less safe now than they did when crime was at its highest.

Trust was crucial to social and political order in the new South Africa, but trust could not be built until the country as a whole began to come to terms with its past. In this respect, the Truth and Reconciliation Commission was central to the emergence of a new, more stable South African polity and society. The TRC made important compromises to advance political reconciliation. It chose to focus only on "gross human rights abuses" rather than examining the social, structural, and institutional foundations of the previous system, and it limited its purview to the period between 1960 and 1994. The commission recognized that an indefinitely drawn out series of hearings would place undue strain on an already fragile polity. Especially important was the decision effectively to eschew the pursuit of justice (that is, prosecution of offenders) in return for as full a telling of the "truth" as possible. The TRC relied on the testimony of thousands of victims and perpetrators alike, using the promise of amnesty to reward perpetrators for "full disclosure" of their misdeeds. This testimony, beamed by radio and television daily and directly into millions of living rooms across the country, represents the TRC's most significant achievement. It laid the groundwork for a common South African history that black and white South Africans might one day understand in the same way.

POLITICS AS USUAL?

Electoral coalitions that have emerged offer perhaps the most telling evidence of South Africa's new political stability. Many of today's political parties are the heirs of political forces that, prior to 1994, engaged in deadly military conflict with each other. Since then parties have engaged in all manner of extraordinary alliances—between Afrikaner and Zulu nationalists, between liberals and conservatives, and, most extraordinary, between the former liberation movement and its former oppressors—at local, regional, and national levels. In this respect, South Africa is well on the way to a reassuringly tedious brand of politics as usual.

The new constitution, adopted in 1996, is an impressive document that has greatly assisted in increasing social and political order. Alongside a comprehensive bill of rights (including social and economic rights), it established proportional representation in the legislature, a relatively centralized government with limited powers devolved to regional and local levels, and an independent judiciary together with a range of other independent agencies (such as a Human Rights Commission) to ensure government responsiveness. Given the extent of the ANC's electoral dominance (in 1999, it won two-thirds of the seats in the national legislature), such institutionalized protections may be vital. South Africa's opposition parties have not for the most part succeeded in broadening their support outside of narrow racially or regionally defined constituencies. Of course, the dominance of a single party is not new in South Africa; in the apartheid era the National Party controlled the country's white legislature for 46 years, hardly a reassuring precedent.

It is striking how quickly South Africa has achieved political and social stability. But that stability is threatened by the growing disorder in neighboring Zimbabwe. The deterioration of food security, political order, and economic growth in that country has profound negative repercussions for South Africa, given the highly integrated nature of the regional economy. (Zimbabwe is South Africa's largest trading partner in Africa, and has consistently ranked in the top 10 of South Africa's international trading partners. South Africa's exports in Zimbabwe also comprise a large share of high-value added and manufactured goods, precisely the kind of exports the country would like to see grow.)

South Africa publicly has yet to demonstrate any decisive leadership with respect to Zimbabwe. This public reticence can be traced to the new South Africa's first rather naïve and heavy-handed attempts to raise issues of human rights on the continent, which were resented by many African states accustomed to a style of diplomacy that prioritized solidarity in public and pressure only behind the scenes. When their early efforts served only to offend other leaders and isolate South Africa on the continent, South Africa's diplomats switched to a brand of foreign policy-making that addressed human rights issues less assertively.

South Africa has attempted to play a diplomatic role in conflict resolution elsewhere on the continent, most notably in the Democratic Republic of the Congo. But solutions to such multifaceted conflicts may be beyond the resources of a single country. The new South Africa is, after all, a mid-sized power with all the ambiguities and ambivalence that this entails for its international role, and it often finds itself regarded with some coolness on its home continent.

The country's leadership in international economic forums, such as the World Trade Organization, the Non-Aligned Movement, and the United Nations Conference on Trade and Development, has been more nuanced and assertive. South Africa has begun to demonstrate some ability to navigate the international political economy, which can be seen, for example, in the successful marketing of the New Economic Partnership for African Development (NE-PAD) to international donors. The real threat to South Africa's prospects has come from a different source.

THE AIDS CRISIS

The greatest challenge facing South Africa in the past decade, as well as the source of the ANC's most important failing, has been the HIV/AIDS epidemic. According to the country's Department of Health, less than 1 percent of the population was infected with HIV in 1990. By the end of the decade that figure had jumped to over 25 percent, one of the fastest growing infection rates in the world. The epidemic is now considered to have reached its "mature" phase: infection rates are slowing, and the next phase—the onset of full-blown AIDS, increased mortality rates, and the creation of large numbers of AIDS orphans—has begun. AIDS is expected to slash the average life expectancy of black South Africans to 40 years, undoing the impact of

other improvements in health and welfare. HIV/AIDS is now the largest single cause of death in the country.

The ANC's failure to respond promptly and adequately to this threat has undoubtedly exacerbated the health crisis and weakened the government's international standing as well as public trust in the government. First as Mandela's deputy and later as the country's president, Thabo Mbeki reacted first with inaction, then denial, then reluctant engagement. His questioning of the link between HIV and the development of AIDS, a view he has since downplayed, and his defensive and hostile interactions with nongovernmental organizations and activists on the AIDS issue have undermined the government's response at every level.

The inadequate response may have resulted from an inability on the part of leading figures in the ANC to comprehend the sheer scale of the problem posed by HIV/AIDS, but this only explains the initial lack of a response. As the dimensions of the crisis became clearer, the epidemic and the question of how to respond to it became highly politicized. The ANC had a long tradition of regarding the state as central to any systematic attempt to combat poverty in South Africa. After decades of struggle against apartheid, democracy had granted the ANC control of the state and seemed to provide the long-awaited opportunity to finally begin to address the backlog of social needs. HIV/AIDS may have been seen as an unwelcome problem that threatened to undo all of the good that the ANC hoped to accomplish. The epidemic threatened also to confirm the trope of Africa as a source of disease and hopeless despair and to displace Mbeki's vision of an "African Renaissance," a resurgence of the continent, powered by home-grown solutions. This may have further confirmed the tendency to ignore or downplay the emerging crisis.

AIDS is expected to slash the average life expectancy of black South Africans to 40 years.

As the ANC and Mbeki cast about for solutions to the growing epidemic, they were predisposed toward and against certain diagnostic and policy options. After decades of attempts by an apartheid government to restrict the growth of the black population through birth control campaigns, they were inclined not to accept the diagnosis of AIDS as a sexually transmitted disease that the use of condoms could curb. As politicians speaking to a conservative popular culture that frowned on the public airing of anything related to sex, they were inclined not to promote a solution that would require a frank and public discussion of matters sexual. By contrast, as progressives and social activists, they were predisposed toward a solution that regarded AIDS as arising out of poverty—the very ill that they were primed to address—and hence they were open to the dissident view that questioned the causal link between HIV and AIDS and pointed instead to the impact of poverty (through chronic malnutrition, a lack of clean water and shelter, and constant exposure to disease and environmental stress) on the immune system.

Governmental inaction did not go unchallenged. The Treatment Action Campaign (TAC), part of South Africa's dynamic civil society, has lobbied energetically for expanded treatment. Most notably, the TAC filed a legal challenge to the government's policy of restricting the availability of nevirapine, an important antiretroviral drug used to treat the HIV infected. In a landmark ruling in 2002, South Africa's constitutional court ordered the government to make the drug available, progressively and within the limits of available resources, in all public health facilities and, likewise, to provide for testing and counseling across the country. But with roughly 5 million South Africans infected with HIV, the annual cost of treatment could run to billions of dollars.

THE INEQUALITY CHALLENGE

The AIDS crisis and the cost of antiretrovirals will inevitably affect the country's ability to deal with another of its challenges—socioeconomic justice—although the two issues are inextricably connected, as the TAC case showed. Beyond HIV/AIDS, the ANC leadership has made a concerted effort to reorient government spending toward the needs of the poorest stratum of society within existing budget constraints, even if this has not always been to the satisfaction of the courts.

The government's chosen instrument in this regard was the Reconstruction and Development Program (RDP). It outlined an exhaustive list of the country's outstanding social needs but failed to prioritize among them or specify how they were to be addressed. The list, moreover, included so many urgent needs—such as access to safe drinking water, basic health care, and housing—that the government's ability to address them would ultimately depend on the overall health of the economy.

Within just a couple of years, the government realized it needed to rethink the strategy. It dismantled the RDP ministry and reassigned core programs to the relevant line ministries. Provision of basic needs has since seen significant improvement. The percentage of households with access to clear water rose from 60 percent in 1996 to 85 percent in 2001, and those with sanitation increased from 49 percent in 1994 to 63 percent in 2003. The proportion of households with electricity rose from 32 percent in 1996 to 70 percent in 2001. The government has made free health care available to pregnant women and children under five years old. Progress on housing was initially less impressive, but 10 years later just over a million homes have been provided.

One of the greatest successes in improving the lives of the poor arose, almost inadvertently, from the racial equalization of pensions for the elderly. In the dying days of apartheid, the National Party, no doubt with an eye to the prospect of democratic elections, raised the level of state pensions for Africans, Indians, and coloreds (mixed race) to the level that whites had long enjoyed. In one stroke this authorized the monthly injection, into hundreds of thousands of households across the country, of $90 a month, roughly twice the median black monthly income. Close to 80 percent of age-qualified Africans now receive a state pension, prorated in terms of household income. Overall, 1.6 million elderly people receive a state pension. Because age in South Africa is closely correlated with household poverty, this has proved a remarkably effective way to target some of the country's poorest households. The state pension has brought dramatic benefits in the health and nutritional status of children, too, because of the prevalence in South Africa of the extended family structure.

The continuing expansion of welfare payments has proved an effective weapon in the battle against poverty. But there are concerns about its fiscal sustainability. If the number of households receiving state pensions continues to expand at the current rate, and if the value of the grant is maintained in real terms, the costs of the program may bankrupt the treasury, especially when added to the costs of providing medical care and welfare to all the South Africans affected by AIDS.

Many of apartheid's most pernicious economic and social effects resulted from the systematic undereducation of black South Africans. Since 1994, the ANC has continued the broadening of educational access that began under the National Party, but—again, as under the National Party—that education is of decidedly uneven quality. Education currently receives almost a quarter of the government's total budget, but that investment goes largely to meeting the salaries of teachers, many of whom are poorly trained and perform badly in the classroom. Combined with weak local governance, the overall result is an education system that continues to produce large numbers of underskilled graduates in an economy that suffers from a dramatic shortage of skilled labor.

THE ECONOMY IN GEAR

In 1994 the ANC inherited an economy that had only just begun to expand again after a recession in the early 1990s. In fact, the malaise was longstanding: the economy had not grown strongly or created enough jobs to keep up with the expanding labor pool since the late 1970s. In addition, while the relatively smooth political transition had stanched the worst of capital flight, capital inflows through the mid-1990s primarily took the form of easily liquidated portfolio flows, rather than direct investment in new productive enterprises. International investors, along with their domestic counterparts, adopted a wait-and-see attitude.

Key leaders within the ANC, including Mbeki, Trade and Industry Minister Alec Erwin, and Finance Minister Trevor Manuel, came to understand the importance of establishing their credibility with the markets. The result, launched in June 1996, was a policy entitled Growth, Employment and

Redistribution (GEAR). Despite its name, GEAR was essentially an orthodox stabilization package, almost indistinguishable in policy content from those imposed by the World Bank and International Monetary Fund in developing economies around the world. GEAR, however, was voluntary, undertaken without any direct pressure or conditionalities from the international financial institutions.

GEAR was based on the premise that in order to grow, South Africa's economy would have to become more competitive and export oriented. At the core of GEAR were requirements to lower government deficits and inflation, reduce tariff barriers, privatize existing state-owned enterprises, and move toward a more dynamic, deregulated market.

Some of these goals proved easier to meet than others. Under the National Party, the budget deficit had reached 8 percent of GDP. The ANC succeeded in lowering that to 2 percent, a significant achievement for a government that had been expected to face irresistible demands to boost spending. It helped that the new South African government had inherited from the apartheid government a relatively efficient and progressive tax system with high rates of compliance. The tax system has since been further strengthened, with happy consequences for the government's deficits.

Inflation has slowly come down, albeit not as dramatically, into single-digit rates. The government also has successfully lowered tariff barriers, though there is still some way to go before South African manufacturing experiences the full extent of competitive pressures. Privatization and commercialization of a panoply of state-owned enterprises—including the national airline, telecommunications, railways, and harbors—have proceeded much more slowly. The effort is complicated by sometimes conflicting objectives, seeking not only to maximize the selling price of state assets and improve their efficiency but also to extend service provision to previously unserved communities and promote black economic empowerment.

The finishing touches were being applied to the GEAR strategy in 1996 when South Africa's domestic currency, which had to that point been remarkably stable, came under attack from international currency speculators. The South African Reserve Bank did not handle the crisis well. In April 1994, the exchange rate had been 3.50 rand to the US dollar. By February 1996, it had plunged to R4.20 to the dollar. In part, the currency crisis was related to technical factors (including an arbitrage gap and the difference between South Africa's inflation rate and that of its major trading partners), but it also reflected financial markets' judgment about overall management of the economy and political stabilization. There were tremors, for example, when Mandela appointed the country's first black minister of finance, Trevor Manuel. The same reaction met the news that the National Party's outgoing governor of the Reserve Bank, Chris Stals, would be replaced by Tito Mboweni, a top-ranking ANC official without extensive financial sector experience. In 1998, international money markets sub-

jected the ANC to a further set of lessons on the power of perceptions of risk, taking the rand down to R6.84 to the dollar. By the end of December 2001, the rand had plummeted to an all-time low of R13.10 to the dollar—an undervaluation by most accounts. Indeed, since that time, the currency has appreciated considerably and for much of 2003 and into 2004 has ranged between R6 and R7 to the dollar.

GEAR has done much to stabilize the South African economy and lay the basis for long-term growth, but it has not significantly shifted the racial basis of economic power. On the tenth anniversary of the new democracy, whites continue to dominate the economy. Black South Africans, who make up 75 percent of the population, are generally considered to own between 2 percent and 7 percent of overall market capitalization in terms of stock shares, hardly a dramatic improvement on 1994's figures. In addition, much of what has been achieved has benefited a tiny black share-owning elite. A sustained transformation of black South Africans' role in the economy has yet to emerge.

BIG BUSINESS, BAD BUSINESS

Economic stabilization aside, the failure of the economy to create large numbers of jobs threatens social stability. The modest revival of growth has been insufficient to generate a requisite increase in employment. Reform of the labor market was initially regarded as a critical part of the GEAR program, but there have been few effective policy innovations on this front. Instead, South Africa's elites have opted to continue with a highly corporatist decision-making model in which the interests of big business and organized labor are privileged over those of small and medium-sized businesses and the unemployed.

The prevalent labor framework aspires toward a high-value-added economy model, staffed by a skilled labor force suitably compensated for its productivity. The trouble is that most of South Africa's current work force is unskilled and not highly productive. The only basis on which this generation of workers conceivably can compete in international labor markets is on price—which, within the current regulatory framework, has not been possible. The result, effectively, has been to shut the unskilled unemployed out of labor markets and to reduce the hiring capacity of small and medium-sized firms.

It is easy to understand why organized labor would press for such a labor market; the trade unions' job after all is to secure the best possible terms for their members. It is less clear why big business would agree to this—at least until one considers the nature of big business in South Africa and, in particular, of its overall cost structure. Since the 1970s, South African big business has tended to be highly capital-intensive, labor costs forming a relatively small proportion of overall costs. Even high wages have not cost big business a huge amount in overall terms. By contrast, the cost of political unrest that could result from unpleasant wrangling over wages, in the form of strikes and industrial action, would be high. The preference of big business invariably has been for a high wage model that would at least ensure minimal disruption to industrial processes.

There is more evidence that big business may be bad for business. South Africa performs poorly in indexes intended to measure levels of entrepreneurship, possibly because of the historically high levels of concentration of ownership and the dearth of real competitiveness in the South African economy. The mark-up in the South African manufacturing sector is roughly twice that found in US manufacturing. The government has attempted to introduce pro-competitive policies to curb collusive behavior within the South African economy but without any great success so far.

A DEMOCRATIC BALANCE-SHEET

South Africans can celebrate with pride a series of significant achievements since 1994. A society that had been organized on the basis of separate and inequitable access to the most basic needs has been restructured into an open and democratic system in which all South Africans enjoy the right to representation and services. The government has made real progress in addressing some of the social welfare inequalities that apartheid created. And its macroeconomic policies have laid the foundation for sustainable long-term economic growth.

With the dawning of the new democracy's tenth anniversary, it is natural for black South Africans to also consider what it is that has *not* changed—or, at least, not for the better: their job prospects and their chances of living a long and healthy life. Both of these outlooks will be shaped by the state of the country's education system, the economy's capacity to create jobs, and the course of the HIV/AIDS epidemic. None of these factors, in origin, flow from the policies of the current government. Today's principals and teachers were, for example, the product of apartheid education; the economy's inability to create enough jobs is related to longstanding capital-to-labor ratios that resulted from business and government decisions made in the 1970s; and the HIV/AIDS epidemic likewise came from elsewhere (both geographically and historically in terms of the basic health care system set up by the National Party).

Yet it is evolving government policy that will shape how these three crises develop from this point. Attempts are being made to address the problems in the education system; in the interim, middle-class children will continue to receive a relatively good standard of education, but it will take some time before reforms are able to improve the quality of education in rural and working class areas. Similarly, the government is attempting to address job creation with public works programs—but here the problem is less amenable to policy tinkering. Only fundamental shifts in the labor relations framework and in the nature of the economy will create the requisite number and kind of jobs for the millions currently unemployed. An effective response to the HIV/AIDS epidemic, by contrast, may well require a change of leadership in the top ranks of the ANC.

ANTOINETTE HANDLEY is a professor of political science at the University of Toronto.

Reprinted by permission from *Current History* Magazine, May 2004, pp. 195–201. Copyright © 2004 by Current History, Inc.

Latin America's Populist Turn

"Latin America's political landscape, highly complex and variegated, defies easy categorization and raises fundamental questions—including whether it might be better to jettison the term 'left' altogether."

MICHAEL SHIFTER AND VINAY JAWAHAR

When Tabaré Vázquez won an impressive victory in Uruguay's presidential election on October 31, 2004, some newspapers could not resist proclaiming that the triumph solidified a wider, regional pattern. "Uruguay Completes the Leftward Realignment of the Southern Cone" was a typical headline. Given Vázquez's pedigree—the new president represents the Broad Front, a coalition of democratic socialists, communists, and former Tupamaro urban guerrillas—he fit right in with the region's other leaders: Ricardo Lagos, Chile's first socialist president in more than three decades; Luiz Inácio "Lula" Da Silva, Brazil's president from the leftist Workers' Party; and Néstor Kirchner, the Argentine president whose political roots can be traced back to the strand of leftism practiced by the Peronist party in the 1970s. The entire lineup now ruling the southern cone of South America exhibits strong "leftist" credentials.

Yet if this is Vázquez's—and the left's—moment, it is unclear just what that moment means. The broad category of "leftist" offers a variety of possibilities. The moderate, even orthodox economic policies that the Southern Cone's "leftist" leaders have recently undertaken contrast sharply with other variants of leftism found in Latin America. Cuban President Fidel Castro has of course largely embodied and practically defined the leftist label over almost half a century. Castro's revolutionary project centered on the radical reordering of Cuba's economy, marked by confiscation and nationalization of private property. And since coming to office in early 1999 as Venezuela's democratically elected president, Hugo Chávez has often been described as a leftist, in part based on his close relationship with Castro and also because of his highly charged rhetoric.

Other so-called leftists, still aspiring to be presidents, include Bolivian indigenous leader Evo Morales and former Salvadoran guerrilla figure Shafik Handal. Few doubt that Nicaragua's Daniel Ortega, who formerly led the Sandinista regime in the 1980s, hopes to return to his old executive office, this time via the ballot. So does Peru's rehabilitated former president, Alan García, who governed in the late 1980s and in the past gladly embraced the leftist label. Curiously, even the president whom García hopes to succeed in 2006,

Alejandro Toledo, has been depicted as a leftist, as has Ecuador's president, Lucio Gutiérrez.

In fact, Latin America's political landscape, highly complex and variegated, defies easy categorization and raises fundamental questions—including whether it might be better to jettison the term "left" altogether. Does "left" actually provide a useful handle for understanding the forces today shaping the region's politics, or for anticipating the policies that a president might pursue once in office? Does it capture what is happening in Latin America, or is it merely an artificial construct that obfuscates more than it illuminates? Are observers confusing a natural concern for acute social conditions, cast in markedly populist rhetoric, with "leftist" agendas? Or are "leftists" simply those who identify themselves as such?

THE REGION AND ITS DISCONTENTS

It is not surprising that Latin America has been undergoing political ferment in recent years. Economic and social progress has been meager, and expectations for a better life have largely not been met. Over the past 25 years the only Latin American country that has witnessed a significant increase in its real per capita income has been Chile. Regionally inflation has successfully been brought under control, and most governments have exercised fiscal discipline. But the results of economic recipes applied throughout Latin America—contained in the so-called Washington consensus and advocating greater privatization and liberalization—have been disappointing.

It is hard to discern coherent proposals and policies that could constitute a viable, alternative approach to the prevailing economic model in Latin America. The region has witnessed greater concern for the urgent social agenda and appeals to the popular sectors of society—traditionally excluded and recently mobilized—not only in Lula's Brazil, Lagos's Chile, and Chávez's Venezuela, but also in Álvaro Uribe's Colombia and Vicente Fox's Mexico. Unsettled national politics have combined with a perceptible tendency across the region to resist pressure from the world's only superpower, the United States, and an attempt to chart a more independent economic and political course. But an effort to tackle thorny social problems with practical solutions, coupled with a more autonomous foreign policy, is a far cry from leftism—at least as generally understood and practiced in Latin America.

It is now common to point out that, consistent with global trends, most Latin Americans are increasingly unhappy with politics as usual and are seeking new political options. Since 1995, Latino-barómetro, the public opinion survey carried out by a Santiago-based organization, has shed light on this public disenchantment. Dissatisfaction with government performance in a number of critical areas, especially the provision of economic and physical security, has grown considerably. Unemployment, crime, and corruption typically top the list of public concerns.

At the same time, there is little evidence that Latin Americans are systematically rejecting either the principles that underpin the democratic system or the market economy. In fact, in 2004 Latino-barómetro reported that some 56 percent of the region's respondents favor the market economy. It is true that the prescriptions associated with the Washington consensus—distinct from, or at least a subset of, the market economy—have yielded unfavorable results for most Latin Americans. Rates of poverty and inequality, long the region's Achilles heel, have remained stubbornly stagnant, or have deteriorated.

Only in 2004—ironically, the same year that the leftist label seemed to acquire greater appeal in Latin America—was overall economic performance reasonably robust. The 5.5 percent growth rate for the year was the region's highest for several decades. Experts attribute the performance mainly to favorable commodity prices, the extraordinarily high demand from China, and the fact that statistics for many of the region's key countries —Argentina and Venezuela, most notably—had risen from an extremely low base. Whether a similar growth rate can be sustained in 2005 and beyond, and whether it will effectively translate into more balanced and equitable development, remains a key question.

If Latin Americans show few signs of being eager to abandon the market economy, they do appear keen to soften the rougher edges of policies commonly associated with the Washington consensus. The substantial scaling back of an array of government functions in the 1990s is widely viewed as having gone too far, and is therefore in need of redress. Perhaps most alarmingly, the privatizations that took place in a variety of sectors, while not objectionable in principle, were accompanied by high levels of corruption and significant social strains. The protests triggered by the attempted privatization of electrical companies in Arequipa, Peru, in 2002, or the popular mobilizations in Bolivia surrounding the fight for control over water in 1999 and gas in 2003, should be construed as demands for honest government and rightful sharing of wealth. But they do not necessarily reflect a defense of state-owned and operated enterprises.

PRESCRIPTIONS AND POLITICS

Indeed, the problem is less the prescriptions for liberalizing economies than the fragile governance structures in most of Latin America that have proved ill-equipped to accommodate and sustain such reforms. To succeed, privatization of state enterprises and assets needs to be well managed politically, and in many of the countries where it has occurred, regulatory frameworks and oversight mechanisms were manifestly deficient. Political parties and their leaders have failed to modernize government structures and properly prepare them to handle these important reforms. As Human Rights Watch and other respected groups have regularly reported, adherence to the rule of law sadly remains more the exception than the norm in much of the region.

Despite the prominence of rule of law in political speeches and on policy agendas, corruption is still a profound and vexing problem. Last year public opinion regarding this issue became highly galvanized in two of Latin America's most unlikely countries: Costa Rica and Chile. Both had long enjoyed reputations as having relatively clean governments, exceptions in a region where corruption is rampant. In Costa Rica, two former presidents, Rafael Ángel Calderón and Miguel Ángel Rodríguez, were charged with having engaged in corrupt practices during their administrations. (Rodríguez was forced to resign as secretary general of the Organization of American States just two weeks after assuming the post, causing enormous embarrassment for the organization and the region.) Similar allegations of corruption were also brought against another former Costa Rican president, José María Figueres.

In perhaps the most notorious case, former Chilean President Augusto Pinochet, long accused of having presided over massive human rights violations, was recently found to have accounts holding millions of dollars at Riggs Bank in Washington, DC. The exposed impropriety further disgraced Pinochet, even among his previous supporters, opening the way for judicial prosecution of past crimes. While these cases highlighted public concern about corrupt activities and the ability of the judicial systems to respond, at least in Costa Rica and Chile, they also seriously tarnished the image of both countries.

Although political and public sector institutions in Latin America are generally held in low regard, polls consistently show that most citizens want democracy to work. A majority of respondents in the region recently agreed with the statement: "A democracy is preferable to any other form of government." A major United Nations Development Program report in 2004 revealed that many in the region could well be tempted by a more authoritarian option if it better addressed their economic and social needs. But it is far from clear whether that is the fundamental choice citizens are likely to confront. (For that matter, it is far from clear whether, faced with a similar choice, citizens of advanced Western democracies would express a significantly different opinion.) The critical, often frustrating task is to devise more effective policies to address complex economic and social problems within the democratic framework.

THE SOUTHERN CONE'S PRAGMATISM

In this regard, the experience of the Southern Cone countries is instructive. Compared with much of Latin America, the political contours of the Southern Cone countries are substantially smoother. This is particularly so in the case of Chile, increasingly in Brazil, and is even evident in Argentina.

Although leaders in all three countries have been called leftists, they are pursuing policies that blend a market economy and democratic politics. At the same time they are attempting to give higher priority to the long-pending and formidable social agenda—at least rhetorically and, when possible, through concrete actions. As in the case of Uruguay's Vázquez, their ascension to the presidency can in part be attributed to the electorate's perception in the three countries that social policies would be at the top of these leaders' agendas. In addition, Lula's Workers' Party, like Vázquez's Broad Front, had ample experience at the local level and had demonstrated its capacity for effective governance.

> *Exaggerating the challenge posed by "leftist" governments would only harm the quality of inter-American relations and prove extremely counterproductive.*

In Latin America generally, Chile is widely regarded as the premiere success story, having forged a broad consensus and fashioned a recipe for democratic stability and relatively broad-based economic growth. Following the end of the Pinochet regime in 1989, Chile has had three successive governments of the Concertación de Partidos por la Democracia, or Coalition of Parties for Democracy, the latest headed by Ricardo Lagos. Under Lagos, the country's first socialist president since Salvador Allende in the early 1970s, Chile's economy has grown admirably and poverty levels have continued to decline. Income inequalities remain a huge problem, although Lagos, building on his Concertación predecessors, has sought to reduce the gap between rich and poor through progressive social policies, particularly education reform.

Sergio Bitar, the education minister in the Lagos administration and the only member of the current cabinet to have also served in the Allende government (he was then minister of energy and mines), perhaps best epitomizes the dramatic evolution of "leftism" in Chile, and in Latin America, over the past three decades. Bitar is directing an ambitious program known as "English Opens Doors" that seeks to make all of Chile's 15 million people fluent in English within a generation. Although the effort has aroused some questions because of its excessive identification with the United States, Bitar has emphasized its democratic character, describing it as "an instrument of equality for all children" in Chile. And as *The New York Times* reported in December 2004, "that argument seems to resonate deeply with working-class families eager to see their children prosper in an increasingly competitive and demanding job market."

LULA'S ACCOMMODATIONS

While a comparable program would be politically unpalatable in Brazil, where anti-American sentiment is more pronounced than in Chile, Lula of the Workers' Party has pursued markedly orthodox market-friendly economic policies as well. In 2004, Brazil's economy grew by over 5 percent, and Lula's

support correspondingly rose—above 65 percent as of December 2004, according to one poll. Lula's policies have been regarded by some as a betrayal of his professed radical stance, and have created fissures within his own party. But Lula has responded pragmatically to a national and global context that leaves little margin for radical policy experimentation, and has tempered his goals accordingly. Midway through his term, Lula has achieved noteworthy political success, and is in a strong position—provided current trends continue—to win re-election in 2006.

Curiously, supporters of the previous, more centrist government of Fernando Henrique Cardoso have criticized the former metal worker for failing to push Brazil's social agenda sufficiently. Lula's social programs, including health policy and agrarian reform, along with the much touted "zero hunger" initiative, have so far failed to generate much enthusiasm or yield important results. In a region anxious for answers to complex social problems, the timidity and moderation of the greatest hope for "leftist" renewal in Latin America's largest country have broad and significant political implications.

Internationally, the Lula government has also been, to the dismay of some, notably accommodating and pragmatic. Lula has devoted considerable energy to going beyond the Southern Cone trade group known as MERCOSUR and seeking to construct a South American trade bloc—to some degree as a counterpoint to the United States. But this effort has been fraught with difficulties and has gained little traction. The Lula government has, however, successfully built on the efforts of the Cardoso administration and has made important inroads in its campaign to secure a permanent seat on the United Nations Security Council. Perhaps most noteworthy has been the Lula government's fine standing with Wall Street, the IMF, and the rest of the international financial community because of its sound macroeconomic management and performance.

In trade policy, too, Washington and Brasilia have been able to work together constructively. Following the sharp disagreements and tensions—chiefly around the issue of agricultural subsidies in the United States—that accompanied the September 2003 trade talks in Cancún, Mexico, the two governments made headway in narrowing their differences in 2004. At the conclusion of the Doha round of global trade negotiations in Geneva in July 2004, US Trade Representative Robert Zoellick praised Brazil's interest in pursuing common ground with the United States. And at the Asia Pacific Economic Cooperation meeting in Santiago in November 2004, Zoellick appeared more sanguine about the prospects for reenergizing the stalled process for creating a Free Trade Area of the Americas—a shift based in part on having developed a better understanding with the Brazilians.

RECOVERY IN ARGENTINA

In contrast with Lagos or Lula, Argentine President Néstor Kirchner has taken a more critical stand toward the prescriptions advocated by the international financial community. In particular, Kirchner has strongly disagreed with the position taken on his country's substantial foreign debt by the IMF and has refused to be rushed into signing a debt-schedule agreement. So far, Kirchner's gamble seems to be working. Argentina's economy has recovered spectacularly from its meltdown in late 2001 (growth in both 2003 and 2004 was 8 percent), and Kirchner has benefited politically from his defiant, independent posture. His resistance to the IMF's demands has been reinforced by other gestures in protecting human rights and fighting corruption that have similarly yielded political dividends.

An effort to tackle thorny social problems with practical solutions, coupled with a more autonomous foreign policy, is a far cry from leftism.

Nonetheless, it would be a mistake to overstate Kirchner's deviation from the policies pursued by Lagos and Lula. Despite strong disagreement on debt rescheduling, Kirchner's management of the economy has been, on the whole, fairly orthodox. He has carefully eschewed any public spending that could risk another bout of high inflation. There are budget surpluses at both the central and provincial levels. And private investment in Argentina—by the Chinese and South Koreans, other Latin Americans, and Argentines themselves—is on the rise. Judged by historical standards, and by some of the rhetoric coming from the Kirchner administration, these policies show considerable pragmatism, moderation, and acceptance of the tenets of the market economy. In neighboring Bolivia, unfortunately, the situation is slightly more unsettled.

BOLIVIA'S TURBULENT POLITICS

Few recent elections in Latin America have so eloquently illustrated the breakdown of ossified and discredited political institutions as did Bolivia's municipal elections in October 2004. In that vote, the country's traditional political parties imploded, and new and independent political forces emerged on the scene.

Morales, the indigenous leader who has successfully extended his support beyond his original base of coca growers, gained some ground in the vote (although his party did not win in any of Bolivia's 10 largest cities). More than any other figure, Morales, who just barely lost the presidential election in 2002, symbolizes the aspirations of the country's majority indigenous population, and underscores Bolivia's highly complicated, fluid, and precarious political landscape.

The man now in charge of the executive office, Carlos Mesa, struggles to maintain order and hold the country together until the next elections, which are scheduled for 2006. Mesa, who has no political party base, had been vice president before moving into his current post following the forced resignation of Gonzalo Sánchez de Lozada in October 2003. The mounting social protests and accompanying violence sparked by the proposed export of Bolivian natural gas—compounded by the fact the gas would be routed through Chile, a country that Bolivia has had a sensitive relationship with for more than a century because of the "War of the Pacific" that stripped Bolivia of access to the sea—highlighted the frustration among many poor Bolivians. The protests also reflected these groups' heightened ability to organize on behalf of their interests, and rendered the Sánchez de Lozada government unsustainable.

Although Mesa gained some breathing space with a national referendum on the gas question in July 2004, the respite has proved short-lived. His ambivalence and tendency to postpone key decisions have generated mounting suspicion with radical national groups and foreign investors alike, with both pressing for more favorable, friendly policies.

Underlying Bolivia's agitated politics and uncertain future is the need to find practical ways to balance a more just distribution of resources with a formula for sustained growth. In key respects, Morales, with his emphasis on social justice, comes closest to the classic definition of a leftist. But in such a transformed context, where Morales is seeking to construct a more hospitable institutional order for more equitable economic development, even he resists that label.

Some analysts believe that Bolivia could split into two separate entities: the overwhelmingly poor, indigenous altiplano; and the more modern and industrial lowland region centered in Santa Cruz. These predictions may be overstated, but some change in the prevailing, highly skewed order is inevitable and desirable, given the mostly legitimate demands of a previously excluded majority. Whether the constitutional assembly planned for 2005 will help find the right mix and satisfy key constituencies is unclear.

VENEZUELA'S DISSENTING VOICE

As Bolivia struggles to find its footing and move forward, Venezuela, after six years under Chávez, has set out on a dramatically different path. Cuba aside, Venezuela is the Latin American country that has most sharply deviated from the regionwide acceptance of the market economy and the principles of liberal democracy. Evidence can be found mainly in Chávez's own rhetoric, with its harsh condemnation of capitalism's ills and free trade and what he calls the "rancid oligarchy" associated with Venezuela's previous civilian, constitutional governments. Chávez has been especially unsparing in his remarks about the nefarious role of US imperialist designs in the world and Washington's presumed determination to impose its own economic and political model on weaker governments and societies. Chávez has also resisted the expanding notion of sovereignty—that setbacks in democratic progress are matters of hemispheric concern—which has gained considerable ground in the region since the end of the cold war.

The Venezuela factor is especially significant regionally. First, the country's social and economic conditions have deteriorated more dramatically than in any other Latin American country over the past two decades. This decline can be attributed to widespread mismanagement and corruption along with excessive reliance on petroleum. As a result, Chávez's charged discourse becomes more compelling, and has wider resonance. Second, with oil prices rising toward $50 per barrel, Chávez has money to spend, and corresponding political muscle, which makes him an important player in hemispheric affairs.

Still, whatever the rhetoric, it is hard to make the case that Chávez is steering a markedly revolutionary or even "leftist" course in the traditional sense of that term. He has welcomed foreign investment in the petroleum sector. Since investors are generally making money, one hears few complaints: provided the oil is flowing, Wall Street is pleased. In this regard, Chávez has been quite shrewd, since he calculates that any political pressure in response to his more authoritarian measures will be tempered by recognition that he is courting foreign investment in a strategically important industry. More radical actions such as land reform have been carried out half-heartedly at best. State-led attempts to instill revolutionary fervor and fashion a "new man" have an anachronistic quality to them, and suggest more posturing and experimentation than a serious effort at sustained, institutional transformation.

It is questionable, moreover, whether Chávez's tightening grip on Venezuelan institutions, and the growing presence and political role of the military, can properly be seen as reflecting a "leftist" orientation. They suggest instead a conscious attempt by the Chávez government to consolidate control and amass power. The president has continued to build on the momentum of a failed April 2002 coup against him (which enabled him to solidify control over the military) and a general strike in early 2003 (which allowed him to further dominate the state petroleum enterprise). In 2004, Chávez emerged stronger than ever after he defeated an August referendum aimed at removing him from office. He has subsequently achieved greater international legitimacy.

With an opposition disoriented and in disarray, Chávez gained further ground in local elections in October 2004, and seems likely to do as well in congressional elections later this year. Also in 2004, Venezuela's Chavista-led legislature passed bills that pack the Supreme Court and authorize the government to determine whether radio and television programs meet standards of "social responsibility." Concerned about the ominous climate in Venezuela, the Inter-American Commission on Human Rights noted that such measures further undermine judicial independence and risk the onset of government censorship of the media. None of this, however, makes Chávez a leftist. (For similar transgressions, Russia's Vladimir Putin is called a rightist.)

The Chávez government's displays of concern for the poor—and occasionally virulent attacks against a popular target like the United States—no doubt resonate among certain sectors throughout the region. Chávez's control of ample resources also makes him highly attractive, and will likely enhance his power and ability to cause mischief in an already unsettled Latin America. But the essential features of the Chávez model, which is likely to continue in force at least until presidential elections in late 2006, hold little appeal for a region searching for viable alternatives and practical solutions to problems.

THE TEST FOR WASHINGTON

Whatever the actual policy orientation that prevails in Latin America, it is undeniable nevertheless that the leftist banner in recent years has been politically wise and effective. This can be attributed in part to the failure of previous governments, many of which could be regarded as "rightist," and the generalized sense of frustration and disappointment that pervades the region. At the same time, the political rhetoric—much of it with a populist flavor—that has recently been heard in Latin America is inseparable from a growing distrust of, and resistance to, the United States.

Although such a reaction to US power has historically been evident, the region's current resentful mood has been compounded by several factors. One is a sense that the United States has been disengaged from Latin American concerns—even by historical standards. The gap between the rhetoric coming from Washington and the actual US commitment to the region is striking. So, too, is the gap between the agendas and priorities of the United States and those of Latin America. Indeed, the disconnect has seldom been greater. In this regard, the climate significantly shaped by Washington's preoccupation with the war on terror—and particularly with what is widely regarded as a disastrous military adventure in Iraq—has only aggravated the strain.

Against the backdrop of a region whose politics are especially sensitive to the words and actions that emanate from Washington, it is crucial to have a nuanced appreciation of Latin America's differentiated political landscape. Exaggerating the challenge posed by "leftist" governments would only harm the quality of inter-American relations and prove extremely counterproductive. Washington's excessive concern about Ortega's potential return to power in Nicaragua, for example, suggests a hangover from the cold war mindset.

To its credit, the United States has generally understood and supported the pragmatism displayed by leaders like Lagos and Lula. One of Lagos's chief accomplishments, after all, was the signing of a free trade agreement with the United States in December 2003. Despite differences over Iraq policy and some attendant friction, relations have been excellent between the Bush administration and Chile's socialist president, who presides over the region's most robust economy and most stable democracy. Similarly, Washington has not treated Lula as the threatening leftist some initially feared he would become. There have been differences, some of them rather sharp, over trade and other matters, but in general a sense of mutual accommodation has dominated. One would expect that a similar relationship may greet the new Uruguayan administration led by Vázquez.

Washington's relationship with Chávez's Venezuela has been far more problematic. Indeed, Chávez poses a vexing policy challenge for the United States. As in other situations throughout the world, the dilemma is how to reconcile a pragmatic relationship that takes into account a vital interest—Venezuela provides about 15 percent of US oil imports—with serious concerns about the erosion of democratic practices and safeguards within the country, and with the leader's support (at a minimum, financially) for political forces that oppose US interests. Washington so far has lacked a thoughtful, strategic approach in dealing with the Chávez government. Its overly reactive posture has resulted in major, costly blunders, such as its initial support of the April 2002 coup against Chávez. Washington needs to do a better job of thinking through how to balance conflicting policy goals and of consulting more systematically and at higher levels with other key Latin American governments. Otherwise, it risks repeating mistakes and further fueling anti-American sentiment throughout the region.

The situation presented by Bolivia's fluid and complicated politics is less clear-cut and poses another severe test for Washington. It may be tempting to regard Morales, who has fiercely opposed US drug policy and has been denied a visa as a result, as a threat to democratic stability in the region. But common sense would suggest an effort to engage such figures and understand their objections to a policy that has destroyed the livelihoods of many coca growers and their families. What is striking in the case of Bolivia is not only that Washington's criticism of Morales has boosted him so much politically but also that the Bush administration failed to support a loyal ally like Sánchez de Lozada when he requested additional aid in 2002. If Washington had been more responsive—and less shortsighted—the collapse of his government might have been averted.

LEAVING "LEFT" BEHIND

Whatever governments are in place in Latin American countries—and whoever is in charge—they must deliver concrete results for broad sectors of the population that have not seen much, if any, improvement in their well-being in recent years. Absent such results, citizens will again become frustrated and will inevitably be drawn to different political banners. (Meanwhile, it is worth noting that none of the five Central American countries could even remotely be considered left.)

There are many obstacles to attaining effective performance. In Latin America, retiring buzzwords such as "left" and focusing instead on practical solutions to difficult problems would be a welcome step forward.

MICHAEL SHIFTER, *a* Current History *contributing editor and adjunct professor at Georgetown University, is vice president for policy at the Inter-American Dialogue.* VINAY JAWAHAR *is a program associate at the Inter-American Dialogue*

UNIT 5

Population, Development, Environment, and Health

Unit Selections

Key Points to Consider

- How do current population trends represent a departure from the past?

- What are the implications of these trends?

- How can conservation efforts succeed in developing countries?

- What are the development implications of the AIDS epidemic?

- What are the constraints on dispensing new drug therapies for diseases in poor countries?

Student Website

www.mhcls.com/online

Internet References

Further information regarding these websites may be found in this book's preface or online.

Earth Pledge Foundation
  http://www.earthpledge.org
EnviroLink
  http://envirolink.org
Greenpeace
  http://www.greenpeace.org
Linkages on Environmental Issues and Development
  http://www.iisd.ca/linkages/
Population Action International
  http://www.populationaction.org
The Worldwatch Institute
  http://www.worldwatch.org

The developing world's population continues to increase at an annual rate that exceeds the world average. The average fertility rate (the number of children a woman will have during her life) for all developing countries is 2.9 while for the least developed countries the figure is 5.1. Although growth has slowed considerably since the 1960s, world population is still growing at the rate of approximately 80 million per year, with most of this increase taking place in the developing world. Increasing population complicates development efforts, puts added stress on ecosystems, and threatens food security. Population migration patterns are increasingly shaped by trends in population growth in both the industrialized countries and the developing world.

World population surpassed 6 billion toward the end of 1999 and, if current trends continue, could reach 9.8 billion by 2050. Even if, by some miracle, population growth was immediately reduced to the level found in industrialized countries, the developing world's population would continue to grow for decades. Approximately one-third of the population in the developing world is under the age of 15, with that proportion jumping to 43 percent in the least developed countries. The population momentum created by this age distribution means that it will be some time before the developing world's population growth slows substantially. Some developing countries have achieved progress in reducing fertility rates through family planning programs, but much remains to be done. At the same time, reduced life expectancy, especially related to the HIV/AIDS epidemic, is having a significant demographic impact in parts of the developing world, especially sub-Saharan Africa.

Over a billion people live in absolute poverty, as measured by a combination of economic and social indicators. As population increases, it becomes more difficult to meet the basic human needs of the developing world's citizens. Indeed, food scarcity looms as a major problem among the poor as population increases, production fails to keep pace, demand for water increases, per capita, cropland shrinks, and prices rise. Larger populations of poor people also place greater strains on scarce resources and fragile ecosystems. Deforestation for agriculture and fuel has reduced forested areas and contributed to erosion, desertification, and global warming. Intensified agriculture, particularly for cash crops, has depleted soils. This necessitates increased fertilization, which is costly and also produces runoff that contributes to water pollution.

Economic development, regarded by many as a panacea, has not only failed to eliminate poverty but has exacerbated it in some ways. Ill-conceived economic development plans have diverted resources from more productive uses. There has also been a tendency to favor large-scale industrial plants that may be unsuitable to local conditions. Where economic growth has occurred, the benefits are often distributed inequitably, widening the gap between rich and poor. If developing countries try to follow Western consumption patterns, sustainable development will be impossible. Furthermore, economic growth without effective environmental policies can lead to the need for more expensive clean-up efforts in the future. Conservation efforts such as Kenya's Green Belt Movement founded by Nobel Peace Prize winner Wangari Maathai are indicative of progress, but much more needs to be done to balance resource exploitation with environmental protection.

Divisions between North and South on environmental issues became more pronounced at the 1992 Rio Conference on Environment and Development. The conference highlighted the fundamental differences between the industrialized world and developing countries over causes of and solutions to global environmental problems. Developing countries pointed to consumption levels in the North as the main cause of environmental problems and called on the industrialized countries to pay most of the costs of environmental programs. Industrialized countries sought to convince developing countries to conserve their resources in their drive to modernize and develop. Divisions have also emerged on the issues of climate and greenhouse gas emissions. The Johannesburg Summit on Sustainable Development, a follow-up to the Rio conference, grappled with many of these issues, achieving some modest success in focusing attention on water and sanitation needs.

Rural-to-urban migration has caused an enormous influx of people to the cities, lured there by the illusion of opportunity and the attraction of urban life. In reality, opportunity is limited. Nevertheless, most choose marginal lives in the cities rather than a return to the countryside. As a result, urban areas in the developing world increasingly lack infrastructure to support this increased population and also have rising rates of pollution, crime, and disease. Additional resources are diverted to the urban areas in an attempt to meet increased demands, often further impoverishing rural areas. Meanwhile, food production may be affected, with those remaining in rural areas having to choose either to farm for subsistence because of low prices or to raise cash crops for export.

Poverty and urbanization also contribute to the spread of disease. Environmental factors account for about one-fifth of all diseases in developing countries and also make citizens more vulnerable to natural disasters. Hazardous waste also represents a serious health and environmental risk. The HIV/AIDS epidemic has focused attention on public health issues, especially in Africa. Africans account for 70 percent of the over 40 million AIDS cases worldwide. Aside from the human tragedy that this epidemic creates, the development implications are enormous. The loss of skilled and educated workers, the increase in the number of orphans, and the economic disruption that the disease causes will have a profound impact in the future. The development and availability of drugs to treat HIV/AIDS and malaria, which kills more children worldwide, is constrained by patent and profitability concerns. Little of the spending on research and development of new drugs goes toward finding therapies for diseases prevalent in developing countries.

The Global Baby Bust

Phillip Longman

THE WRONG READING

YOU AWAKEN to news of a morning traffic jam. Leaving home early for a doctor's appointment, you nonetheless arrive too late to find parking. After waiting two hours for a 15-minute consultation, you wait again to have your prescription filled. All the while, you worry about the work you've missed because so many other people would line up to take your job. Returning home to the evening news, you watch throngs of youths throwing stones somewhere in the Middle East, and a feature on disappearing farmland in the Midwest. A telemarketer calls for the third time, telling you, "We need your help to save the rain forest." As you set the alarm clock for the morning, one neighbor's car alarm goes off and another's air conditioner starts to whine.

So goes a day in the life of an average American. It is thus hardly surprising that many Americans think overpopulation is one of the world's most pressing problems. To be sure, the typical Westerner enjoys an unprecedented amount of private space. Compared to their parents, most now live in larger homes occupied by fewer children. They drive ever-larger automobiles, in which they can eat, smoke, or listen to the radio in splendid isolation. Food is so abundant that obesity has become a leading cause of death.

Still, both day-to-day experience and the media frequently suggest that the quality of life enjoyed in the United States and Europe is under threat by population growth. Sprawling suburban development is making traffic worse, driving taxes up, and reducing opportunities to enjoy nature. Televised images of developing-world famine, war, and environmental degradation prompt some to wonder, "Why do these people have so many kids?" Immigrants and other people's children wind up competing for jobs, access to health care, parking spaces, favorite fishing holes, hiking paths, and spots at the beach. No wonder that, when asked how long it will

take for world population to double, nearly half of all Americans say 20 years or less.

Yet a closer look at demographic trends shows that the rate of world population growth has fallen by more than 40 percent since the late 1960s. And forecasts by the UN and other organizations show that, even in the absence of major wars or pandemics, the number of human beings on the planet could well start to decline within the lifetime of today's children. Demographers at the International Institute for Applied Systems Analysis predict that human population will peak (at 9 billion) by 2070 and then start to contract. Long before then, many nations will shrink in absolute size, and the average age of the world's citizens will shoot up dramatically. Moreover, the populations that will age fastest are in the Middle East and other underdeveloped regions. During the remainder of this century, even sub-Saharan Africa will likely grow older than Europe is today.

FREE FALLING

THE ROOT CAUSE of these trends is falling birthrates. Today, the average woman in the world bears half as many children as did her counterpart in 1972. No industrialized country still produces enough children to sustain its population over time, or to prevent rapid population aging. Germany could easily lose the equivalent of the current population of what was once East Germany over the next half-century. Russia's population is already contracting by three-quarters of a million a year. Japan's population, meanwhile, is expected to peak as early as 2005, and then to fall by as much as one-third over the next 50 years—a decline equivalent, the demographer Hideo Ibe has noted, to that experienced in medieval Europe during the plague.

Although many factors are at work, the changing economics of family life is the prime factor in discouraging childbearing. In nations rich and poor, under all forms of government, as more and more of the world's population moves to urban areas in which children offer little or no economic reward to their parents, and as women acquire economic opportunities and reproductive control, the social and financial costs of childbearing continue to rise.

In the United States, the direct cost of raising a middle-class child born this year through age 18, according to the Department of Agriculture, exceeds $200,000—not including college. And the cost in forgone wages can easily exceed $1 million, even for families with modest earning power. Meanwhile, although Social Security and private pension plans depend critically on the human capital created by parents, they offer the same benefits, and often more, to those who avoid the burdens of raising a family.

Now the developing world, as it becomes more urban and industrialized, is experiencing the same demographic transition, but at a raster pace. Today, when Americans think of Mexico, for example, they think of televised images of desperate, unemployed youths swimming the Rio Grande or slipping through border fences. Yet because Mexican fertility rates have dropped so dramatically, the country is now aging five times faster than is the United States. It took 50 years for the American median age to rise just five years, from 30 to 35. By contrast, between 2000 and 2050, Mexico's median age, according to UN projections, will increase by 20 years, leaving half the population over 42. Meanwhile, the median American age in 2050 is expected to be 39.7.

Those televised images of desperate, unemployed youth broadcast from the Middle East create a similarly misleading impression. Fertility rates are falling faster in the Middle East than anywhere else on earth, and as a result, the region's population is aging at an unprecedented rate. For example, by mid-century, Algeria will see its median age increase from 21.7 to 40, according to UN projections. Postrevolutionary Iran has seen its fertility rate plummet by nearly two-thirds and will accordingly have more seniors than children by 2030.

Countries such as France and Japan at least got a chance to grow rich before they grew old. Today, most developing countries are growing old before they get rich. China's low fertility means that its labor force will start shrinking by 2020, and 30 percent of China's population could be over 60 by mid-century. More worrisome, China's social security system, which covers only a fraction of the population, already has debts exceeding 145 percent of its GDP. Making demographics there even worse, the spreading use of ultrasound and other techniques for determining the sex of fetuses is, as in India and many other parts of the world, leading to much higher abortion rates for females than for males. In China, the ratio of male to female births is now 117 to 100—which implies that roughly one out of six males in today's new generation will not succeed in reproducing.

All told, some 59 countries, comprising roughly 44 percent of the world's total population, are currently not producing enough children to avoid population decline, and the phenomenon continues to spread. By 2045, according to the latest UN projections, the world's fertility rate as a whole will have fallen below replacement levels.

REPAYING THE DEMOGRAPHIC DIVIDEND

WHAT IMPACT will these trends have on the global economy and balance of power? Consider first the positive possibilities. Slower world population growth offers many benefits, some of which have already been realized. Many economists believe, for example, that falling birthrates made possible the great economic boom that occurred in Japan and then in many other Asian nations beginning in the 19608. As the relative number of children declined, so did the burden of their dependency, thereby freeing up more resources for investment and adult consumption. In East Asia, the working-age population grew nearly four times faster than its dependent population between 1965 and 1990, freeing up a huge reserve of female labor and other social resources that would otherwise have been committed to raising children. Similarly, China's rapid industrialization today is being aided by a dramatic decline in the relative number of dependent children.

Over the next decade, the Middle East could benefit from a similar "demographic dividend." Birthrates fell in every single Middle Eastern country during the 1990s, often dramatically. The resulting "middle aging" of the region will lower the overall dependency ratio over the next 10 to 20 years, freeing up more resources for infrastructure and industrial development. The appeal of radicalism could also diminish as young adults make up less of the population and Middle Eastern societies become increasingly dominated by middle-aged people concerned with such practical issues as health care and retirement savings. Just as population aging in the West during the 1980s was accompanied by the disappearance of youthful indigenous terrorist groups such as the Red Brigades and the Weather Underground, falling birthrates in the Middle East could well produce societies far less prone to political violence.

Declining fertility rates at first bring a "demographic dividend." That dividend has to be repaid, however, if the trend continues. Although at first the fact that there are fewer children to feed, clothe, and educate leaves more for adults to enjoy, soon enough, if fertility falls beneath replacement levels, the number of productive workers drops as well, and the number of dependent elderly increase. And these older citizens consume far more resources than children do. Even after considering the cost of education, a typical child in the United States consumes 28 percent less than the typical working-age adult, whereas elders consume 27 percent more, mostly in health-related expenses.

Largely because of this imbalance, population aging, once it begins creating more seniors than workers, puts severe strains on government budgets. In Germany, for example, public spending on pensions, even after accounting for a reduction in future benefits written into current law, is expected to swell from an already staggering 10.3 percent of GDP to 15.4 percent by 2040—even as the number of workers available to support each retiree shrinks from 2.6 to 1.4. Meanwhile, the cost of government health-care benefits for the elderly is expected to rise from today's 3.8 percent of GDP to 8.4 percent by 2040.

Population aging also depresses the growth of government revenues. Population growth is a major source of economic growth: more people create more demand for the products capitalists sell, and more supply of the labor capitalists buy. Economists may be able to construct models of how economies could grow amid a shrinking population, but in the real world, it has never happened. A nation's GDP is literally the sum of its labor force times average output per worker. Thus a decline in the number of workers implies a decline in an economy's growth potential. When the size of the work force falls, economic growth can occur only if productivity increases enough to compensate. And these increases would have to be substantial to offset the impact of aging. Italy, for example, expects its working-age population to plunge 41 percent by 2050—meaning that output per worker would have to increase by at least that amount just to keep Italy's economic growth rate from falling below zero. With a shrinking labor supply, Europe's future economic growth will therefore depend entirely on getting more out of each remaining worker (many of them unskilled, recently arrived immigrants), even as it has to tax them at higher and higher rates to pay for old-age pensions and health care.

Theoretically, raising the retirement age could help to ease the burden of unfunded old-age benefits. But declining fitness among the general population is making this tactic less feasible. In the United States, for example, the dramatic increases in obesity and sedentary lifestyles are already causing disability rates to rise among the population 59 and younger. Researchers estimate that this trend will cause a 10–20 percent increase in the demand for nursing homes over what would otherwise occur from mere population aging, and a 10–15 percent increase in Medicare expenditures on top of the program's already exploding costs. Meanwhile, despite the much bally-hooed "longevity revolution," life expectancy among the elderly in the United States is hardly improving. Indeed, due to changing lifestyle factors, life expectancy among American women aged 65 was actually lower in 2002 than it was in 1990, according to the Social Security Administration.

The same declines in population fitness can now be seen in many other nations and are likely to overwhelm any public health benefits achieved through medical technology. According to the International Association for the Study of Obesity, an "alarming rise in obesity pre-

sents a pan-European epidemic." A full 35 percent of Italian children are now overweight. In the case of European men, the percentage who are overweight or obese ranges from over 40 percent in France to 70 percent in Germany. And as Western lifestyles spread throughout the developing world so do Western ways of dying. According to the World Health Organization, half of all deaths in places such as Mexico, China, and the Middle East are now caused by noncommunicable diseases related to Western lifestyle, such as cancers and heart attacks induced by smoking and obesity.

GLOBAL AGING AND GLOBAL POWER

CURRENT POPULATION TRENDS are likely to have another major impact: they will make military actions increasingly difficult for most nations. One reason for this change will be psychological. In countries where parents generally have only one or two children, every soldier becomes a "Private Ryan"—a soldier whose loss would mean overwhelming devastation to his or her family. In the later years of the Soviet Union, for example, collapsing birthrates in the Russian core meant that by 1990, the number of Russians aged 15–24 had shrunk by 5.2 million from 25 years before. Given their few sons, it is hardly surprising that Russian mothers for the first time in the nation's history organized an antiwar movement, and that Soviet society decided that its casualties in Afghanistan were unacceptable.

Another reason for the shift will be financial. Today, Americans consider the United States as the world's sole remaining superpower, which it is. As the cost of pensions and health care consume more and more of the nation's wealth, however, and as the labor force stops growing, it will become more and more difficult for Washington to sustain current levels of military spending or the number of men and women in uniform. Even within the U.S. military budget, the competition between guns and canes is already intense. The Pentagon today spends 84 cents on pensions for every dollar it spends on basic pay. Indeed, except during wartime, pensions are already one of the Pentagon's largest budget categories. In 2000, the cost of military pensions amounted to 12 times what the military spent on ammunition, nearly 5 times what the Navy spent on new ships, and more than 5 times what the Air Force spent on new planes and missiles.

Of course, the U.S. military is also more technically sophisticated than ever before, meaning that national power today is much less dependent on the ability to raise large armies. But the technologies the United States currently uses to project its power—laser-guided bombs, stealth aircraft, navigation assisted by the space-based Global Positioning System, nuclear aircraft carriers—are all products of the sort of expensive research and development that the United States will have difficulty affording if the cost of old-age entitlements continues to rise.

The same point applies to the U.S. ability to sustain, or increase, its levels of foreign aid. Although the United States faces less population aging than any other industrialized nation, the extremely high cost of its health care system, combined with its underfunded pension system, means that it still faces staggering liabilities. According to the International Monetary Fund (IMF), the imbalance between what the U.S. federal government will collect in future taxes under current law and what it has promised to pay in future benefits now exceeds 500 percent of GDP. To close that gap, the IMF warns, "would require an immediate and permanent 60 percent hike in the federal income tax yield, or a 50 percent cut in Social Security and Medicare benefits." Neither is likely. Accordingly, in another 20 years, the United States will be no more able to afford the role of world policeman than Europe or Japan can today. Nor will China be able to assume the job, since it will soon start to suffer from the kind of hyper-aging that Japan is already experiencing.

AGING AND THE PACE OF PROGRESS

EVEN IF there are fewer workers available to support each retiree in the future, won't technology be able to make up the difference? Perhaps. But there is also plenty of evidence to suggest that population aging itself works to depress the rate of technological and organizational innovation. Cross-country comparisons imply, for example, that after the proportion of elders increases in a society beyond a certain point, the level of entrepreneurship and inventiveness begins to drop. In 2002, Babson College and the London School of Business released their latest index of entrepreneurial activity. It shows that there is a distinct correlation between countries with a high ratio of workers to retirees and those with a high degree of entrepreneurship. Conversely, in countries in which a large share of the population is retired, the amount of new business formation is low. So, for example, two of the most entrepreneurial countries today are India and China, where there are currently roughly five people of working age for every person of retirement age. Meanwhile, Japan and France are among the least entrepreneurial countries on earth and have among the lowest ratios of workers to retirees.

This correlation could be explained by many different factors. Both common sense and a vast literature in finance and psychology support the claim that as one approaches retirement age, one usually becomes more reluctant to take career or financial risks. It is not surprising, therefore, that aging countries such as Italy, France, and Japan are marked by exceptionally low rates of job turnover and by exceptionally conservative use of capital. Because prudence requires that older investors take fewer risks with their investments, it also stands to reason that as populations age, investor preference shifts toward safe bonds and bank deposits and away from speculative stocks and venture funds. As populations age further, ever-higher shares of citizens begin cashing out their investments and spending down their savings.

Also to be considered are the huge public deficits projected to be run by major industrialized countries over the next several decades. Because of the mounting costs of pensions and health care, government spending on research and development, as well as on education, will likely drop. Moreover, massive government borrowing could easily crowd out financial capital that would otherwise be available to the private sector for investment in new technology. The Center for Strategic and International Studies has recently calculated that the cost of public benefits to the elderly will consume a dramatically rising share of GDP in industrialized countries. In the United States, such benefits currently consume 9.4 percent of GDP. But if current trends continue, this figure will top 20 percent by 2040. And in countries such as France, Germany, Italy, Japan, and Spain, somewhere between a quarter and a third of all national output will be consumed by old-age pensions and health care programs before today's 30-year-olds reach retirement age.

Theoretically, a highly efficient, global financial market could lend financial resources from rich, old countries that are short on labor to young, poor countries that are short on capital, and make the whole world better off. But for this to happen, old countries would have to contain their deficits and invest their savings in places that are themselves either on the threshold of hyper-aging (China, India, Mexico) or highly destabilized by religious fanaticism, disease, and war (most of the Middle East, sub-Saharan Africa, Indonesia), or both. And who exactly would buy the products produced by these investments? Japan, South Korea, and other recently industrialized countries relied on massive exports to the United States and Europe to develop. But if the population of Europe and Japan drops, while the population of the United States ages considerably, where will the demand come from to support development in places such as the Middle East and sub-Saharan Africa?

Population aging is also likely to create huge legacy costs for employers. This is particularly true in the United States, where health and pension benefits are largely provided by the private sector. General Motors (GM) now has 2.5 retirees on its pension rolls for every active worker and an unfunded pension debt of $19.2 billion. Honoring its legacy costs to retirees now adds $1,800 to the cost of every vehicle GM makes, according to a 2003 estimate by Morgan Stanley. Just between 2001 and 2002, the U.S. government s projected short-term liability for bailing out failing private pension plans increased from $11 billion to $35 billion, with huge defaults expected from the steel and airline industries.

An aging work force may also be less able or inclined to take advantage of new technology. This trend seems to be part of the cause for Japan's declining rates of productivity growth in the 19903. Before that decade, the aging of Japan's highly educated work force was a weak but positive force in increasing the nation's productivity, according to studies. Older workers learned by doing, developing specialized knowledge and craft skills and the famous company spirit that made Japan an unrivaled manufacturing power. But by the 1990s, the continued aging of Japan's work force became a cause of the country's declining competitiveness.

Population aging works against innovation in another way as well. As population growth dwindles, so does the need to increase the supply of just about everything, save health care. That means there is less incentive to find ways of making a gallon of gas go farther, or of increasing the capacity of existing infrastructure. Population growth is the mother of necessity. Without it, why bother to innovate? An aging society may have an urgent need to gain more output from each remaining worker, but without growing markets, individual firms have little incentive to learn how to do more with less—and with a dwindling supply of human capital, they have fewer ideas to draw on.

IMPORTING HUMAN CAPITAL

IF HIGH-TECH isn't the answer, what about immigration? It turns out that importing new, younger workers is at best only a partial solution. To be sure, the United States and other developed nations derive many benefits from their imported human capital. Immigration, however, does less than one might think to ease the challenges of population aging. One reason is that most immigrants arrive not as babies but with a third or so of their lives already behind them—and then go on to become elderly themselves. In the short term, therefore, immigrants can help to increase the ratio of workers to retirees, but in the long term, they add much less youth to the population than would newborn children.

Indeed, according to a study by the UN Population Division, if the United States hopes to maintain the current ratio of workers to retirees over time, it will have to absorb an average of 10.8 million immigrants annually through 2050. At that point, however, the U.S. population would total 1.1 billion, 73 percent of whom would be immigrants who had arrived in this country since 1995 or their descendants.

Just housing such a massive influx would require the equivalent of building another New York City every 10 months. And even if the homes could be built, it is unclear how long the United States and other developed nations can sustain even current rates of immigration. One reason, of course, is heightened security concerns. Another is the prospect of a cultural backlash against immigrants, the chances of which increase as native birthrates decline. In the 1920s, when widespread apprehension about declining native fertility found voice in books such as Lothrop Stoddard's *The Rising Tide of Color Against White World-Supremacy*, the U.S. political system responded by

shutting off immigration. Germany, Sweden, and France did the same in the 1970s as the reality of population decline among their native born started to set in.

Another constraint on immigration to the United States involves supply. Birthrates, having already fallen well below replacement levels in Europe and Asia, are now plummeting throughout Latin America as well, which suggests that the United States' last major source of imported labor will dry up. This could occur long before Latin nations actually stop growing—as the example of Puerto Rico shows. When most Americans think of Puerto Rico, they think of a sunny, overcrowded island that sends millions of immigrants to the West Side of New York City or to Florida. Yet with a fertility rate well below replacement level and a median age of 31.8 years, Puerto Rico no longer provides a net flow of immigrants to the mainland, despite an open border and a lower standard of living. Evidently, Puerto Rico now produces enough jobs to keep up with its slowing rate of population growth, and the allure of the mainland has thus largely vanished.

For its part, sub-Saharan Africa still produces many potential immigrants to the United States, as do the Middle East and parts of South Asia. But to attract immigrants from these regions, the United States will have to compete with Europe, which is closer geographically and currently has a more acute need for imported labor. Europe also offers higher wages for unskilled work, more generous social benefits, and large, already established populations of immigrants from these areas.

Even if the United States could compete with Europe for immigrants, it is by no means clear how many potential immigrants these regions will produce in the future. Birthrates are falling in sub-Saharan Africa as well as in the rest of the world, and war and disease have made mortality rates there extraordinarily high. UN projections for the continent as a whole show fertility declining to 2.4 children per woman by mid-century, which may well be below replacement levels if mortality does not dramatically improve. Although the course of the AIDS epidemic through sub-Saharan Africa remains uncertain, the CIA projects that AIDS and related diseases could kill as many as a quarter of the region's inhabitants by 2010.

A FUNDAMENTAL PROBLEM

SOME BIOLOGISTS now speculate that modern humans have created an environment in which the "fittest," or most successful, individuals are those who have few, if any, children. As more and more people find themselves living under urban conditions in which children no longer provide economic benefit to their parents, but rather are costly impediments to material success, people who are well adapted to this new environment will tend not to reproduce themselves. And many others who are not so successful will imitate them.

So where will the children of the future come from? The answer may be from people who are at odds with the modern environment— either those who don't understand the new rules of the game, which make large families an economic and social liability, or those who, out of religious or chauvinistic conviction, reject the game altogether.

Today there is a strong correlation between religious conviction and high fertility. In the United States, for example, fully 47 percent of people who attend church weekly say that the ideal family size is three or more children, as compared to only 27 percent of those who seldom attend church. In Utah, where 69 percent of all residents are registered members of the Church of Jesus Christ of Latter Day Saints, fertility rates are the highest in the nation. Utah annually produces 90 children for every 1,000 women of childbearing age. By comparison, Vermont—the only state to send a socialist to Congress and the first to embrace gay civil unions—produces only 49.

Does this mean that the future belongs to those who believe they are (or who are in fact) commanded by a higher power to procreate? Based on current trends, the answer appears to be yes. Once, demographers believed that some law of human nature would prevent fertility rates from remaining below replacement level within any healthy population for more than brief periods. After all, don't we all carry the genes of our Neolithic ancestors, who one way or another managed to produce enough babies to sustain the race? Today, however, it has become clear that no law of nature ensures that human beings, living in free, developed societies, will create enough children to reproduce themselves. Japanese fertility rates have been below replacement levels since the mid-1950s, and the last time Europeans produced enough children to reproduce themselves was the mid-1970s. Yet modern institutions have yet to adapt to this new reality.

Current demographic trends work against modernity in another way as well. Not only is the spread of urbanization and industrialization itself a major cause of falling fertility, it is also a major cause of so-called diseases of affluence, such as overeating, lack of exercise, and substance abuse, which leave a higher and higher percentage of the population stricken by chronic medical conditions. Those who reject modernity would thus seem to have an evolutionary advantage, whether they are clean-living Mormons or Muslims, or members of emerging sects and national movements that emphasize high birthrates and anti-materialism.

SECULAR SOLUTIONS

How CAN secular societies avoid population loss and decline? The problem is not that most people in these societies have lost interest in children. Among childless Americans aged 41 years and older in 2003, for example, 76 percent say they wish they had had children, up from 70 percent in 1990. In 2000, 40-year-old women in the United States and in every European nation told surveys that they had produced fewer children than they intended. Indeed, if European women now in their 40s had been able to produce their ideal number of children, the continent would face no prospect of population loss.

The problem, then, is not one of desire. The problem is that even as modern societies demand more and more investment in human capital, this demand threatens its own supply. The clear tendency of economic development is toward a more knowledge-based, networked economy in which decision-making and responsibility are increasingly necessary at lower levels. In such economies, however, children often remain economically dependent on their parents well into their own childbearing years because it takes that long to acquire the panoply of technical skills, credentials, social understanding, and personal maturity that more and more jobs now require. For the same reason, many couples discover that by the time they feel they can afford children, they can no longer produce them, or must settle for just one or two.

Meanwhile, even as aging societies become more and more dependent on the human capital parents provide, parents themselves get to keep less and less of the wealth they create by investing in their children. Employers make use of the skills parents endow their children with but offer parents no compensation. Governments also depend on parents to provide the next generation of taxpayers, but, with rare exception, give parents no greater benefits in old age than non-parents.

To change this pattern, secular societies need to rethink how they go about educating young adults and integrating them into the work force, so that tensions between work and family are reduced. Education should be a lifetime pursuit, rather than crammed into one's prime reproductive years. There should also be many more opportunities for part-time and flex-time employment, and such work should offer full health and pension benefits, as well as meaningful career paths.

Governments must also relieve parents from having to pay into social security systems. By raising and educating their children, parents have already contributed hugely (in the form of human capital) to these systems. The cost of their contribution, in both direct expenses and forgone wages, is often measured in the millions. Requiring parents also then to contribute to payroll taxes is not only unfair, but imprudent for societies that are already consuming more human capital than they produce.

To cope with the diseases of affluence that make older workers less productive, rich societies must make greater efforts to promote public health. For example, why not offer reduced health care premiums to those who quit smoking, lose weight, or can demonstrate regular attendance in exercise programs? Why not do more to discourage sprawling, automobile-dependent patterns of development, which have adverse health effects including pollution, high rates of auto injuries and death, sedentary lifestyles, and social isolation? Modern, high-tech medicine, even for those who can afford it, does little to promote productive aging because by the time most people come to need it, their bodies have already been damaged by stress, indulgent habits, environmental dangers, and injuries. For all they spend on health care, Americans enjoy no greater life expectancy than the citizens of Costa Rica, where per capita health expenditure is less than $300.

In his 1968 bestseller *The Population Bomb*, Paul Ehrlich warned, "The battle to feed all of humanity is over. In the 1970s the world will undergo famines—hundreds of millions of people are going to starve to death in spite of any crash programs embarked upon now." Fortunately, Ehrlich's prediction proved wrong. But having averted the danger of overpopulation, the world now faces the opposite problem: an aging and declining population. We are, in one sense, lucky to have this problem and not its opposite. But that doesn't make the problem any less serious, or the solutions any less necessary.

PHILLIP LONGMAN is Senior Fellow at the New America Foundation and author of the forthcoming *The Empty Cradle* (Basic Books, 2004), from which this article is adapted.

Development in Africa:
The Good, the Bad, the Ugly

"Whether HIV/AIDS and civil conflict are tamed or left unrestrained is primarily in the hands of Africans and, above all, African leaders—and so is the region's future economic performance."

CAROL LANCASTER

This is the year of African development. The eyes of the world will be on the region in 2005 with a Group of Eight meeting in Scotland that is to focus on African development and a special United Nations session to assess progress toward achieving the Millennium Development Goals agreed to by all UN member states in 2000. This year has also seen the publication of a UN Millennium Project report and one by the Commission for Africa, organized by British Prime Minister Tony Blair.

Sub-Saharan Africa is at the center of these discussions because it is the region with the most intractable development problems and constitutes the core of the worldwide development challenge for the foreseeable future. China and India—the world's two largest countries—have achieved rapid rates of growth over the past decade and have reduced poverty significantly. Most Latin American countries have been making steady progress over the past 10 years. Yet the average per capita income in Africa is no better than it was three decades ago. Nor does it seem likely that the countries of sub-Saharan Africa will reach the Millennium Development Goals during the coming decade. This is the bad news about the state of Africa's economies. There is also good news, and news that can only be termed ugly.

THE GOOD NEWS

Some bright spots emerge within the region's overall disappointing growth performance. First there is Botswana, which enjoyed an average annual rate of growth of 11 percent during the 1980s and more than 5 percent in the 1990s. Mauritius is also a bright spot with an average annual growth rate over the two de-

cades of 5.5 percent. These performances would be admirable in any part of the world. They remind us that economic success over an extended period is achievable in Africa as well, despite the difficulties of geography, climate, history, and the vagaries of the international economy.

Economic performance for a number of African countries has improved considerably in the past 15 years. Benin, Ethiopia, Mali, Mauritania, Mozambique, Namibia, Sudan, and Uganda all have seen substantial increases in their growth rates that exceeded their rates of population increase. Another group of countries has made credible economic progress (with annual growth of gross domestic product at 4 percent or more), including Burkina Faso, Eritrea, Ghana, and Guinea. In 2003, buoyed by high prices for many of Africa's primary product exports, several nations experienced exceptional growth rates, including Mozambique (7.6 percent), Burkina Faso (6.5 percent), and Benin, Ghana and Uganda, with growth rates in excess of 5 percent.

There is also modest good news on the economic management front in Africa. Many reforms adopted in the 1980s and 1990s remain in place, and economic management has improved moderately—or at least has not deteriorated. Currency adjustments, regarded as politically dangerous at the beginning of the reform period in the 1980s, have now been widely implemented and maintained. Government budget deficits have been reduced, with inflation averaging below 10 percent in the region. Commodity boards have been virtually eliminated, and prices for most agricultural products have been left to the market to determine.

While foreign direct investment into sub-Saharan African countries remains a small portion of total global FDI flows (1.6 percent in 2003), it has increased. In 1998, FDI in the region amounted to $6 billion; by 2003, it had risen to $9 billion. Much of this investment has gone to oil-producing countries, but some is clearly in response to improved economic performance in, for example, Kenya, Madagascar, Mali, Uganda, and Tanzania.

Meanwhile, the burden of debt servicing in the region has fallen—from 14 percent of exports in 1999 to 10 percent in 2003. The drop is in part a result of debt-relief measures, including the World Bank's initiative for highly indebted poor countries (HIPC). This program considerably eased the debt stock and servicing of 23 participating countries in 2004.

The HIPC initiative has also led to the introduction of a new process in Africa to help focus governments' attention on using their resources effectively for poverty reduction and making development plans in consultation with their peoples. To gain HIPC debt relief and large-scale concessional lending from international financial institutions, governments must produce a Poverty Reduction Strategy Paper (PRSP) in consultation with their civil society groups, and this paper must be approved by the boards of the World Bank and International Monetary Fund. As of February 2005, some 30 African countries had produced either full-fledged or "interim" PRSPs.

It is difficult to determine at this point whether the PRSPs are good news or no news for Africa. The idea behind them is a good one: persuade governments to adopt an overall strategy for reducing poverty and to consult with citizens in forming the strategy. But because PRSPs are a requirement for certain kinds of aid and debt relief, providing evidence for the seriousness and "ownership" of the PRSPs can turn into game playing—preparing another report in order to get the money. Some have argued that the PRSP differs from earlier structural adjustments linked to conditional lending in that aid is conditioned now on changing processes rather than policies. A recent World Bank evaluation recognized past problems with the PRSP process but was optimistic that, with some tweaking, it could be valuable for all concerned.

Another piece of good news for African development has been an increase in aid flows to the region from $12.7 billion in 1999 to $23.2 billion in 2003 (in net disbursement terms). Most countries not involved in civil conflict have benefited from increased aid flows. In addition, around $4 billion in remittances flowed to Africa in 2003, albeit the lowest level of remittances to developing countries.

There are other sources of good news. A few important social indicators have improved steadily if modestly since the 1990s: infant mortality has fallen from 110 to 103 per 1,000 births; literacy has increased from half to nearly two-thirds of the population. Social indicators gauge not only the quality of life, but also the extent to which the foundations of growth are in place. Without a pool of ed-

---

Millennium Development Goals For 2015

- Halve the proportion of people whose income is less than $1 per day
- Achieve universal primary education
- Eliminate gender disparities at all levels of education
- Reduce child mortality by two-thirds
- Halt and reverse the spread of HIV/AIDS, malaria, and other diseases
- Reduce by three-fourths the maternal mortality rate
- Halve the proportion of people without sustainable access to safe drinking water
- Develop a global partnership for development

---

ucated and healthy workers, investors will not find it easy to establish enterprises, create jobs, and spur growth.

THE BAD NEWS

Despite the hopeful economic and social progress in some areas, the overall economic news is not so good. The number of people living in poverty in Africa—already large at the end of the twentieth century, at just over 315 million—is projected to increase to over 400 million by 2015. Significant and sustained economic growth is required to raise people out of poverty. But economic growth in the region has not been on the average robust enough to raise per capita incomes over the past 30 years. And Africa is the only major developing region where this has been the case.

Behind this disappointing performance are a number of proximate causes. One is rapid population growth. Even though this rate has been declining, it still averaged 2.5 percent per year in the late 1990s, which produced nearly a doubling of Africa's population between 1980 and 2003—from 380 million to nearly 700 million. Other contributory factors include low rates of saving and investment. (The gross rate of capital formation in sub-Saharan Africa was 18 percent of gross national product in 2003—up from the past but still the lowest of any region.) And 40 percent of African wealth is held outside the region, in banks or investments in Europe, the United States, and elsewhere, a form of capital flight that reduces the savings in Africa available to finance investment there.

In many poorly performing African countries, such as Zimbabwe and Zambia, the economy relies on low-productivity agriculture (with little increase in the value added over recent years) and the export of a few primary products like cotton, cocoa, sugar, coffee, and copper that are subject to the vagaries of international price movements. Efforts to diversify exports—for example, through various kinds of manufacturing—have had little impact on the overall profile. As a result, the share of African exports in world trade has fallen over the past 25 years from

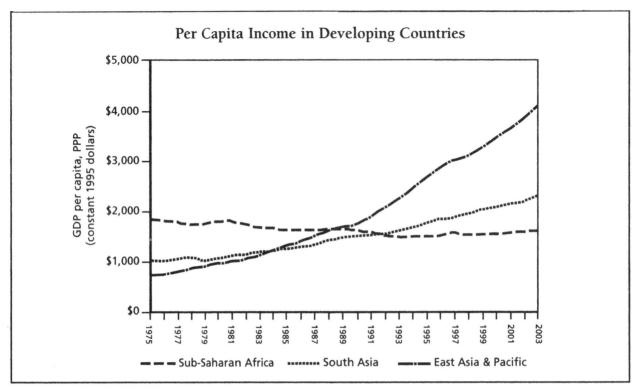

Per Capita Income in Developing Countries

Source: Commission for Africa, *Our Common Interest* (March 2005), p. 99
<http://www.commissionforafrica.org/English/report/thereport/cfafullreport.pdf>.

nearly 4 percent in 1980 to just 1.5 percent in 2002, leading to further economic marginalization of the region.

AND THE UGLY

There are two main sources of "ugly" in sub-Saharan Africa: HIV/AIDS and civil conflict. No one can regard the HIV/AIDS pandemic that has struck the region as anything but catastrophic. By 2004, 25 million Africans were living with HIV/AIDS (57 percent of them women), resulting in more than 2 million deaths, 3 million new infections, and 12 million AIDS orphans per year. The disease is especially intense in southern Africa, with 5 million in South Africa alone infected with the virus and a quarter of the total population of Botswana.

Even as the rate of increase in Africans living with HIV/AIDS has slowed considerably, the high overall rate of infection continues, as does the devastation it wreaks on human lives, families, economies, and entire countries. One illustration of its impact is the collapse in life expectancy in many countries—Botswana saw it drop from 57 years in 1990 to 38 years in 2002. In recent years, economic growth in Botswana has remained several percentage points below its 7 percent annual average for the previous several decades. This decline may have more to do with decreases in the prices of Botswana's exports than the impact of HIV/AIDS, but most analysts expect countries with high rates of the disease eventually to experience slower economic growth.

HIV/AIDS typically attacks individuals in the prime of their economically productive lives. Whether they are farmers, civil servants, or factory workers, as they sicken their productivity falls or they leave the labor force entirely. Their family incomes thus fall, while more of their remaining income is absorbed by the costs of the illness. And all of these effects ripple throughout society when HIV/AIDS reaches epidemic proportions as it has in parts of Africa.

Conflict—primarily civil conflict within states—is the other source of ugly news for Africa. According to the World Bank, one in four African countries currently suffers the effects of armed conflicts. The total number of casualties from conflicts in Africa surpasses that of all other regions in the world combined. The violence has forced approximately 15 million Africans from their homes, nearly a third of whom have sought refuge in neighboring countries.

Fueling this death, destitution, and destruction are the discontent of excluded, disadvantaged, or repressed groups; the greed of warlords; and the external support of commercial and political interests that are often corrupt and manipulative. African governments have attempted to deal with some of these conflicts through the creation of various peacekeeping forces (as in Sudan's Darfur, for example) while troops from non-African countries have also sought to keep the peace in Liberia, the Ivory Coast, and elsewhere.

Although the number of conflicts has diminished in recent years, several (such as the nearly decade-long war in the Democratic Republic of the Congo) drag on. And the fragility of a number of African states (Guinea, Sierra Leone, Burundi, the Republic of the Congo) leaves plenty of room for concern about future outbreaks of violence. Conflicts not only kill and displace people and make them destitute, but also destroy assets such as roads, housing, factories, and health and education facilities. It often takes years for a poor country to recover economically from the impact of prolonged civil war. Indeed, the African countries suffering from conflict are almost always those with the poorest growth and development performance: Sierra Leone, the Ivory Coast, the Democratic Republic of the Congo, Burundi, and others.

THE CURSES OF GOVERNANCE AND LOCATION

The economic news from sub-Saharan Africa speaks of stagnation—not the widespread deterioration experienced in the 1980s to be sure, but not the long hoped-for renaissance either. What explains this disappointing outcome?

The two culprits most blamed for Africa's lagging economic performance are poor governance and difficult geography. Good governance is usually regarded as predictable, accountable, and transparent decision-making on the part of a government. This means implementing the rule of law and ensuring political representation, respect for human rights, the independence of media and civil society organizations, and the absence of corruption. The concept is sometimes extended to mean capable and competent economic and political management as well.

The average per capita income in Africa is no better than it was three decades ago.

It is widely recognized that governance has been poor in much of sub-Saharan Africa, and that poor governance has inhibited the investment and growth required to support economic and social progress and reduce poverty. Poor governance almost always involves corruption, insecurity, and the absence of the rule of law. Investors—apart from those involved in natural resources extraction, which often provides large and quick profits—are reluctant to risk their monies in such an environment. And with low investment, growth and development typically are low. Kenya under President Daniel arap Moi, the former Zaire under Mobutu Sese Seko, and Cameroon today are examples of the impact of state malfeasance on economic progress.

Some interesting attempts have been made to measure governance, itself an inherently vague concept. One recent effort by the Economic Commission for Africa (based on public perceptions in 28 countries) found greater progress in implementing the formal aspects of electoral democracy in Africa than in combating corruption or building the effectiveness of institutions and the state itself. These findings are not significantly different from those of the World Bank. And they are echoed in the two major reports published in 2005 that focus some or all of their attention on development in sub-Saharan Africa: the Millennium Project report (led by Columbia University professor Jeffrey Sachs) entitled *Investing in Development* (http://www.unmillenniumproject.org/) and the Commission for Africa's report, Our Common Interest. Both reports acknowledge the major problems in the past with governance while urging a substantial increase in aid to the region.

What explains the pattern of poor governance? There is no consensus on the answer to this question. Some argue it is an outcome of the experience of colonial domination and exploitation, though this argument becomes less convincing as the period of independence grows. Development technocrats emphasize the lack of adequate training and management capacity on the part of government officials. But even those governments with a measure of capacity—like that of Kenya during the 1980s—have experienced poor governance.

Others argue that the ethnically divided, poorly integrated states of independent Africa have prevented the development of disciplined national governments focused on the good of the entire country because the "politics of the belly" have predominated. Aid donors have sought to strengthen civil society in Africa, reflecting the view that there are not enough institutions of accountability in the region to ensure good governance. Critics of aid have argued that relatively large amounts of foreign assistance have made it easy for political elites to act with little accountability.

Clearly, at the heart of the governance problem is the weakness of Africa's political institutions— their inability to constrain the behavior of rapacious, capricious, or repressive politicians. But we still do not understand well why this pattern exists or what can be done about it.

Some analysts of Africa's underdevelopment emphasize the disadvantages of geography—the impact of the location and climate of countries on their economic performance and prospects. Countries such as Mali, Niger, Chad, and Malawi that are landlocked and far from a coast confront major transport costs for commodities they export or import, reducing their competitiveness internationally and, therefore, their ability to lift their growth through trade.

Countries with a tropical climate—that is, nearly all of sub-Saharan Africa except South Africa—often have a larger disease burden. Afflictions such as malaria, schistosomiasis, various forms of sleeping sickness, and parasitic worms are not suppressed by frosts (as in many temperate latitudes), and the cures are typically nonexistent, expensive, or painful. Granted, Malaysia, Brazil, Vietnam, and Costa Rica, all in tropical climates, have enjoyed healthy growth. But the tropics can make the health

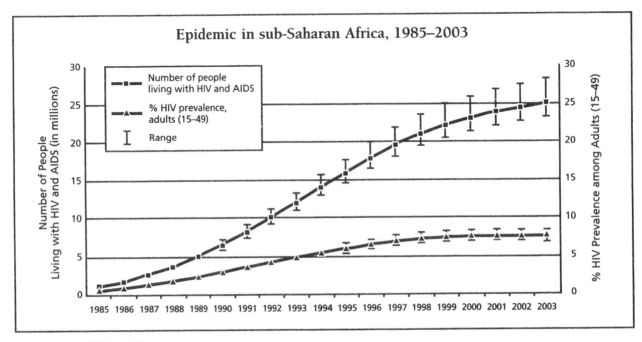

Source: UNAIDS/WHO, 2004.

challenges of development greater. The large disease load reduces the health and productivity of the population and so the performance of their economies.

While location and climate do not doom countries to poverty, they make overcoming poverty significantly more difficult. Much of sub-Saharan Africa struggles with one or both of these geographic misfortunes.

PROSPECTS FOR GROWTH

What do the good, the bad, and the ugly tell us about Africa's future economic prospects? Their basic message suggests slow and modest progress in economic management, in governance, and in social gains. The gains are not uniform: some countries are doing better than others. And the gains are not always sustained over time: some countries performing better yesterday have slipped backward today, especially, as in the case of the Ivory Coast, if they have dissolved into civil conflict.

There is, in short, no evidence of an African economic renaissance. And it remains unclear, even in the best of circumstances, what would drive such a renaissance. Trade is important, but manufacturing led growth seems limited since labor costs are relatively high and labor productivity is low—the success of Mauritius notwithstanding. In most of Africa, labor laws and the fact that labor is not as plentiful as in other regions continue to raise labor costs, while low levels of education and technology continue to impair productivity. The advantages of cheap labor that China has so effectively exploited do not seem within Africa's grasp at present.

The US African Growth and Opportunity Act of 2000 has eased restrictions on imports from eligible African countries on items such as textiles and apparel. But the act appears to have stimulated only a small amount of exports from Africa to the United States thus far. And even with easier entry to US markets, it is not clear how competitive African textiles will be in light of the removal this year of barriers to Chinese textile exports with the termination of the international Multifiber Agreement.

The export of services (software and back-office services, for example) that has driven growth in India and Costa Rica does not at present appear to offer much potential for Africa, given the limitations of communications infrastructure and labor market skills. The expansion of call centers in South Africa—with its well-developed communications infrastructure and its pool of educated labor—is an example of the possibilities but also the limits of this approach for Africa. Tourism and high-value agricultural products do hold promise, but the full exploitation of these opportunities has yet to be realized. Again, problems of governance and security as well as infrastructure impose obstacles.

As a result, the production and export of natural resources—minerals, food, fiber, and beverages—seem likely to remain the major source of growth for sub-Saharan Africa for the immediate future. Unfortunately, the world prices of many of these commodities have declined over recent decades relative to the prices of manufactured goods; this is one of the reasons Africans have wanted to diversify their economies.

The challenge for the few lucky countries exporting high-priced minerals—primarily petroleum and mainly from Angola, Nigeria, Gabon, Chad, and Equatorial Guinea—is to use their windfall profits to invest in the foundations for long-term development beyond mineral production. The past experience of the "resource curse" in Africa—the tendency for resource-rich governments to become less rather than more accountable to their people and so to govern less rather than more effectively for development —is, however, not reassuring.

THE BIG IFS

Several factors could brighten or dim these prospects for Africa's future economic performance. On the bright side is the prospect of increased economic assistance—especially if even a portion of the increases recommended by the Blair Commission and Millennium Project reports are realized. The latter report found that much of Africa is not progressing rapidly enough to meet the Millennium Development Goals, and that a doubling in aid would likely be necessary if the goals are to be achieved by 2015. If such aid is directed to countries that can make good use of it, it could help ease obstacles to more rapid economic progress—for example, by expanding and strengthening infrastructure, education, and health services. It also could ease the burden of HIV/AIDS, if it is used effectively to provide anti-retroviral drugs. And it could reduce the incidence of malaria if some of it is spent on significantly increasing the availability of pesticide-impregnated bed nets, as recommended in the UN report.

Also on the bright side is an initiative undertaken by Africans to improve governance in the region. In 2001, member states of the African Union agreed to the creation of the New Economic Partnership for African Development (NEPAD) that proposed the development of a "peer review" mechanism to assess the quality of governance in African countries. For each government that volunteered to have the quality of its governance assessed, a panel of eminent Africans would be set up to conduct the assessment and, eventually, to publish its findings.

Eight countries so far have signed up to be reviewed: Ghana, Mauritius, Rwanda, Kenya, Angola, Lesotho, Tanzania, and Malawi. Work has begun on several assessments, but no peer reviews have yet been published. If this mechanism turns out to be fair and probing and if the assessments provide incentives for African governments to improve the quality of their governance, NEPAD could make an important contribution to the continent's future. But these remain big "ifs."

The ugly side—HIV/AIDS and conflict—also looms large. If governments in Africa take an energetic stand against the spread of HIV/AIDS, including prevention and (with help from abroad) treatment, the disease could be controlled and reduced. If little is done beyond today's efforts, the death toll from HIV/AIDS could quadruple, with ensuing problems of deepening poverty and perhaps insecurity.

As for conflict, it will continue to be closely linked to the quality of governance. Where governance is oppressive or exclusionary, conflict almost always has been the result. And there is no shortage of country candidates for conflict in the future, in addition to those already plagued by violence. Whether HIV/AIDS and civil conflict are tamed or left unrestrained is primarily in the hands of Africans and, above all, African leaders—and so is the region's future economic performance.

CAROL LANCASTER *is an associate professor of foreign service at Georgetown University and a former deputy administrator of the US Agency for International Development.*

Malaria, the child killer

A special report by the BBC Focus on Africa on the devastating effects of the disease on Africa's population and economy.

WE are driving through the desert of Sudan's Dafur region when we spot in the distance two girls searching for water. They are orphans of the war, and have well developed malaria. Using all their energy signalling to us, they flail their arms, and with each rotation of the shoulder it seems they may fall over.

Death by malaria begins with cyclical attack of fever, shaking chills, sweating, dizziness, vomiting, diarrhoea and muscle pain. In the final states the liver and kidneys fail and the person enters a comma and eventually dies.

Debilitated by malaria shaking, the girls are gaunt, desiccated, visibly destroyed by the disease, literally dying in the Darfur sun the girls quiver with fever. Their faces are empty, heavy-eyed, absent of even the most basic human wonder.

This process of dying happens to someone in Africa every 30 seconds, according to the United Nation Children's agency, UNICEF.

Africa bears the vast majority of the human and economic cost of malaria worldwide, between 2 million to 3 million people die from the disease every year, about 90 per cent of which are in Africa, mostly young children and pregnant women.

At least one in five infants on the continent is born with the malaria parasite and the disease is the leading cause of death among children under five.

While HIV/AIDS kills more adults, malaria is the biggest killer of children worldwide. Yet work to find an AIDS vaccine receives seven times more funding than Malaria vaccine research.

In April 2000, African leaders signed the Abuja Declaration, committing their countries to specific malaria control achievements by this year. The declaration stated: "By 2005, at least 60 per cent of those suffering from malaria will have prompt access to be able to use correct, affordable and appropriate treatment". It also said: "At least 60 per cent of those at risk of malaria should benefit from in-secticide-treated mosquito nets." But most African nations still struggle with ineffective treatments, and the World Health organisation (WHO) reports that today mosquito nets are used by less than five per cent of Africans at risk.

Annually malaria costs African countries between US$10 billion to US$12 billion in lost domestic product even though it could be controlled, perhaps even eradicated, for a fraction of that amount.

A cross Africa, as in much of the developing world, poor people suffer the most from malaria. It is both a disease of poverty and a cause of poverty. Economic growth in countries with high level of malaria transmission is significantly lower than countries without malaria. African families at risk from malaria are estimated to spend up to 25 per cent of their income on direct malaria prevention and treatment. Although the cost of a mosquito net is low between Western standards (about US$2.33), the price is beyond the reach of Africa's poorest residents.

Poverty aside, African countries have been further plagued by a strong resurgence of the disease over the last ten years, in large part due to mosquitoes developing parasite developing resistance to many antimalarial drugs.

Resistance to Chloroquine, the cheapest and most widely used antimalaria, is now common throughout Africa. Yet international donors still subsidise and fund chloroquine therapy programmes. "Donors must stop wasting their money funding drugs that don't work, medicines Sans Frontiers says in a report.

Resistance to sulfadoxine-pyrimethamine, often the first and least expensive alternative to choloroquine is also increasing in east and southern Africa.

International organisations are urging Western governments to support multi-drug combinations based on artemisinin, a herb extract used in China for centuries but little know in the West.

But artemisinin is not yet produced in large enough quantities to be reasonably priced and many African nations are stuck, unable to afford the only drug combinations prevent to be effective.

Western donors have been hesitant to promote the treatment claiming a lack of funds—artemisinin is currently ten times more expensive than chloroquine. But advocacy groups argue that international funding priorities must change to recognise the forgotten epidemic, and are urging Western donors to create subsidies for antimalarial like artemisinin. Without subsidies, they argue, in areas with insufficient government and private purchasers, a large proportion of residents at risk from malaria will be unable to afford appropriate courses of antimalaria therapies.

International efforts to contain or even eradicate the disease have received a boost in recent years with a US$4.7 billion pledge by donors over five-years to the UN global fund for AIDS, tuberculosis and malaria.

The drugs company GloxoSmithKline claims it is close to producing a malaria vaccine in a joint venture with the malaria Vaccine initiative. The vaccine prevented almost 60 per cent cases in children during trials in Mozambique. But the earliest a vaccine is expected to be ready for sale in 2010.

Malaria has been a non-issue in North America and Europe since world War II, when DDT came into use as a cheap and highly effective insecticide.

In a massive public health campaign, tiny amounts of DDT were sprayed on the inside walls of houses. This indoor Residual Spraying (IRS) method was so successful that malaria was eradicated in Europe and the US within a few years.

Following success in the West, in the 1950s the WHO launched a programme to eradicate malaria using DDT. But the sheer resources, manpower and transport needed to mount serious campaigns in Africa mean that it was never sincerely target. Funding for eradication wanted as the US and Europe became malaria-free, but by the 1970s rates of malaria began to rise again in Africa.

Ever since there has been a raging DDT debate. In many countries environmentalist concerns about the consequence of extensive DDT use in agriculture led to its banning for all uses, including public health. As a consequence, poorer African countries that want to but cannot fund their own DDT-based programme are reliant on foreign donors, who now prefer funding insecticide-treated nets to DDT. But critics say mosquito nets show limited success in reducing malaria when compared to DDT programmes, and argue it is a question of quantity: the amount of DDT used to control pests on a large farm may suffice to control mosquitoes in a small country. South Africa almost eradicated the disease through a mosquito-control programme using DDT.

Many African nations are increasing resources for malaria control. Conscious of the drain on their economies, the first against malaria is now seen to be an important element of national poverty reduction strategies for malaria-endemic countries. Many African nations have reduced taxes on insecticides, mosquito nets and the materials use to make them.

Malaria advocacy efforts are even beginning to make in roads into pop culture, with a two-day superstar-packed concert held in Senegal in March in support of battle to eradicate the disease.

There is even an on-line malaria game, created to help educate children. On www.nobelprize.org you can run around with Speedy Ann the mosquito, who sport a digital human smell detection meter to help her find humans to bite. "It's not my fault", says Speedy Ann in the games introduction, making it clear that she did not create the epidemic." I only spread the disease.

The Price of Life

Rachel Glennerster, Michael Kremer, and Heidi Williams

The pharmaceutical industry spends billions of dollars each year developing drugs to fight disease. For the most part, though, the major drug companies respond to rich-country markets and neglect diseases concentrated in poor countries.

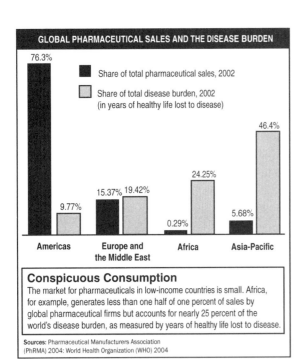

GLOBAL PHARMACEUTICAL SALES AND THE DISEASE BURDEN

■ Share of total pharmaceutical sales, 2002

☐ Share of total disease burden, 2002 (in years of healthy life lost to disease)

Americas: 76.3%, 9.77%
Europe and the Middle East: 15.37%, 19.42%
Africa: 0.29%, 24.25%
Asia-Pacific: 5.68%, 46.4%

Conspicuous Consumption
The market for pharmaceuticals in low-income countries is small. Africa, for example, generates less than one half of one percent of sales by global pharmaceutical firms but accounts for nearly 25 percent of the world's disease burden, as measured by years of healthy life lost to disease.

Sources: Pharmaceutical Manufacturers Association (PhRMA) 2004: World Health Organization (WHO) 2004

Selective Killers

Cause	Deaths per day in poor countries	Deaths per day in rich countries
Lower respiratory infections	7830	923
HIV/AIDS	5861	56
Diarrheal diseases	4210	16
Malaria	3409	< 1
Tuberculosis	3000	41

Note: Data are for 2002.

The Pill Bottleneck

Little private research and development (R&D) aims to solve health problems such as malaria or tuberculosis, which are concentrated in poor countries. Of all the new medicines licensed between 1975 and 1997, only a handful were for the deadliest tropical diseases.

1,223 New Medicines licensed worldwide between 1975 and 1997

Tropical disease medicines resulting from pharmaceutical industry R&D activities **4**

Source: *Journal of the American Medical Association*

Death Taxes

People in poor countries face more dangerous disease environments than those in rich countries because of their geography, climate, and limited health systems. Infectious and parasitic diseases—among the world's leading causes of death—account for about 33 percent of the disease burden in poor countries, but only 2.5 percent in rich countries.

Source: *WHO Global Burden of Disease Project, 2004*

Shot in the Arm

One way to encourage the development of products that low-income countries need is for international organizations, national governments, and private foundations to guarantee a market for desired vaccines. Companies would then have an incentive to make the modest investments needed to develop vaccines for common diseases. The costs per year of savings a life would be dramatically lower than those for other major diseases.

Sources: *Health Affairs;* Center for Global Development; Global Health Policy Research Network

India and Goliath

With major drug companies focusing on rich-world diseases, it might seem that the developing world could direct its own research at the diseases that kill its people. Several developing countries have pharmaceutical industries, but their financial resources are minuscule compared to those for large pharmaceutical firms. The Indian pharmaceutical industry is the world's fourth largest, but its resources are small and its research is often directed at Western markets.

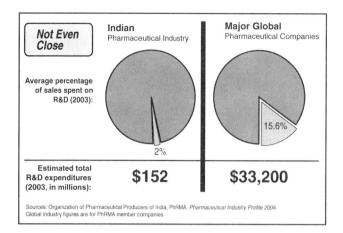

Rachel Glennerster is executive director of the Poverty Action Lab at the Massachusetts Institute of Technology. Michael Kremer is the Gates professor of developing societies at Harvard University. **Heidi Williams** *is a doctoral candidate at Harvard University.*

Why we owe so much to victims of disaster

At the G8 summit, Brown and Blair should think of our debts to Africans, not theirs to us. We have stolen their share of the planet's resources.

Andrew Simms

If you want to know how to tackle global warming, try the simple wisdom of Wilkins Micawber in Dickens's David Copperfield. "Annual income twenty pounds, annual expenditure nineteen pounds nineteen and six, result happiness," he said. "Annual income twenty pounds, annual expenditure twenty pounds ought and six, result misery."

It is rarely understood this way, but climate change is really a problem of debt. Not a cash debt, but an ecological one. Environmentally, we're living way beyond our means, spending more than the bank of the earth and the atmosphere can replace in our accounts. It is this debt—not the hole in the nation's public spending plans—that ought to have been the subject of the election campaign. And it is this debt—not the financial debts of poor nations to rich—that should guide the thinking of the Chancellor and other western leaders as they approach the G8 summit in July.

Gordon Brown and Tony Blair have set Africa and global warming as the summit's key themes. Yet newly released documents reveal one of the government's more embarrassing oversights. It was agreed at an international summit, more than three years ago, to create a special pot of money to help poor countries cope with climate change. Britain, alone among major European aid donors, has failed to contribute to the "Least Developed Countries Fund".

For years, we have been pilfering from the natural resource accounts of the rest of the world. When the people of Asia, Africa and Latin America decide they want to spend their fair share of nature's equity, either it won't be there or we could be on the verge of a crash in its already overstretched banking system. If the whole world wanted to live like people in the UK, we would need the natural resources of three more planets. If the US were the model, we would need five.

It's not just that we owe these countries for our profligate use of the planet's resources. It is also that they suffer the worst effects of our overuse. The most vulnerable people in the poorest countries—particularly children and women—are in effect paying the interest on our ecological debts. According to the World Disasters Report, the number of mostly climate-related disasters rose from just over 400 a year in

1994-98 to more than 700 a year in 1999-2003, with the biggest rise in the poorest countries.

The sight of a Mozambican woman giving birth in a tree during the great storms of 2000 is seared into the world's consciousness. Mozambique was desperately poor and burdened with debt payments. The floods were the worst for 150 years. Not only had its potential to develop been mismanaged by western creditors, Mozambique was left more vulnerable because it had to choose between preparing for disasters or spending its meagre resources on health and education. Now, in a warming world, Africa's rainfall, so crucial to its farming, is about to become even more erratic.

The story is similar outside Africa. In the mid- to late 1990s, at the height of the Jubilee 2000 debt cancellation campaign, nearly half the Jamaican government's spending went on debt service. The island is rich in natural resources, but it was getting harder for it to earn a living from exporting crops such as sugar and bananas. Yet, under pressure from the IMF and the World Bank, the money available for social programmes in Jamaica was halved.

Angela Stultz-Crawle, a local woman who ran a project in Bennetlands, Kingston to provide basic health and education services, saw the consequences at first hand: reductions in health programmes, in education, in road repairs, in lights. "Just walking around," she said, "you see people living in dirt yards, scrap-board houses. It is repaying. Every day you hear the government come out and say, 'Oh, we have met our IMF deadlines, we have paid,' and everyone claps." Again, Jamaica is particularly vulnerable to the extreme weather that climate change will make more frequent. Last year alone, two major hurricanes, Ivan and Charley, skirted its shores.

So across the developing world, the poorest people suffer from two crises, to neither of which they contributed: financial debt (which their governments are repaying) and ecological debt (which our governments aren't repaying).

In case after case—the IMF-approved kleptocracy of Mobutu's Zaire, the collusion with corruption, asset-stripping and violence in Nigeria's oilfields—the responsibility for financial debts lies at least as much in western capitals as in developing countries in the south. Yet, to win paltry debt relief, poor countries had to swallow the economic-policy equivalent of horse pills. Even the Financial Times commented that the IMF "probably ruined as many economies as they have saved". Yet we still expect poor countries to repay most of their debts, despite the effects on their people's lifestyles. Rich countries, faced with ecological debt, will not even give up the four-wheel-drive school run.

The widening global gap in wealth was built on ecological debts. And today's economic superpowers soon became as successful in their disproportionate occupation of the atmosphere with carbon emissions as they were in colonial times with their military occupation of the terrestrial world. Until the Second World War, they managed this atmospheric occupation largely through exploiting their own fossil-fuel reserves. But from around 1950 they became increasingly dependent on energy imports. By 1998, the wealthiest fifth of the world was consuming 68 per cent of commercially produced energy; the poorest fifth, 2 per cent.

In 2002, many rich countries were pumping out more carbon dioxide per person than they were a decade earlier, when they signed the UN Framework Convention on Climate Change. Now, with Africa and climate change at the top of the G8 summit agenda, there couldn't be a better time for a little paradigm shift. If Blair and Brown want to show leadership, they could relabel the G8 as the inaugural meeting of the ecological debtors' club, and start discussing how to pay back their creditors down south.

But is there any chance that the advanced industrial economies could make the cuts in consumption needed to clear their debts? Perhaps we should ask the women recently seen reminiscing about VE Day, women who during the world war had to keep house under severe constraints. After all, global warming is now described as a threat more serious than war or terrorism.

Drawing on articles in Good Housekeeping, and on guides with such titles as Feeding Cats and Dogs in Wartime or Sew and Save, they enormously reduced household consumption—use of electrical appliances, for example, dropped 82 per cent—while at the same time dramatically improving the nation's health.

The ecological debt problem of climate change, if it is to be solved, will still require a proper global framework, eventually giving everybody on the planet an equal entitlement to emit greenhouse gases, and allowing those who under-emit to trade with those who wish to over-emit. But such efforts will be hollow unless the argument to cut consumption can be won at household level.

To refuse the challenge would be the deepest hypocrisy. We have demanded that the world's poorest countries reshape their economies to pay service on dodgy foreign debts. It would be an appalling double standard now to suggest that we couldn't afford either to help developing countries adapt to climate change, or to cut our emissions by the 80-90 per cent considered necessary.

The language of restraint on public spending permeates our public discourse, yet the concept of living within our environmental means still escapes mainstream economics. That will have to change. "Balancing nature's books" could be the simple language that enables the green movement to resonate with the public. Imagine opening a letter from the bank over breakfast to learn that, instead of your usual overdraft, you had an ecological debt that threatened the planet. I wouldn't want to be there when the bailiffs called for that one.

Andrew Simms's Ecological Debt: the health of the planet and the wealth of nations is published this month by Pluto Books ([pounds sterling]12.99 from www.plutobooks.com)

UNIT 6
Women and Development

Unit Selections

Key Points to Consider

- What is the payoff from enhancing women's rights?

- What prevents wider opportunities for women in the developing world?

- What factors have contributed to progress in women's rights in the Middle East?

- How did the genocide in Rwanda change women's roles in society?

- Why might a larger role for women help to resolve conflicts?

Student Website
www.mhcls.com/online

Internet References
Further information regarding these websites may be found in this book's preface or online.

WIDNET: Women in Development NETwork
   http://www.focusintl.com/widnet.htm
WomenWatch/Regional and Country Information
   http://www.un.org/womenwatch/

There is widespread recognition of the crucial role that women play in the development process. Women are critical to the success of family planning programs, bear much of the responsibility for food production, account for an increasing share of wage labor in developing countries, are acutely aware of the consequences of environmental degradation, and can contribute to the development of a vibrant civil society and good governance. Despite their important contributions, however, women lag behind men in access to health care, nutrition, and education while continuing to face formidable social, economic, and political barriers.

Women's lives in the developing world are invariably difficult. Often female children are valued less than male offspring, resulting in higher infant and child mortality rates. In extreme cases, this undervaluing leads to female infanticide. Those females who do survive face lives characterized by poor nutrition and health, multiple pregnancies, hard physical labor, discrimination, and perhaps violence.

Clearly, women are central to any successful population policy. Evidence shows that educated women have fewer and healthier children. This connection between education and population indicates that greater emphasis should be placed on educating women. In reality, female school enrollments are lower than those of males because of state priorities, family resources that are insufficient to educate both boys and girls, female socialization, and cultural factors. Although education is probably the largest single contributor to enhancing the status of women and thereby promoting development, access to education is still limited for many women. Sixty percent of children worldwide who are not enrolled in school are girls. Education and higher status for women also has benefits in terms of improved health, better wages, and greater influence in decision making.

Women make up a significant portion of the agricultural workforce. They are heavily involved in food production from planting to cultivation, harvesting, and marketing. Despite their agricultural contribution, women frequently do not have adequate access to advances in agricultural technology or the benefits of extension and training programs. They are also discriminated against in land ownership. As a result, important opportunities to improve food production are lost when women are not given access to technology, training, and land ownership commensurate with their agricultural role.

The industrialization that has accompanied the globalization of production has meant more employment opportunities for women, but often these are low-tech, low-wage jobs. The lower-labor costs in the developing world that attract manufacturing facilities are a mixed blessing for women. Increasingly, women are recruited to fill these production jobs because wage differentials allow employers to pay women less. On the other hand, expanding opportunities for women in these positions contributes to family income. The informal sector, where jobs are smaller scale, more traditional, and labor-intensive, has also attracted more women, because these jobs are often their only source of employment, due to family responsibilities or discrimination. Clearly, women also play a critical role in economic expansion in developing countries. Nevertheless, women are often the first to feel the effects of an economic slowdown.

The consequences of the structural adjustment programs that many developing countries have had to adopt have also fallen disproportionately on women. As employment opportunities have declined because of austerity measures, women have lost jobs in the formal sector and faced increased competition from males in the informal sector. Cuts in spending on health care and education also affect women, who already receive fewer of these benefits. Currency devaluations further erode the purchasing power of women.

Because of the gender division of labor, women are often more aware of the consequences of environmental degradation. Depletion of resources such as forests, soil, and water are much more likely to be felt by women who are responsible for collecting firewood and water and who raise most of the crops. As a result, women are an essential component of successful environmental protection policies but they are often overlooked in planning environmental projects.

Enhancing the status of women has been the primary focus of several international conferences. The 1994 International Conference on Population and Development (ICPD) focused attention on women's health and reproductive rights and the crucial role that these issues play in controlling population growth. The 1995 Fourth World Conference on Women held in Beijing, China, proclaimed women's rights to be synonymous with human rights. These developments represent a turning point in women's struggle for equal rights and have prompted efforts to pass legislation at the national level to protect women's rights. International conferences have not only focused attention on gender issues, but also provided additional opportunities for developing leadership and encouraging grassroots efforts to realize the goal of enhancing the status of women. A recent review of the women's agenda established at the 1994 ICPD showed mixed results in achieving access to contraceptives and sex education for young women and improving reproductive health.

There are some indications that women have made progress in certain regions of the developing world. In the Middle East, the 2002 Arab Human Development Report highlighted the extent to which women in the region lagged behind their counterparts in other parts of the world. The influence of religious conservatives threatens to limit women's political participation in Iraq, raising further concerns about whether that country can be a model for democratic reform in the region. While there has been some progress recently, the gap in gender equality between the Middle East and the rest of the world remains wide. While there remains a wide divergence in the status of women worldwide, the recognition of the valuable contributions they can make to society is increasing the pressure to enhance their status.

The Payoff
From Women's Rights

Isobel Coleman

THE COST OF INEQUALITY

OVER THE PAST DECADE, significant research has demonstrated what many have known for a long time: women are critical to economic development, active civil society, and good governance, especially in developing countries. Focusing on women is often the best way to reduce birth rates and child mortality; improve health, nutrition, and education; stem the spread of HIV/AIDS; build robust and self-sustaining community organizations; and encourage grassroots democracy.

Much like human rights a generation ago, women's rights were long considered too controversial for mainstream foreign policy. For decades, international development agencies skirted gender issues in highly patriarchal societies. Now, however, they increasingly see women's empowerment as critical to their mandate. The Asian Development Bank is promoting gender-sensitive judicial and police reforms in Pakistan, for example, and the World Bank supports training for female political candidates in Morocco. The United States, too, is increasingly embracing women's rights, as a way not only to foster democracy, but also to promote development, curb extremism, and fight terrorism, all core strategic objectives.

Women's status has advanced in many countries: gender gaps in infant mortality rates, calorie consumption, school enrollment, literacy levels, access to health care, and political participation have narrowed steadily. And those changes have benefited society at large, improving living standards, increasing social entrepreneurship, and attracting foreign direct investment.

Yet significant gender disparities continue to exist, and in some cases, to grow, in three regions: southern Asia, the Middle East, and sub-Saharan Africa. Although the constraints on women living in these areas—conservative, patriarchal practices, often reinforced by religious values—are increasingly recognized as a drag on development, empowering women is still considered a subversive proposition. In some societies, women's rights are at the front line of a protracted battle between religious extremists and those with more moderate, progressive views. Deep tensions are evident in Saudi Arabia, Iraq, and Afghanistan, for example, and to a lesser extent in Nigeria, Pakistan, and Indonesia. Their resolution will be critical to progress in these countries, for those that suppress women are likely to stagnate economically, fail to develop democratic institutions, and become more prone to extremism.

Washington appreciates these dangers, but it has struggled to find an appropriate response. Since September 11, 2001, largely thanks to growing awareness of the Taliban's repression of Afghan women, gender equality has become a greater feature of U.S. policy abroad. But the Bush administration's policies have been inconsistent. Although Washington has linked calls for democracy with increased rights for women, especially in the Middle East, it has done too little to enforce these demands. It has supported women's empowerment in reform-oriented countries such as Morocco, but it has not promoted it in countries less amenable to change such as Saudi Arabia. Although women's rights feature prominently in U.S. reconstruction plans for Afghanistan and Iraq, Washington has not done enough to channel economic and political power to women there.

Given the importance of women to economic development and democratization—both of which are key U.S. foreign policy objectives Washington must promote their rights more aggressively. In particular, it must undertake, consistently and effectively, more programs designed to increase women's educational opportunities, their control over resources, and their economic and political participation. With overwhelming data now showing that women are critical to development, good governance, and stable civil life, it is time that the United States does more to advance women's rights abroad.

A HIGH-RETURN INVESTMENT

GENDER DISPARITIES hit women and girls the hardest, but ultimately all of society pays a price for them. Achieving gender equality is now deemed so critical to reducing poverty and improving governance that it has become a development objective in its own right. The 2000 UN Millennium Development Goals, the international community's action plan to attack global poverty, lists gender equality as one of its eight targets and considers women's empowerment essential to achieving all of them. Nobel Prize-winning economist Amartya Sen has argued that nothing is more important for development today than the economic, political, and social participation of women. Increasingly, women, who were long treated as passive recipients of aid, are now regarded as active promoters of change who can help society at large. And various studies specifically show that the benefits of promoting women are greatest when assistance focuses on increasing their education, their control over resources, and their political voice.

Although there is no easy formula for reducing poverty, many argue that educating girls boosts development the most. Lawrence Summers, when he was chief economist at the World Bank, concluded that girls' education may be the investment that yields the highest returns in the developing world. Educated women have fewer children; provide better nutrition, health, and education to their families; experience significantly lower child mortality; and generate more income than women with little or no schooling. Investing to educate them thus creates a virtuous cycle for their community.

Educating women, especially young girls, yields higher returns than educating men. In low-income countries, investing in primary education tends to pay off more than investing at secondary and higher educational levels, and girls are concentrated at lower levels of the education system than are boys. So closing the gender gap in the early years of schooling is a better strategy than promoting other educational reforms that allow gender gaps to remain. Similarly, children benefit more from an increase in their mother's schooling than from the equivalent increase in their father's. Educating mothers does more to lower child mortality rates, promote better birth outcomes (for example, higher birth weights) and better child nutrition, and guarantee earlier and longer schooling for children.

Educating girls is the single most effective way to boost economic progress.

Girls' education also lowers birth rates, which, by extension, helps developing countries improve per capita income. Better-educated women bear fewer children than lesser-educated women because they marry later and have fewer years of childbearing. They also are better able to make informed, confident decisions about reproduction. In fact, increasing the average education level of women by three years can lower their individual birth rate by one child. Studies show that in India, educating girls helps lower birth rates even more effectively than family planning initiatives.

Female education also boosts agricultural productivity. World Bank studies indicate that, in areas where women have very little schooling, providing them with at least another year of primary education is a better way to raise farm yields than increasing access to land or fertilizer usage. As men increasingly seek jobs away from farms, women become more responsible for managing the land. Because women tend to cultivate different crops than their husbands do, they cannot rely on men for training and need their own access to relevant information. As land grows more scarce and fertilizers yield diminishing returns, the next revolution in agricultural productivity may well be driven by women's education.

It is no coincidence, then, that in the last half-century, the regions that have most successfully closed gender gaps in education have also achieved the most economically and socially: eastern Asia, southeastern Asia, and Latin America. Conversely, regions with lagging growth—southern Asia, the Middle East, and sub-Saharan Africa—have also lagged in their investments in girls' education. Today, illiteracy among adult females is highest in southern Asia (55 percent), the Arab world (51 percent), and sub-Saharan Africa (45 percent). Simulation analyses suggest that, had these three regions closed their gender gaps in education at the same rate as eastern Asia did from 1960 to 1992, their per capita income could have grown by up to an additional percentage point every year. Compounded over three decades, that increase would have been highly significant.

Robust democracy is exceedingly rare in societies that marginalize women.

Giving women more control over resources also profits the community at large because women tend to invest more in their families than do men. Increases in household income, for example, benefit a family more if the mother, rather than the father, controls the cash. Studies of countries as varied as Bangladesh, Brazil, Canada, Ethiopia, and the United Kingdom suggest that women generally devote more of the household budget to education, health, and nutrition, and less to alcohol and cigarettes. For example, increases in female income improve child survival rates 20 times more than increases in male income, and children's weight-height measures improve about 8 times more. Likewise, female borrowing has a greater positive impact on school enrollment, child nutrition, and demand for health care than male borrowing.

These differences help explain why extending microfinance (small-scale lending with little or no collateral) to women has become such a powerful force for development. Mohammed Yunus founded Grameen Bank in Bangladesh—and launched the microfinance wave—by reasoning that if loans were granted to poor people on appropriate and reasonable terms, "millions of small people with their millions of small pursuits [could] add up to create the biggest development wonder." Yunus deliberately promoted microfinancing for women for reasons of equity: women are generally poorer and more credit-constrained than men, and they have limited access to the wage labor market and an inequitable share in decision-making at home. But he was also moved by sound economics: women pay their loans back more consistently than men.

Microfinance has been lauded for alleviating poverty in a financially sustainable way. But its greatest long-term benefit could be its impact on the social status of women. Women now account for 80 percent of the world's 70 million microborrowers. And studies show that women with microfinancing get more involved in family decision-making, are more mobile and more politically and legally aware, and participate more in public affairs than other women. Female borrowers also suffer less domestic violence—a consequence, perhaps, of their perceived value to the family increasing once they start to generate income of their own.

Allowing women to participate in politics also benefits democracy—and not only because it advances their civil rights. Intriguing new studies suggest that women in power make different policy choices than their male counterparts, with profound implications for the local allocation of public resources and, thus, for development. Esther Duflo, an economist at the Massachusetts Institute of Technology, has examined the impact of a constitutional amendment that India passed in 1993, which requires states to devolve more power over expenditures to *panchayats* (local councils) and reserve a third of council leadership positions for women. Duflo found that when women are in charge, the *panchayat* invests more in infrastructure that is directly relevant to women's needs. This is not to say that women's priorities are somehow better than men's, only that they are different and that in countries in which women are neglected, putting them in charge may begin to redress the imbalance.

The research of Steven Fish, a political scientist at the University of California at Berkeley, into why Muslim countries are generally less democratic than other countries reveals other benefits of female political participation. Fish has found that robust democracy is exceedingly rare in societies that display a large gender gap in literacy rates and a skewed gender ratio (usually a marker of inferior nutrition and health care for girls and infanticide or sex-selective abortion). He argues that societies that marginalize women generally count both fewer anti-authoritarian voices in politics and more men who join fanatical religious and political brotherhoods—two factors that stifle democracy.

ONE AT A TIME

GIVEN THE IMPORTANCE of women to economic and political development, it is no surprise that they are on the front line of modernization efforts around the world. But empowering women is rarely easy: it produces tensions everywhere, because it often collides with the twin powers of culture and religion.

Today, much scrutiny is given to the impact of Islam on women, often as evidence of a deep cultural rift between the West and conservative Muslim societies. But the real cultural rift may be within the Muslim world: between highly traditional rural populations and their more modernized urban compatriots or between religious fundamentalists and more moderate interpreters of Islam. Such tensions can be felt in countries ranging from Nigeria to Indonesia, but nowhere are they starker than in the Middle East.

Mustafa Kemal Atatürk, the founder of modern Turkey, may be the best-known leader to have pushed his country into modernity by transforming the role of women in society. After the abolition of the Ottoman caliphate in 1924, Atatürk promoted an aggressive program of secularization, replacing sharia with European constitutional law, prohibiting traditional Muslim dress, abolishing religious schools, and turning education into a state monopoly. Believing that women are intrinsically important to society, he launched many reforms to give them equal rights and more opportunities. A new civil code abolished polygamy and recognized the rights of women to inherit, divorce, and get custody of their children. Segregation in education was ended, and women were given full political rights. By the mid-1930s, Turkey had 13 female judges and 18 female parliamentarians. It was the first country in the world to appoint a female justice to its highest court, and in the mid-1990s, a woman was elected prime minister.

Similarly, when Tunisia won its independence in 1956, President Habib Bourguiba adopted an authoritarian, top-down approach to empower women as part of broader efforts to modernize the country. In his first year, he adopted a revolutionary Code of Personal Status that greatly enhanced women's rights: it banned polygamy, required a girl's consent for marriage, raised the minimum marriage age to 17, and allowed women to request divorce. At the time, these were progressive measures not only for Tunisia, but also for the world. And they stood in especially stark contrast to the laws then in force in Morocco, which gained independence from France at the same time but adopted a highly restrictive personal status law (*moudawana*) that institutionalized many conservative constraints on women.

Tunisia's enlightened policies toward women have contributed to its markedly superior record on developing human capital and economic growth. Today, the country's overall literacy rate is 70 percent (80 percent for men and 60 percent for women), compared to only 48 percent in

Morocco (60 percent for men and 35 percent for women). Tunisia's better-educated work force has helped the country attract more foreign direct investment. And tens of thousands of Tunisian women have brought their families into the middle class by working in export-oriented light manufacturing and foreign service centers. Not surprisingly, Tunisia's population growth rate has been notably lower than Morocco's, which accounts in part for its stronger gains in per capita income.

The aggressive promotion of women's rights has not come without a significant backlash. Because the notion of female empowerment is often strongly associated with secularism and Western values, it has generated widespread resistance in certain societies, among both men and women. To appeal to religious conservatives, leaders throughout the Arab world have long given them significant influence over women, usually by letting them oversee family law and personal status codes. But now that the importance of women to economic and political development is becoming increasingly clear, several young, Western-educated reformist leaders—King Mohamed VI of Morocco, King Abdullah of Jordan, and Sheik Hamad of Qatar—are reclaiming control over these areas. These leaders are engaged in the delicate exercise of pushing women forward without alienating their still highly conservative constituencies. Their efforts were boosted by the groundbreaking *Arab Human Development Report 2002*, which attributed the Arab world's economic and political stagnation in part to gender inequality.

Women in Morocco have made some remarkable advances in recent years. In the mid-1990s, with the support of the World Bank, Morocco launched a program promoting women's participation in development by increasing girls' education, health care for mothers and their children, and economic and political opportunities for women. It guaranteed that women would get 10 percent of the seats in the lower house of parliament in the 2002 elections. This quota helped raise the number of female representatives from 2 to 35—a notable achievement in the Arab world, which has the lowest percentage of women parliamentarians anywhere (about 3 percent). Several international organizations, including the National Democratic Institute and the UN Development Fund for Women, helped train the female candidates.

Women's groups have also been encouraged to play a more active role in Moroccan politics. In recent years, they have lobbied hard to reform the *moudawana*, and despite vehement opposition from fundamentalists, Mohammed VI established a "royal consultative committee" to assist their efforts. In January, the Moroccan parliament enacted one of the most progressive women's rights laws in the region, allowing women to marry without their father's consent, initiate divorce, and share with their husbands responsibility over family matters. The minimum marriage age was raised from 15 to 18, and the practice of polygamy severely restricted.

In conservative societies, debate over gender equality is often a proxy for more difficult debates about religious liberties and human rights.

Similarly, in Jordan, King Abdullah is improving the education of women and increasing their participation in the work force and in politics. The government has eliminated any gender gap in primary school enrollment, and girls now outnumber boys in secondary and tertiary education. Queen Rania, Abdullah's wife, has actively promoted microfinance initiatives, and under her patronage, in late 2003, Jordan hosted the region's first microfinance conference. The government has also implemented limited electoral quotas, reserving 6 out of 110 seats in parliament for women.

Nowhere, however, is women's reform more startling than in tiny Qatar, an otherwise highly conservative Wahhabi state. Sheik Hamad has launched a number of political reforms, including the country's first popular elections in 1999, in which both men and women were allowed to vote and run for office. Hamad and his wife, Sheika Mouza, have also encouraged educational reform. The government hired the Rand Corporation to advise on restructuring the country's educational system, and it launched the Education City Initiative, which has invited several American universities to set up local branches in Qatar (including the Virginia Commonwealth School of Arts and the Weill Cornell Medical College). Women now make up nearly 70 percent of the country's university students. Although Qatar's population is less than a million, the effects of its reforms are likely to ripple beyond its borders.

These reforms have not gone unnoticed in neighboring Saudi Arabia, for example, where religious conservatives still maintain strict control over women's access to public life. Saudi society is nearly completely segregated: in health care, education, and the work force. Women are treated as minors: they must have a male chaperon in public, they are not allowed to drive, and they need permission from their closest male relative to travel. The Saudi government has recently agreed to issue women identity cards, but only with the permission of a male guardian. The notorious *mutaween* (religious police) patrol malls to ensure that women are fully covered in public. In a tragic incident in Mecca in 2002, 15 schoolgirls were killed in a fire after *mutaween* allegedly forced them back inside the burning building because they were not appropriately covered.

But the Mecca fire prompted a national debate over religious extremism, after which control over the education of Saudi girls was transferred from religious authorities to the Ministry of Education. And that controversy helped revive long-standing calls for change. Female literacy in Saudi Arabia has risen from 2 percent in the mid-1960s, when universal female education was introduced (over vehement protest from the religious authorities), to more than 70 percent today. Women now account for nearly 60 percent of all university students, and they increasingly question the constraints on their lives. In January 2003, Saudi women signed a petition demanding that the government recognize their legal and civil rights. These efforts are beginning to pay off. The government, risking the wrath of religious conservatives, recently offered to let women take part in elections scheduled for later this year.

The demands of Saudi women may be helped along by new economic circumstances that are fueling the pressure for change. (A joke circulating around Riyadh says that the woman most sought after these days is the one with a job.) As GNP per capita has plunged from $25,000 in 1984 to roughly $8,500 today, more Saudis are wondering why half the country's human capital should be so severely handicapped. Indeed, a World Bank study on labor markets in the Middle East indicates that the increased participation of women in the work force can raise the average household income by as much as 25 percent without raising unemployment. So it is an encouraging sign that 10 percent of private Saudi businesses today are believed to be run by women.

Still, the Saudi debate over women's rights is often a proxy for difficult and dangerous debates over civil liberties, religious extremism, and human rights more generally, which are largely stifled. Talk of women's liberation encounters powerful resistance and recurring backlashes. Women played a major role in the recent Jeddah Economic Forum, which featured a keynote address by Lubna al Olayan, a prominent businesswoman, calling for greater economic and political rights for women. But soon after the conference, Saudi Arabia's grand mufti, or highest religious authority, issued a *fatwa* denouncing the public role of women. (It could not have helped that al Olayan also appeared on the front page of several leading Saudi newspapers with her head uncovered.)

Fundamentalists draw such a close link between women's empowerment and Western decadence that reformists such as Crown Prince Abdullah must be exceedingly careful when they endorse the former not to appear to be condoning the latter. For now, the role of women continues to be a line in the sand between those who want to modernize the country and those who seek to impose a harsh, medieval version of Islam in the kingdom.

A FAIRER FUTURE

THE BUSH ADMINISTRATION appears to recognize the importance of women to development. Women's rights have been a prominent element of its nation-building efforts in Afghanistan and Iraq and a central motif of its project to promote democracy in the Middle East. U.S. policymakers have been instrumental in pushing for women's rights in the new constitutions of Afghanistan and Iraq. Their influence has helped bridge deep cultural rifts in both countries, and it was critical in securing a 25 percent electoral quota for women there.

Although that accomplishment is significant, it was no more than a fragile first step. For Afghan women to benefit from the quota—or from any political or economic opportunity—they will need to become much better educated. Female literacy (defined as the ability to read a newspaper and write a letter) is well below 20 percent in Afghanistan. The U.S. Agency for International Development (USAID) has stated that Afghan girls' education is one of its priorities, and already more girls attend school than ever before in the country's history. But USAID has committed only $100 million to all education initiatives in Afghanistan over the next two years. That is grossly insufficient funding for the most pressing educational need of Afghan girls—the training of female teachers—especially since a large part of it is earmarked for school construction.

Women's progress since the fall of the Taliban has been significant, but it may be short-lived. Several of the powerful warlords who effectively control large swaths of the country fiercely oppose women's rights. Their militias have burned down girls' schools and pressured village leaders to prevent women from registering to vote in upcoming elections. And the United States supports these leaders, tacitly and explicitly, using their help in the hunt for terrorists. This marriage of convenience threatens Washington's policy of advancing women's rights, which will have to be pursued for another generation before the status of Afghan women improves substantially.

Washington says it is committed to women's rights abroad, but it has compromised on them in Afghanistan and Iraq.

Washington has also compromised on women's issues in Iraq. On the one hand, it has placed women's rights high on its reconstruction agenda: U.S. officials meet frequently with female Iraqi leaders, emphasize the importance of women's rights, and have channeled several million dollars to local women's groups. (In March, Secretary of State Colin Powell announced the Iraqi Women Democracy Initiative, which earmarks $10 million for leadership, political advocacy, and media training for women.) On the other hand, Washington has bowed to pressure from Shia leaders, backing down from appointing several female judges and designating only three women to the Iraqi Governing Council (IGC) and none to the 24-member Constitutional Committee.

The Coalition Provisional Authority (CPA) has also vacillated on the sensitive issue of sharia. Over the past year, many Iraqis have expressed deep concern that the rights women enjoyed under Saddam Hussein's secular Baathist regime would be significantly eroded if the new government adopted sharia wholesale. Concerns intensified in December when the IGC canceled Iraq's relatively liberal Personal Status Law, placing several aspects of family law, such as marriage, divorce, child custody, and inheritance, under the control of religious authorities. (Critics have argued that the move proves the CPA should have appointed more women to the IGC.) During deliberations on the interim constitution, Iraqi women organized protests and U.S. civilian administrator Paul Bremer said he would veto any draft that tried to impose a rigid version of sharia on Iraq. As a compromise, the interim constitution states that Islam will be an important source of future legislation but not the only one, as Shia leaders demanded. How strenuously the United States will continue to push this issue is unclear and still a cause for deep concern to many Iraqis.

The U.S. experiences in Afghanistan and Iraq show that advancing women's rights is a complicated and delicate task, particularly in Muslim societies. But just as the United States promoted human rights even when doing so conflicted with its other strategic objectives (arms control in the Soviet Union or economic relations with China), it should now wholeheartedly promote women's rights. Washington should make sure its policy on women's rights is consistent, it should generously fund the projects it undertakes, and, where necessary, it should condition its aid to developing countries on their efforts to close gender gaps.

For the sake of consistency and credibility, the United States must promote international efforts intended to advance the role of women worldwide. It should lead the implementation of UN Resolution 1325, unanimously passed by the General Assembly in 2000, which committed the UN to giving women a greater role in peacekeeping and post-conflict transitions. More important, the United States should finally endorse the 1981 Convention to Eliminate Discrimination Against Women, the only global treaty that deals exclusively with the rights of women, which 175 countries, including every industrialized democracy but the United States, have ratified. U.S. critics of the treaty have called it "antifamily," even though nothing in it contradicts traditional family values. They have also argued that the United States does not need it, since U.S. laws go well beyond what it recommends. So why not ratify it? By failing to support the agreement, Washington undermines its professed commitment to women's rights and exposes itself to charges of hypocrisy.

The United States should also dramatically increase funding to improve the status of women in regions where gender gaps are widest and assistance is most needed: the Middle East, southern Asia, and sub-Saharan Africa. Given the demonstrated high returns on investments in girls' education, the United States should, as its top development priority, work to eliminate gender gaps in primary education (USAID support for basic education is roughly $250 million). Likewise, it should expand support for microfinancing beyond its current level of roughly $200 million. Women's health and family planning deserve more funding too, particularly in countries such as Afghanistan where maternal mortality rates are alarmingly high. Adequate primary, maternal, and reproductive health care is critical to women's empowerment, especially in areas with high rates of adolescent marriage. (In those countries, the Bush administration's emphasis on abstinence is unrealistic.) Finally, the United States should use the Middle East Partnership Initiative, a $150 million program to advance democracy in the Middle East, which promotes programs for female literacy and health, as well as business and political training, as a model for more activist, pro-women policies in other parts of the world.

With the Millennium Challenge Account, the United States is now undertaking the largest expansion of foreign development aid in more than a generation. It should seize this opportunity to leverage its aid for the benefit of women's rights, by incorporating specific gender measures into the criteria that determine eligibility for funds. None of the 16 current criteria specifically takes account of women's status, but these could easily be adjusted. A country's maternal mortality ratio and its female primary school completion rate are both good indicators of gender equality there.

Similarly, the United States could promote respect for women's rights by making adherence to them a more explicit condition for U.S. military and economic aid. The State Department should be tasked with writing country reports tracking worldwide progress on key gender measures such as girls' literacy, maternal health, gender ratios, and political participation, much as it already does on human rights. Funding data collection on gender disparities is also important. Such information is lacking in many countries, and improving it could, by itself, help close gender gaps resulting from neglect.

"The worldwide advancement of women's issues is not only in keeping with the deeply held values of the American people," Powell has said, "it is strongly in our national interest as well." The United States has advocated women's rights as a moral imperative or as a way to promote democracy. In so doing, it might have compounded the difficulty of its task, by irking conservative religious forces or the authoritarian regimes it otherwise supports. But now Washington can also make an economic case for women's rights, which may be more acceptable to traditionalists. Promoting women's rights because they spur development and economic growth is a powerful way for the United States to advance its foreign policy in the future while minimizing the ideological debates that have frustrated it in the past.

ISOBEL COLEMAN is a Senior Fellow on U.S. Foreign Policy at the Council on Foreign Relations.

Iraq's excluded women

Swanee Hunt; Cristina Posa

Building democracy in Iraq will prove impossible without immediate leadership from the country's forsaken majority: its women. But while the Bush administration trumpets women's rights in the Middle East, it neglects to back words with action. The failure to empower women would condemn Iraq to the fate of its Arab neighbors—autocracy, economic stagnation, and social malaise.

It was August 2003 in the Iraqi city of Najaf—long before the holy city's takeover by Muslim cleric Moktada al-Sadr—and U.S. Marine Lt. Col. Christopher Conlin faced a dilemma. Arriving at a swearing-in ceremony for Nidal Nasser Hussein, Najaf's first female lawyer and Conlin's selection for a judgeship on the local court, he encountered a gaggle of demonstrators protesting her appointment. Despite their relatively small number (about 30 in a city of more than half a million), Conlin relented and delayed Hussein's appointment indefinitely.

Sadly, this episode of sacrificing Iraqi women's political participation to pacify vocal minorities is hardly anomalous. Although the administration of U.S. President George W. Bush extols women's advancement as a centerpiece of its Iraq strategy, good intentions have seemingly substituted coherent policy. The administration devoted millions of dollars to women's professional training via the Coalition Provisional Authority (CPA), the U.S. entity created to run Iraq until the sovereignty transfer on June 30, 2004. But after Bush ended major combat operations in May 2003, the CPA undermined its own good work by allowing Iraqi women to become a bargaining chip in political negotiations with powerful religious parties.

The United States thus made the classic mistake of sacrificing long-term stability for political expediency. Failing to include women in Iraq's government notifies other countries in the region that women's political engagement is not, in fact, the pillar of democracy the West portrays. Ultimately, such failure could undermine support for the U.S. mission in Iraq by reinforcing the notion that Washington used human rights as a pretext for war rather than committing to it out of principle. Moreover, it condemns Iraq to the fate suffered by its Arab neighbors: autocracy, economic stagnation, and social malaise

caused by sidelining half—or 60 percent, in Iraq's case—of the population.

The political and religious climate in Iraq practically guarantees that if women are frozen out of a nascent Iraqi government, their chances of breaking through later are slim to none. For Iraqi women, it's now or never.

CHECKERED PAST

Both supporters and critics of the war in Iraq politicize the history and status of Iraqi women. In the absence of weapons of mass destruction, the Bush administration deployed the rhetoric of "rape rooms" to justify the ouster of Iraqi President Saddam Hussein. Meanwhile, critics of the invasion note that women in Iraq had considerably better educational and professional opportunities under Hussein than did many women in the region.

Conditions for Iraqi women have certainly deteriorated since the Persian Gulf War of 1991. Today, mothers who can read have daughters who cannot, and the older generation often displays more modern views than the younger. Those who recall pre-Hussein Iraq remember women's political activism. The Iraqi Women's League was founded in 1952 but forced underground by Hussein soon after the Baath Party took over in 1968. Members kept in touch as exiles and recently reconstituted the league in Baghdad with the aim of maintaining women's involvement in the new government. Although the Baathists usurped Iraqis' political freedom, women's advancement fit the party's secular, nationalist agenda, and it established a Soviet-style General Federation of Iraqi Women in 1969 with branches throughout the country. Women's educational and professional prospects improved, particularly in the fields of education, medicine, and engineering, and women became breadwinners when their husbands left for the battlefields of the Iran-Iraq war in the 1980s.

Many of these gains were lost during the economic depression that followed international sanctions in the 1990s. Men took priority in the shrinking job market. Families pulled girls out of school to work at home, and female literacy plummeted. Iraqis increasingly turned to religion for solace, sharpening the divide between the country's Shiite Muslims (who constitute roughly 60 percent of the population), and Sunni Muslims (who account for about 35 percent). Saddam Hussein, a Sunni, launched a "Faith Campaign" in the early 1990s that attempted to co-opt the support of conservative religious leaders while eradicating Shiite leadership, rolling back women's legal protections in the process. Nevertheless, Shiite Islam's influence grew steadily throughout the 1990s, chiefly because its focus on social justice attracted the poor and oppressed and also because Hussein's crackdowns strengthened Shiite solidarity. This growing religious divide mirrored an increasing gap between rich and poor Iraqis that radical Shiite leaders such as Sadr would later successfully exploit.

AFFIRMATIVE REACTION

Following the U.S. invasion in March 2003, the CPA attempted to elevate Iraqi women's status through educational programs, job-skills training, rights awareness, and networking—no easy feat in a country isolated for more than a decade. But women in Iraq today need immediate political representation to secure those long-term efforts.

Unlike other post-conflict situations—wherein international bodies such as the United Nations dispatch civilian experts to handle political negotiations—Iraq has seen coalition forces take the lead in establishing local governing councils and appointing local administrators, particularly in rural areas. Nongovernmental organizations have helped Iraqis build political coalitions and take other steps toward democracy. But when it came to political appointments, the military was usually the only game in town.

In Iraq's volatile security climate, women's empowerment did not concern military personnel on the front lines of nation building. "We didn't give special considerations to engaging the women," recalls Lt. Col. Carl E. Mundy III, now a federal executive fellow at the Brookings Institution in Washington. Mundy commanded a Marine battalion during the invasion of Iraq and handled post-conflict operations in the Al Qadisiya province. "My concern was not stepping where I shouldn't step, or dragging a woman in there that would anger the local men," he explains. "Maybe once security has been established to a certain degree and most people are back to work, then you can start working around the edges at fair representation of both sexes."

The problem with this approach is that involving women becomes considerably more difficult after U.S. forces entrench conservative religious clerics and tribal leaders in positions of power. "Women may have greater voices in urban areas, but they remain under the hold of tribal leaders and religious clerics in rural areas," observes Hind Makiya, founding director of the U.K.-based Iraqi Women's Foundation.

The United States set a disastrous precedent even before the invasion of Iraq by creating a government-in-waiting led by an exile group almost entirely bereft of women. The CPA appointed only three women to the 25-member Iraqi Governing Council, and Minister of Public Works Nesreen Berwari is the only female in the Cabinet of Ministers. Although more than 80 women serve on city, district, and neighborhood councils in Baghdad, far fewer serve in the 18 Iraqi provinces, and none has been appointed a provincial governor. Worse, no women were appointed to the 24-member constitutional drafting committee, which produced the document currently serving as the interim constitution.

The drafting process guaranteed women's underrepresentation. The CPA declined to demand a mandatory number of female-held seats in the future National Assembly, despite support by Iraqi women's groups and Sunni statesman Adnan Pachachi (an influential member of the Governing Council), because the Bush administration didn't want to contradict its anti-affirmative action policy back home. Safia al-Souhail, a leader of the Bani Tamim tribe in central Iraq, scoffs at such squeamishness and dismisses non-Iraqis' concerns that enforcing women's representation violates indigenous culture. "They're forcing a lot of changes on this society. Why not force this as well?" she said. "Suddenly, women's rights are the red line?"

UNHOLY ALLIANCES

In February 2004, then council president and Shiite cleric Abdelaziz al-Hakim introduced Resolution 137, which would have abrogated Iraq's 1959 Uniform Civil Code for family law, giving religious courts jurisdiction over matters such as inheritance, marriage, and divorce. Women organized protests in Baghdad and urged American CPA Administrator L. Paul Bremer to pressure the Governing Council to repeal the resolution. Council member Dr. Raja Habib Khuzai, a former hospital director and founder of the Women's Health Center in Baghdad, led the fight. Although she won support from moderate Islamic leaders and convinced the majority of the council to overturn the resolution, a CPA staffer later told her that she had "chosen the wrong time" to pick a fight with the council's Islamists.

Bremer tried to reassure concerned members of the U.S. Congress. "Women's rights in Iraq will be protected by Iraqis, men and women," he wrote in a March 2004 letter to Rep. Carolyn Maloney of New York. "Some elements here wish to make Islamic law the law of the land, but they are neither a majority nor close to constituting one." He cited a poll finding that two thirds of Iraqis support equal legal rights for women.

Yet the will of the majority is hardly reassuring in a country traditionally tyrannized by the minority. According to Makiya, self-declared moderates and conservative council members alike have skipped council meetings to prevent the necessary quorum from voting on issues affecting women's lives and political participation. "Meetings behind closed doors and unholy affiliations are examples of men desperate for power as an end in itself, rather than as a means for social and economic reconstruction," Makiya says.

These maneuvers have compelled many women to turn to clerics for protection, Makiya adds. "We have to rely on a moderate religious leader such as Grand Ayatollah Ali al-Sistani to fight for our rights, as so-called Iraqi liberals barter away our rights amongst themselves," she says. Some Iraqi women have sought and received pronouncements from Sistani (Iraq's most respected Shiite cleric) supporting their participation in government.

By cultivating the support of true moderates such as Sistani, U.S. and Iraqi leaders will be able to preempt a backlash against women's reforms. But if they take the shortsighted approach of delaying women's progress and participation to avoid provoking extremists such as Sadr, they'll set a precedent that will be hard to overcome. Warns American Islamic Congress Executive Director Zainab Al-Suwaij, "If we don't push for women's rights and representation right now, then forget about it later on."

TIMING IS EVERYTHING

Iraqi women are reconciling the principles of their faith with the chance for new political, economic, and professional identities. Even women from the most conservative regions of the country are enthusiastic about becoming involved in public life, says Zainab Salbi, who runs women's centers across Iraq through her U.S.-based nonprofit organization Women for Women International. A group of Shiite women seeking funding for a women's center in their hometown of Karbala recently told potential donors that they wanted room for a prayer space alongside that for computer terminals and English lessons.

Ultimately, greater political participation by women could provide Iraq with a stabilizing force needed to stave off the potentially disastrous division of the country into ethnic states. Despite Iraq's diversity, "women are not a monolith, and neither are Sunni women or Shiite women or Kurdish women," says Susan Kupperstein, senior program officer for the U.S.-based National Democratic Institute's Iraq program. However, women have a shared stake in their economic and social development that impels them to transcend regional, ethnic, and religious divides. "I detect a great spirit of unity among Iraqi women," observes Charlotte Ponticelli, senior coordinator of the U.S. State Department's Office of International Women's Issues.

It is not too late to secure meaningful political representation for Iraqi women. U.N. Special Envoy Lakhdar Brahimi's plan to install a technocratic caretaker government until democractic elections are held could bypass the dangers of a rushed vote that might solidify the base of extremists most likely to limit women's rights. Although the situation in Iraq is constantly changing, Brahimi's plan currently places governance in the hands of a president, two vice presidents, and a prime minister.

Khuzai is pushing for the appointment of at least one woman to this quartet to ensure that the executive branch has the same 25 percent representation prescribed by the constitution. "This is the only way we can encourage women to participate. Otherwise they'll think it's only promises," Khuzai says. "The time is now."

Founding Mothers?

Many Iraqi women are defying cultural conservatives and danger outside their doors to participate in Iraq's incipient government and civil society. Here are some who are taking the risk:

Songul Chapouk represents the Turkoman population on the Iraqi Governing Council, An engineer by training, Chapouk founded the Kirkuk-based Iraqi Women's Organization, which trains women in computer skills, agriculture, and literacy.

Dr. Raja Habib Khuzai was one of three women appointed to the Iraqi Governing Council, which governed Iraq prior to June 30, 2004. Khuzai previously directed a hospital in the southern Iraqi city of Diwaniah while teaching at the local medical college, Her current project is a national women's health initiative for cancer prevention. Khuzai is also president of the Women's Organization in Diwaniah and founder of the Women's Health Center in Baghdad.

Nesreen Berwari was minister of public works in the provisional Iraqi government and minister of reconstruction and development for the Kurdistan Regional Government in Northern Iraq. She was also a member of the economy and infrastructure working group at the U.S. State Department's Future of Iraq project. Berwari previously worked with the International Organization for Migration and the United Nations Department of Humanitarian Affairs.

Hind Makiya is founding director of the Baghdad Women's Foundation and the Iraqi Women's Foundation, a U.K.-based nongovernmental organization that supports women's participation in a democratic Iraq. One of only five women appointed to the Iraqi Reconstruction and Development Council, she has also worked with various administrative bodies to plan a national strategy for Iraq's post-conflict education system.

Siham Hattab Hamdan serves on several committees of the Baghdad City Advisory Council, including the public affairs committee and the legal affairs and human rights committee. A lecturer in English literature at Mustansiriya University in Baghdad, she previously served as the council's vice chairperson, representing the Sadr City district. She is currently working to establish women's centers in Sadr City.

Ala Talabani, a former vice president of the Kurdistan Women's Union, fled Iraq in 1991 for the United Kingdom after she was fired from engineering and teaching positions for her Kurdish ethnicity and for not being a member of the ruling Baath Party. In 2003, Talabani co-founded Women for a Free Iraq and the Iraqi Women's High Council, which drafted policies on the role of women in Iraq's post-conflict reconstruction.

Want to Know More?

For profiles of women leaders in Iraq, visit the Web site of Women Waging Peace, launched by Harvard University's John F. Kennedy School of Government and overseen by the nonprofit Hunt Alternatives Fund. See also "Winning the Peace Conference Report: Women's Role in Post-Conflict Iraq," which highlights the findings of an April 2003 conference by Women Waging Peace and the Woodrow Wilson International Center for Scholars.

Charles Tripp's A History of Iraq, 2nd ed. (Cambridge: Cambridge University Press, 2000) provides a detailed timeline stretching from the country's establishment by the British to the run-up to the recent U.S.-led war. See also Kanan Makiya's Republic of Fear: The Politics of Modern Iraq, rev. ed. (Berkeley: University of California Press, 1998).

The Web site of Women War Peace, a project of the United Nations Development Fund for Women, features in-depth information on the current conflict's impact on Iraqi women. The Web site of Women for a Free Iraq offers links to press reports on women in Iraq. The "Arab Human Development Reports" (New York: United Nations) of 2002 and 2003 identify women's low status as a cause—not a symptom—of the Arab world's economic

and political challenges. Marina Ottaway questions the link between women's empowerment and democratization in "Women's Rights and Democracy in the Arab World" (Washington: Carnegie Papers, No. 42, February 2004). The Web site of the U.S. Department of State's Office of International Women's Issues documents deposed Iraqi leader Saddam Hussein's brutality toward women and promotes U.S. efforts to assist them. See also Human Rights Watch's "Climate of Fear: Sexual Violence and Abduction of Women and Girls in Baghdad" (New York: Human Rights Watch, Vol. 15, No. 7, July 2003).

Swanee Hunt and Cristina Posa discuss women's ability to bridge divides in "Women Waging Peace" (FOREIGN POLICY, May/June 2001). For insight on how women around the world are simultaneously gaining political clout while bearing the brunt of poverty, read Jane S. Jaquette's "Women in Power: From Tokenism to Critical Mass" (FOREIGN POLICY, Fall 1997) and Mayra Buvinic's "Women in Poverty: A New Global Underclass" (FOREIGN POLICY, Fall 1997).

For links to relevant Web sites, access to the FP Archive, and a comprehensive index of related FOREIGN POLICY articles, go to www.foreignpolicy.com.

Swanee Hunt, former U.S. ambassador to Austria, is lecturer at Harvard University's John F. Kennedy School of Government and founder of the nonprofit organization Women Waging Peace. Cristina Posa recently worked in Iraq as a legal and political advisor to Oxfam International.

Ten years' hard labour

More money and less ideology could improve the reproductive health of millions

A DECADE ago, the world's leaders met in Cairo at the International Conference on Population and Development (ICPD). There, they crafted a plan to achieve "reproductive health and rights for all" by 2015. That plan was wide-ranging—from more contraception and fewer maternal deaths to better education for girls and greater equality for women. But more than just setting targets, the ICPD plan also aimed to change the way those at the sharp end of making policy and delivering services thought about reproduction. It wanted to move away from a focus on family planning (and, by extension, government policies on population control) towards a broader view of sexual health, and systems and services shaped by individual needs.

Over the past week, hundreds of government officials, public-health experts and activists met in London to mark the anniversary of the ICPD and to take stock of progress towards achieving its goals. On paper, that progress has been impressive. Governments around the world have introduced legislation that reflects the ICPD's aims. But when it comes to turning policy into practice, "mixed success" is the verdict of a report card just released by Countdown 2015, a coalition of voluntary bodies involved in the field.

Take contraception, for example. According to the United Nations' Population Fund (UNFPA), 61% of married couples now use contraception, an 11% increase since 1994. This has helped push global population growth down from 82m to 76m people a year over the past decade. But in some places—particularly in sub-Saharan Africa and parts of Asia—birth rates remain high (see chart). That has spurred some governments to offer incentives to those who have fewer children, and others to inflict penalties on those who do not.

Sometimes, a high birth rate is a result of people wanting large families. But often it is due to a lack of affordable contraception. UNFPA estimates that 137m women who want to use contraception cannot obtain it. As Amare Bedada, the head of the Family Guidance Association of Ethiopia, points out, "We don't need to tell our clients about contraception. They see their plots of land diminishing, and they tell us they want to limit their family size."

Maternal health is another area where much more needs to be done. Poor women still die in huge numbers from the complications of pregnancy and childbirth. According to UNFPA, 920 women die for every 100,000 live births in sub-Saharan Africa. In Europe, by contrast, the figure is 24 (see the chart on the last page of this article). However, these numbers are, at best, only rough estimates gleaned from hospital statistics. Many women go uncounted because they never reach the health-care system for treatment in the first place.

Plenty of studies have shown what it takes to reduce maternal sickness and death. Good ante-natal health care is vital. So are cheap and simple drugs, such as oxytocin, to prevent haemorrhaging during birth. Trained midwives (or "birth attendants" as they are known in medical parlance)

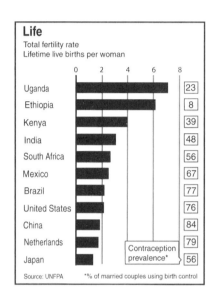

help, too. And so do local emergency obstetric centres that can handle complicated deliveries. Some countries, such as Sri Lanka, have managed to cut maternal mortality by careful spending on such measures. The challenge is to translate these successes to other places.

Yet another subject that needs to be tackled more effectively is youth sex. The largest generation of teenagers in history—a whopping 1.3 billion 10-19-year-olds—is now making its sexual debut. How it behaves, and what it learns, is crucial.

The ICPD plan was the first international agreement to acknowledge the sexual and reproductive rights of teenagers. A few countries, such as Panama, have introduced laws to safeguard some of these. In many others, youth-friendly programmes have sprung up to offer advice and assistance on thorny issues such as unwanted pregnancy and sexually

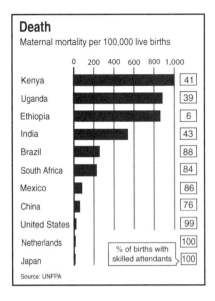

Death
Maternal mortality per 100,000 live births

Kenya	41
Uganda	39
Ethiopia	6
India	43
Brazil	88
South Africa	84
Mexico	86
China	76
United States	99
Netherlands	100
Japan	100

% of births with skilled attendants

Source: UNFPA

transmitted diseases—now soaring worldwide at 340m infections a year.

Such programmes, of course, are complicated by fierce—if probably futile—battles in many countries over whether young people should be having sex at all. These play out in international skirmishes over abstinence versus condoms for the young, parental consent to contraception and abortion, and what, if any, sex education should be provided by the state.

Sex and money

One significant obstacle to tackling these problems is money, or rather the lack of it. Ten years ago, the ICPD estimated the cost of implementing its recommended programmes at $18.5 billion by 2005—or $23.7 billion in today's dollars. The goal was to mobilise one-third of that money from rich donors, and the rest from developing countries themselves. But current spending is well below the mark.

Few poor countries have earmarked enough of their budgets to meet their citizens' reproductive-health needs. Nor have donors lived up to expectations. In 2003, they spent an estimated $3.1 billion on reproductive health. Although contributions have increased over recent years, with a few European countries, such as the Netherlands, chipping in more, and private donors,

such as the Gates Foundation, entering the field, this is still far off even the inflation-devalued $6.1 billion expected from donors by 2005.

Reproduction, it seems, is no longer a sexy subject. As Steve Sinding, the head of the International Planned Parenthood Federation (IPPF), points out, donor interest in the past was stimulated largely by fears of a population crisis. When the Cairo Conference reframed the issues in terms of women's health and reproductive rights, that demographic rationale was lost, taking funding with it.

Moreover, there are other causes competing for international funding, most notably AIDS. At the time of the Cairo Conference, 20m people were infected with HIV, the virus that causes AIDS. Today, that number has doubled. Indeed, AIDS threatens to derail the ICPD strategy. For, although billions of dollars are now pouring in to fight the disease, much of this money is going into AIDS-specific programmes that do not address reproductive health more broadly.

As Nafis Sadik, a former head of UNFPA and now the UN secretary-general's special envoy for HIV in Asia, observes, ten years ago those working in family planning shied away from the field of HIV, with its heavy burden of social stigma. Today, the roles are reversed, as reproductive health is engulfed in a storm of religious and political controversy. One consequence is that organisations concerned with fighting AIDS are failing to make use of valuable infrastructure and expertise already on the ground in places where the disease hits hardest. Given that more than half of HIV infections in sub-Saharan Africa are among women, and that for many African women family-planning services are their main contact with the formal health-care system, such services need to be drafted into the wider battle against HIV. Many family-planning clinics already offer HIV testing and counselling, as well as condoms (against the double whammy of un-

wanted pregnancy and HIV infection), and also a broad based message of sexual health.

What the field of reproductive health lacks in resources, however, it makes up in ideology. Over the past ten years, battles have broken out between contending views of sexuality, pitting religious conservatives—primus inter pares, the Vatican—against social liberals. The fight has become particularly fierce since the election of George W. Bush as America's president. Mr Bush's socially conservative views are reflected in the way America, the world's leading donor for reproductive health, spends its money at home and abroad.

Breeding trouble

The main battles are over abortion. Austin Ruse, the president of the Catholic Family and Human Rights Institute (C-FAM), an American Christian lobby group, argues that the shift in talk from fertility control to reproductive rights and services is just code for making abortion universally available. He regards this as wrong, and believes that the ICPD plan of action and those agencies which support it—particularly UNFPA—should be opposed at every turn by a growing coalition of "pro-family" groups worldwide. "Over the next five years, I see everything coming our way, especially on the question of abortion," says Mr Ruse.

UNFPA, not surprisingly, has a different view. Thoraya Obaid, its head, reckons that those who oppose the ICPD plan of action are not just against legalising abortion, but are fighting against women's rights in general. She points to the text of the plan, which states that abortion should never be promoted as a form of family planning and that women should be helped to avoid abortion through better access to contraception. (It also says that those who have sought abortions are entitled to the best possible medical treatment to deal with the complications.)

All sound stuff, but trouble lies in the plan's statement that abortion policy should be up to national governments to decide. Since 1994, more than a dozen countries have liberalised their laws on abortion (with a couple of countries tightening them up). But none of this comes without a fight, often led by the Catholic Church. Kenya has seen a particularly nasty debate over the past six months. There have been street protests, graphic television "docudramas" showing the perils of abortion, and even the arrests of health-care workers who are alleged to have performed more than a dozen abortions whose fetuses recently ended up in a ditch outside Nairobi. The government, which was looking at its abortion laws as part of a broader constitutional review, has made no changes to the current provision, which bans abortion unless the mother's life is at stake.

In many developing countries, Christian anti-abortion groups such as America's Human Life International—a sister organisation to C-FAM—have been pitching in to help organise resistance to changes in abortion laws. But American officials have entered the fray as well. Delegates to regional meetings held in Latin America during the past year to re-affirm their commitment to the ICPD plan of action have complained about pressure from American officials to reject the plan's calls for broad-based reproductive rights and services.

While pressure by the Catholic Church and other opponents of legal abortion can shape official policy, Tim Black, the head of Marie Stopes International (MSI), a voluntary organisation providing reproductive services, argues it does little to stop women seeking abortions, legal or illegal. Surveys from hospitals in Ethiopia, Uganda and Kenya suggests that anywhere from 20-50% of maternal deaths are due to complications resulting from unsafe backstreet abortions. But these numbers are challenged by the opponents of abortion, who argue that it is a rare phenomenon in the developing world, and that legalising it will make it more common.

The American government's views on abortion are expressed in the Mexico City Policy, which was re-introduced by Mr Bush in 2001. This policy, first implemented by Ronald Reagan in 1984, forbids American government funding of foreign organisations which in any way promote, endorse or advocate abortion. American law has banned foreign assistance for the direct performance of abortions since 1973. But the Mexico City Policy, or "Global Gag Rule" as its critics often refer to it, means that groups which want to perform abortions with money from other sources must also toe the United States' line, or else forfeit American assistance.

Opponents of this policy argue that it imposes on foreigners restrictions which are unconstitutional in America. Indeed, Frances Kissling, the head of Catholics for a Free Choice, an American voluntary organisation which opposes banning abortion, argues that Mr Bush is flexing his conservative muscles abroad—and therefore appeasing his supporters at home—precisely because he cannot deliver a domestic anti-abortion agenda. Last week, for example, a court in New York declared unconstitutional a ban on so-called partial-birth abortion that Mr Bush signed into law in 2003.

Several prominent family-planning organisations, such as MSI and the IPPF, have refused to agree to the Mexico City Policy, saying it compromises their ability to offer women in poor countries the full range of services available in the rich world. In Ethiopia, for example, these groups have had to trim their services and shelve expansion plans as a result of losing both money and contraceptive supplies from the American government.

Such arguments, however, cut little ice with Jeanne Head, the United Nations representative for National Right to Life, an American anti-abortion group. As she puts it, "if they

refuse these funds and they can't keep functioning, then they don't care about these women, they only care about abortion. I think the blame lies on the organisations, not on the US government."

John Kerry has said he will rescind the Mexico City Policy if elected. The Democratic candidate has also promised to restore American funding to UNFPA. This has been withheld by the Bush administration for the past three years under a piece of legislation called the Kemp-Kasten amendment. This amendment authorises the president to restrict funding to any group that "supports or participates in the management of a programme of coercive abortion or involuntary sterilisation."

The White House accuses UNFPA of abetting coercive reproductive practices in China—a claim that UNFPA denies. Several international delegations, including ones from Britain's parliament and the American State Department, have investigated UNFPA's activities in China and failed to find evidence to support such allegations. On the contrary, they argue that where UNFPA operates, policies in China are improving. But these findings are contested. The Bush administration says UNFPA has yet to mend its ways, and refuses to pay the $34m appropriated by Congress. The agency says it has managed to fill the gap this year, from big donors such as Britain, which is raising its annual contribution to £20m ($36m), and tiny ones such as Afghanistan, which chipped in $100.

But making up the money is the easy part. Today's battles over abortion, abstinence and condoms are casting a pall over the field, and complicating what is already a formidable task. Making sex safer and reproduction less risky in the 21st century requires all the tools to hand. Policies that restrict people's choices should not be a fact of life.

Index

A

Abbas, Mahmoud, 107, 108
abortion, 214–215
Abuja Declaration, 294
Access to Information and Protection of Privacy Act, 161
accountability, of governments, 64, 164
Addis Ababa agreement, 120
Afghanistan, elections in, 151–156
Africa, 88; changes in governance in, 70; democracy and, 160–165; development in, 188–193; economic performance in, 188; the IMF and, 67; war in, 111–114
African National Congress (ANC), 166, 168, 170
aging, population and, 183–185
agricultural protectionism, 38–39
agriculture, market access and, 45, 53
aid, category of the CDI, 74–75; United Nations and, 83–86
AIDS, 214; crisis, in South Africa, 168–169, 193; Sub-Saharan Africa and, 17
al-Banna, Gamal, 143, 144
Amish, the, insurance and, 24–25
Angola, foreign investors in, 76
Annan, Kofi, 104, 117
Arabs. @i[See] Muslims
Arafat, Yasir, 107, 108
Argentina, 129, 130, 173–175
Asia, 88; economic growth and, 17–18
Asian crisis in 1997–98, 48–49
Asia-Pacific Economic Cooperation Forum, 57
Association of South East Asian Nations, 53
Atatürk, Mustafa Kemal, 145, 205
Avon, 27

B

Baathist Regime, 207–208; survival of, 157–159
Bhagwati, Jagdish, 38
Bhalla, Surjit, 3
bilateral assistance, U.S., 6–7
"Birth Rate Crisis," 31–32
Blair, Tony, 198–199
Bolivia, politics and, 175
Bonn Accord, 152
Book and the Koran: A Contemporary Reading, The (Shahrour), 143
Botswana, diamonds and, 97
Boujnourdi, Mohammad, 149–150
Bourguiba, Habib, 205
Bourguignon, Francis, 3
brand names, marketing and, 27
Brown, Gordon, 198–199
Brundtland Commission, on health, 33
Burkina Faso legislature, 162
Bush, George W., 8, 9, 68, 108, 136, 139, 140, 141, 158, 159, 207, 208, 209

C

caliphate, 145
Cambodia, 52, 53–54
Canada, 38, 42, 76; technology and, 78
Cancun, WTO talks in, 39, 42–43
Cardoso, Fernando Henrique, 174
Caux Round Tables's Principles for Business, 27
cellular technology, 27
Central American Free Trade Agreement, 57
CGD/FP Commitment to Development Index, 73–79; rankings of, 75; seven categories of, 74. See also specific categories
Chávez, Hugo, 172, 175, 176
children: development and, 21; security and, 77
China, 18, 71, 120, 215; foreign investors in, 76; the IMF and, 66; poor people in, 2, 4; population of, 182; trade policies of, 37
Chirac, Jacques, 97, 104
Civic Forum, 140
civil liberties, 130, 131, 132
civil society, role of the strength of, 137, 138
civil war: in Africa, 190; Darfur and, 118; market for, 95–99
Coalition Provisional Authority (CPA), 207–208, 209, 210
Code of Personal Status, 205
Collier, Paul, on the economic civil war, 95–99
commercial finance, 88
Commitment to Development Index (CDI). See CGD/FP Commitment to Development Index
communication technology, 90
competitiveness, governance and, 11–12
computerization, microfinance and, 88–89
conditionality, of the IMF, 49
Constant Battles (LeBlanc), 101
constitution, conversations on, 163
Convention to Eliminate Discrimination Against Women, 208
corruption, 113, 122, 173; in African government, 164; governance and, 11–12
credit rating systems, 12
credit scoring, microfinance and, 88–89
criminalization, of the state, 106
cross-border supply of services, market access and, in the South, 45
cross-country inequality, 2
cultural imperialism, 21–22

D

Darfur, Sudan: crime in, 118–119; genocide and, 115–121
De Klerk, F. W., 166, 167
death taxes, 196
Deaton, Angus, 2–4
"debt forgiveness grants," 6
"debt overhang," 80
debt relief: calculating the benefits of, 80–82; the IMF and, 66, 67

Democracy in America (Tocqueville), 137
democracy, 136–139; in Afghanistan, 152–153; global capital mobility and, 55–60; women and, 205, 209–212
demographic dividend, population growth and, 182–183
Denmark, adoption of the Commitment to Development Index, 73–79
Development Assistance Committee (DAC), 5, 6
development: elements of, 74–79; Western model of modernity and, 21–25
diamonds, economy and, 96
Diaspora, economic recovery and, 98
digital divide, mobile phones and, 90–91
"Digital Solidarity Fund," 90
discriminatory trade preferences, 62
disease, in Africa, 191–192, 194–195
Doha Round, 39, 42, 44, 45, 46, 59; four-pronged strategy for new approach to, 52–53; three views on trading in, 51–54
Dollar, David, 71
domestic industry, restrictive trade policies on the development of, 61
domestic reforms, 62
domestic regulatory policies, 61
donor aid agencies, 164

E

Easterly, William, 86
"e-Choupals," 26
Economic Commission for Africa, 191
Economic Community of West African States, 162
economic development, 18; in impoverished countries, 5–10; role of institutions in, 17
economic inequality, 2–4
economic performance, 14
economic society, 23–23
Economic Support Funds (ESF), 6
educational reforms, gender gaps and, 204, 206
Egypt: democracy and, 140; Gaza and, 109
Egyptian National Assembly, 139
Ehrlich, Paul, 187
elections: in Afghanistan, 151–156; in Africa, 161–162; in Saudi Arabia, 140
electoral coalitions, 167
electoral commission, 111
electoral democracy, impact of, on the stability of states, 106
"Elusive Quest for Growth, The" (Easterly), 86
emerging-market economies, debt of, 67
"English Opens Doors" program, 174
Enhanced HIPC initiatives, 80
Enhanced Structural Adjustment Facility, 47
entrenched poverty, civil conflict and, 95–99
entrepreneurs, 37
environment, category of the CDI, 74, 76–77
Europe, 102, 124; immigrants and, 185; the IMF and, 66

Index

Mbeki, Thabo, 168
mercantilism, 61
MERCOSUR, 174
Mexico City Policy, 215
Mexico: foreign investors in, 76; NAFTA and, 58–59; population of, 182
microfinance institutions (MFIs), 87, 88, 89
microfinance, 204–205; and the poor, 87–89
Migrant Workers Convention, ratification of, 33
migration, 19, 31–33; globalization and, 31–32
military rule, 144
Millennium Challenge Account (MCA), 7, 9, 11, 12, 208
Millennium Declaration, 5, 67
Millennium Development Goals (MDG), 7–9, 19, 67, 80, 83, 188, 193, 204
modernizing distribution channels, 28
Moi, Daniel arap, 163
Monterrey Consensus, 6, 8, 67
Montreal Protocol, 77
Morocco, women in, 206
Morrisson, Christian, 3
most-favored-nation (MFN) principle, 40, 62
Mueller, John, 100–101, 102–103
mujahideen, factions of, 151
mujtahid, 143
multilateral trade liberalization, 44
multinational corporations, and the developing world, 28–29
Multinational Force and Observers (MFO), 108, 109, 110
Muslim Brotherhood, 141, 142, 143, 144
Muslim World Outreach Policy Coordinating Committee, 141
Muslims, 117, 119, 148, 149

N

Nasser, Gamal Abdul, 144
national income, and liberty, 129–135
national interests, 55; migration and, 32
National Islamic Front (NIF), 115
National Security Council, 141
National Union for the Total Independence of Angola (UNITA), 96, 97
nationalism, 59
NATO, 109, 110
natural resources, 198; civil conflict and, 95–99
neocons, 158
Nestlé Group, 27
Netherlands, adoption of the Commitment to Development Index, 73–79
New Democracy, The (al-Banna), 143
New Economic Partnership for African Development (NEPAD), 193
non-core principles, 62, 63
North American Free Trade Agreement, 43, 57–59, 60
Norway, 76

O

official development assistance (ODA), 5
oil, 120; in Africa, 164
Omar, Mullah, 152

open trade policies, 37
opium wars, 155–156
Organization for Economic Cooperation and Development, 5

P

Party of Davos, 56–57
Party of Porto Alegre, 56, 59
Pashtuns, 151, 155
Peace and Conflict report, 100, 101
peacekeepers, 102
Peoples Democratic Party of Afghanistan, 153
peshmerga, 141
pharmaceutical industry, price of life and, 196–197
physical harassment, 122
Pinochet, Augusto, 173
polarization, war and, 119
political freedom, 14
political goods, governance and, 12–14
political leaders, 33, 160
Political Order in Changing Societies (Huntington), 137
political rights, 130, 131
politics: in Latin American, 173; in South Africa, 166, 167–168
poor, gap between the rich and, 2; selling products to, 26–30
Population Bomb, The (Ehrlich), 187
population growth, 181–187, 213
"Positive Agenda for Trade Negotiations," 51
Poverty Reduction and Growth Facility (PRGF), 47, 49, 81
Poverty Reduction Strategy Paper (PRSP), 47, 49, 189
poverty: politics and, 132; reduction in, 47
pragmatism, in the Southern Cone, 173–174
Preachers Not Judges, 144
preferential access dimension, of SDT, 62
President's Emergency Plan for AIDS Relief (PEPFAR), 7
Program on International Policy Attitudes (PIPA), 6
"protectionism," 57
proxy indicators, governance and, 13–14
"pull-up" effect, 38

R

radicalism, tyranny and, 144
rating systems, for components of governance, 11–12
rebellion, low-income countries and, 96
Reconstruction and Development Program (RDP), 169
reformation, fundamentalism and, 144
refugees: Afghan, 154; and global security, 122–125. *See also* migration
regional integration, 19
regulatory disciplines, 61
religion, freedom of, 148–149
Remnants of War, The (Mueller), 100, 102
reproductive health, money and, 213–215
Resolution 242, of the U.N. Security Council, 108

resource endowments, in development, 17–20
Retreat from Doomsday (Mueller), 100
Revolutionary United Front (RUF), 96
rich nations, ranking of, 73–79
Ricupero, Rubens, on the Doha Round global trade talks, 51–54
Rising Tide of Color Against White World-Supremacy, The (Stoddard), 185
rituals, 147
Robertson, Dennis, 24
Robinson, Joan, 39
Rodrik, Dani, 37, 59, 71
Ruggiero, Renato, 55
rule of law, 71, 137; political goods and, 13
rural telecentres, 90
Rwanda, genocide in, 112

S

Sachs, Jeffrey, 83
Sala-i-Martin, Xavier, 3
Sanusi, Mobolaji, 160
SARS (Severe Acute Respiratory Syndrome), 18
Saudi Arabia, women in, 206–207
SDT (special and differential treatment), 61, 62
secular societies, 186–187
security: in Afghanistan, 153;category of the CDI, 74, 77–78; political goods and, 13
Seko, Mobutu Sese, 112
Sen, Amartya, 204
Senegal, agriculture and, 38
sex, money and, 214
Shahrour, Muhammad, 143, 144–148
Sharia, 145, 146, 147, 149
Sharon, Ariel, 107
Shia parties, 140, 141
Shiites, 157
shrimp farming, 40–41
Sierra Leone, diamonds and, 97
Simon, Julian, 103
Siphana, Sok, on the Doha Round global trade talks, 51–54
Smith, Adam, 17–18, 56
smuggling, of human beings, 31
social contract, 55
social security system, of China, 182
socially oriented financial institutions (SOFIs), 88
socioeconomic and cultural structure, role of, 137
South Africa, 166–171; economy and, 169–170
South Asia, 21
Southern African Development Community (SADCC), 161
Sovereign Debt Restructuring Mechanism, 67
Soviet Union, 100
speech, freedom of, 146
Stern, Nicholas, 39
Stiglitz, Joseph, 37
Stravinsky, Igor, 103
subindicators, governance and, 13
Sub-Saharan Africa, 7, 17, 19, 21; globalization and, 4; Millennium Development Goals in, 83, 84, 188; poverty in, 48

Test Your Knowledge Form

We encourage you to photocopy and use this page as a tool to assess how the articles in *Annual Editions* expand on the information in your textbook. By reflecting on the articles you will gain enhanced text information. You can also access this useful form on a product's book support Web site at *http://www.mhcls.com/online/*.

NAME: DATE:

TITLE AND NUMBER OF ARTICLE:

BRIEFLY STATE THE MAIN IDEA OF THIS ARTICLE:

LIST THREE IMPORTANT FACTS THAT THE AUTHOR USES TO SUPPORT THE MAIN IDEA:

WHAT INFORMATION OR IDEAS DISCUSSED IN THIS ARTICLE ARE ALSO DISCUSSED IN YOUR TEXTBOOK OR OTHER READINGS THAT YOU HAVE DONE? LIST THE TEXTBOOK CHAPTERS AND PAGE NUMBERS:

LIST ANY EXAMPLES OF BIAS OR FAULTY REASONING THAT YOU FOUND IN THE ARTICLE:

LIST ANY NEW TERMS/CONCEPTS THAT WERE DISCUSSED IN THE ARTICLE, AND WRITE A SHORT DEFINITION:

We Want Your Advice

ANNUAL EDITIONS revisions depend on two major opinion sources: one is our Advisory Board, listed in the front of this volume, which works with us in scanning the thousands of articles published in the public press each year; the other is you—the person actually using the book. Please help us and the users of the next edition by completing the prepaid article rating form on this page and returning it to us. Thank you for your help!

ANNUAL EDITIONS: Developing World 06/07

ARTICLE RATING FORM

Here is an opportunity for you to have direct input into the next revision of this volume.
We would like you to rate each of the articles listed below, using the following scale:

1. **Excellent: should definitely be retained**
2. **Above average: should probably be retained**
3. **Below average: should probably be deleted**
4. **Poor: should definitely be deleted**

Your ratings will play a vital part in the next revision.
Please mail this prepaid form to us as soon as possible.
Thanks for your help!

RATING	ARTICLE	RATING	ARTICLE
	1. More or Less Equal?		23. The End of War?
	2. The Development Challenge		24. The Failed States Index
	3. Strengthening Governance: Ranking Countries Would Help		25. Gaza: Moving Forward by Pulling Back
	4. Institutions Matter, but Not for Everything		26. Africa's Unmended Heart
	5. Development as Poison		27. Sudan's Darfur: Is it Genocide?
	6. Selling to the Poor		28. Blaming the Victim: Refugees and Global Security
	7. The Challenge of Worldwide Migration		29. National Income and Liberty
	8. International Trade		30. The Democratic Mosaic
	9. Trade Secrets		31. Keep the Faith
	10. Why Prospects for Trade Talks are Not Bright		32. Voices within Islam: Four Perspectives on Tolerance and Diversity
	11. Why Developing Countries Need a Stronger Voice		33. First Steps: The Afghan Elections
	12. Why Should Small Developing Countries Engage in the Global Trading System		34. The Syrian Dilemma
	13. Without Consent: Global Capital Mobility and Democracy		35. Africa's Democratization: A Work in Progress
	14. Making the WTO More Supportive of Development		36. The New South Africa, a Decade Later
	15. How to Run the International Monetary Fund		37. Latin America's Populist Turn
	16. A Regime Changes		38. The Global Baby Bust
	17. Ranking the Rich 2004		39. Development in Africa: The Good, the Bad, the Ugly
	18. Calculating the Benefits of Debt Relief		40. Malaria, The Child Killer
	19. Recasting the Case for Aid		41. The Price of Life
	20. Microfinance and the Poor		42. Why We Owe So Much to Victims of Disaster
	21. The Real Digital Divide		43. The Payoff From Women's Rights
	22. The Market for Civil War		44. Iraq's Excluded Women
			45. Ten Years' Hard Labour

(Continued on next page)

ANNUAL EDITIONS: DEVELOPING WORLD 06/07

BUSINESS REPLY MAIL
FIRST CLASS MAIL PERMIT NO. 551 DUBUQUE IA

POSTAGE WILL BE PAID BY ADDRESEE

McGraw-Hill Contemporary Learning Series
2460 KERPER BLVD
DUBUQUE, IA 52001-9902

Ialıludlllludlaaallldadalullaadadll

- -

ABOUT YOU

Name   Date
_____

Are you a teacher? ❑   A student? ❑
Your school's name
_____

Department
_____

Address   City   State   Zip
_____

School telephone #
_____

YOUR COMMENTS ARE IMPORTANT TO US!

Please fill in the following information:
For which course did you use this book?
_____

Did you use a text with this ANNUAL EDITION? ❑ yes ❑ no
What was the title of the text?
_____

What are your general reactions to the *Annual Editions* concept?
_____

Have you read any pertinent articles recently that you think should be included in the next edition? Explain.
_____

Are there any articles that you feel should be replaced in the next edition? Why?
_____

Are there any World Wide Web sites that you feel should be included in the next edition? Please annotate.
_____

May we contact you for editorial input? ❑ yes ❑ no
May we quote your comments? ❑ yes ❑ no